Methods in Protein Sequence Analysis

Edited by

Marshall Elzinga

Brookhaven National Laboratory, Upton, New York

Humana Press · Clifton, New Jersey

Organization and Program Committee:
Marshall Elzinga, Chairman
Lowell H. Ericsson
Agnes Henschen
Marcus J. Horn
Richard A. Laursen
John H. Walker

The organizers are grateful to the following for providing financial support for this Conference:
National Science Foundation
AAA Laboratory
Beckman Instruments
Dionex Corporation
Hewlett-Packard Company
MCB Manufacturing Chemists
Pierce Chemical Company
Sequemat/Genetic Design, Inc.
Waters Associates, Inc.

Library of Congress Cataloging in Publication Data
Main entry under title:

Methods in protein sequence analysis.
(Experimental biology and medicine)
"Augmented record of the IVth International Conference on Methods in Protein Sequence Analysis held September 25–25, 1981 at the Brookhaven National Laboratory, Upton, New York"—
Includes index.
1. Amino acid sequence—Congresses.   2. Proteins—Analysis—Congresses. I. Elzinga, Marshall. II. International Conference on Methods in Protein Sequence Analysis (4th: 1981: Brookhaven National Laboratory) III. Series: Experimental biology and medicine (Humana Press) [DNLM: 1. Proteins—Analysis.   2. Amino acid sequence.   QU 60 M592]
QP551.M393   574.19′245   82-80733
ISBN 0-89603-038-5   AACR2

Methods in Protein Sequence Analysis

Experimental Biology and Medicine

PREFACE

Methods in Protein Sequence Analysis contains an intensely practical account of all the new methodology available to scientists carrying out protein and peptide sequencing studies. Many of the striking advances in fields as diverse as immunology, cell motility, and neurochemistry have in fact been fueled by our ever more powerful ability to determine the sequences and structures of key proteins and peptides. It is our hope that the rich array of techniques and methods for sequencing proteins discussed in this volume—methods that generate much of the information crucial to progress in modern biology—will now become accessible to all who can benefit from them.

The papers of the present volume constitute the Proceedings of the IVth International Conference on Methods in Protein Sequence Analysis, which was held at Brookhaven National Laboratory, Upton, NY, September 21–25, 1981. It was the most recent in a series of biennial conferences, the previous one having been held in Heidelberg, GFR, in 1979. The series was originated by Richard Laursen, and initially dealt with one aspect of the field, solid-phase sequencing.

The scope of the meeting was very broad and among the many aspects of protein sequencing discussed were: instrumentation, strategy, chemicals, mass spectrometry, cleavage of proteins and separation of peptides, and solid, liquid, manual, and even "gas-phase" sequencing.

The goals of the methods described here are essentially the same as they were 30 years ago: one wants to know the order of the amino acids in a given protein or peptide because the amino acid sequence is information that is fundamental to understanding its properties. Thus many of the papers describe improvements of the traditional methods used to determine the complete sequences of pure proteins. Especially notable in this regard are the important recent developments that have taken place in the area of peptide separation: here HPLC (high performance liquid chromatography) is rapidly replacing all other column and paper methods.

v

   Similarly, the enormous advances being made in the technology of RNA and DNA sequencing have forced a re-evaluation of many studies that involve protein sequencing. It is now clear that protein and nucleic acid sequencing are complementary techniques, each with advantages in addressing different aspects of many questions about structures and sequence. Thus, if one can now prepare the right piece of nucleic acid, information regarding the amino acid sequence of the protein(s) for which it codes can often be obtained very quickly by sequencing the nucleic acid. However, questions regarding gene expression, posttranslational processing, sidechain modification, and so on generally still require protein sequence data. In modern molecular biology, crucial experiments sometimes require limited amounts of sequence information from specific parts of proteins that are available in only extremely small quantities. The design of such experiments and the development of the necessary micromethods for sequencing at picomole levels are discussed in several of the papers.

   The scientific program of the conference included both oral and poster presentations. In general, papers dealing primarily with methods were presented orally and appear in this volume as full manuscripts, while new sequences were presented as posters and are published as "Communications." Most of the papers provide up-to-date accounts of the authors' experiences in the development of procedures or the improvement of techniques. This book should therefore be useful to practitioners of the protein sequencing art who want to learn about new procedures and alternative methods that are potentially applicable to their own sequencing problems, as well as to those who work in peripheral areas and wish to familiarize themselves with developments in this field.

April, 1982   **Marshall Elzinga**

Memorial

Erhard Gross, 1928–1981, In Memoriam

Christian Birr

Max-Planck-Institut fur Medizinische Forschung, Jahnstrasse 29, Heidelberg, F. R. Germany

The international natural sciences community has lost one of its most active, eager, busy and great members. On the 12th of September, 1981, my friend and mentor, Dr. Erhard Gross, died in a tragic traffic accident in Germany, ten days after the 53rd anniversary of his birthday. The cruelty of fate has thus taken him from his wonderful wife, Gertrud, and his gentle and talented sons, Johannes and Christoph. I am privileged by the organizer of this symposium to briefly commemorate the academic life of Dr. Erhard Gross.

He graduated in chemistry 1956 from the University of Frankfurt, F. R. Germany, at the Institute of Organic Chemistry headed by Theodor Wieland. In those early days Erhard Gross developed a close relationship with his scientific teacher and friendly mentor. This relationship matured into a trustful friendship with Theodor Wieland, which spanned Erhard's too short life. It was a characteristic of Erhard to infuse his fellow-scholars, as well as later students of Theodor Wieland, with his warm feelings for his teacher. In this way, I received the honor and pleasure of Dr. Gross' constant friendship and helpful advice.

In 1958 he passed the Ph.D. examinations with the presentation of his thesis on the synthesis of a bicyclic peptide similar in structure to the mushroom toxin Phalloidin, which he had studied under the auspices of Th. Wieland at the University of Frankfurt. As we shall see, there are three common threads that connect the scientific efforts and results of the natural scientist and organic chemist Dr. Gross: his biomedical interest, the peptide bond, and thioethers, three characteristics of his doctoral thesis.

Just after his marriage in 1958 Erhard left Frankfurt, Germany, for the United States of America. Actually, this was not his first experience with the American people. In 1945 he became an American Prisoner of War for a short period after having had to take part as a youngster in the Second World War in 1944.

In 1958 Dr. Gross started to work in the prestigious chemistry laboratory of B. Witkop at the Institute of Arthritis and Metabolic Diseases of the National Institute of Health (NIH) in Bethesda, Maryland. He was charged with the search for nonenzymatic methods in the cleavage of peptides and proteins. It was during this time that Erhard Gross developed the world-famous cyanogen bromide method for the selective cleavage of methionine peptide bonds, which was published in 1961 together with B. Witkop. To date this specific reaction remains the most useful chemical procedure in the arsenal of methods available to protein sequencers. Without the specificity of the cyanogen bromide cleavage we should not have the great successes in the industrial biosynthesis in bacteria of peptide hormones by the recombinant DNA technique. There, methionine is incorporated between the carrier protein and the target peptide as a specific detachment site.

In the mid sixties Dr. Gross investigated reaction mechanisms of enzymes, and detected aspartic acid in the active center of pepsin. He elucidated the structure of uncommon amino acids and discovered dehydro-alanine as a natural constituent in proteins.

In 1968 Erhard Gross became Chief of the Molecular Structure Section at the Laboratory of Biomedical Sciences of the National Institute of Child Health and Human Development at the NIH. His research in cell biology was concentrated on the isolation and characterization of cell organelles. He investigated placental lysosomes and those from polymorphous leucocytes. His chemical work was centered on the structural elucidation and synthesis of membrane active peptides for the induction of fetal resorption. For this purpose he investigated the gramicidins and published in 1969–1971 the syntheses of Gramicidin A, C and B. His most recent research work was focused upon the conformational characteristics of microbial peptides that lead to formation of channel aggregates and thus allow the translocation of cations across lipid membranes.

In 1971 Erhard described the pentacyclic heterodetic nature of the polypeptide Nisin, an antibiotic from *Streptococcus lactis* which contains 5 thioether bridges. In 1972 he correlated these structural elements with those of the other pentacyclic thioether polypeptide Subtilin, a microbial product from *Bacillus subtilis*. Since then a great deal of his research efforts were devoted to the task of synthesizing those very difficult polycyclic thioether peptides. Though he and his coworkers succeeded in the preparation of a monocyclic element of the antibiotics, by his untimely death the total syntheses of Nisin and Subtilin have to remain a challenge for Erhard's scientific heirs.

Erhard Gross often demonstrated his talents in scientific management, which culminated in his brilliant organization of the most successful Sixth American Peptide Symposium in 1979 at Georgetown University in Washington, D.C. Together with his friend and colleague Johannes Meienhofer, Erhard most efficiently edited, in an intellectual marathon, more than a thousand pages of the proceedings volume, which was published within less than six months. Most recently, the same team of authors and editors received great international recognition and admiration through the publication of their series "THE PEPTIDES, Analysis, Synthesis, Biology." Three excellent volumes have already appeared and a fourth is in preparation. By these monuments of work our memorial to Erhard Gross will remain immortal.

His scientific contributions were honored by the Humboldt Award presented to him in 1978.

In my mind's eye I can see Erhard Gross with his famous trade marks, the bow-tie, his red-blond crest and his finger pointing as an impulsive discussant. His enthusiasm and activity would have helped greatly to make a meeting like this one a success. Our best memorial to him will be to remember these distinguishing characteristics, and use them as an inspiration for a happy and fruitful symposium.

CONTENTS

Manual Sequencing

Sequencing by Mass Spectrometry

Cleavage of Proteins

Coordination of Protein and DNA Sequencing

Use of High Performance Liquid Chromatography (HPLC) For the Separation of Proteins, Peptides, and Amino Acids

Contents

Identification of Phenylthiohydantoins

Communications: Sequence Methods

List of Participants

SEETHARAMA A. ACHARYA • *Rockefeller University, New York, NY*

NICHOLAS ALONZO • *Brookhaven National Laboratory, Upton, NY*

ANTHONY D. AUFFRET • *University of Leeds, Leeds, England*

CARL D. BENNETT • *Merck Sharp and Dohme, West Point, PA*

HANS BENNICH • *Uppsala University, Uppsala, Sweden*

K. BEYREUTHER • *University of Cologne, Cologne, W. Germany*

R. BHIKHABHAI • *Uppsala University, Uppsala, Sweden*

AJIT S. BHOWN • *University of Alabama, Birmingham, AL*

KLAUS BIEMANN • *Massachusetts Institute of Technology, Cambridge, MA*

JEROLD A. BIETZ • *USDA, Northern Regional Research Laboratory, Peoria, IL*

STEVEN BIRKEN • *Columbia University, New York, NY*

CHRISTIAN BIRR • *Max Planck Institute for Med. Res., Heidelberg, W. Germany*

R. BLACHER • *Hoffman-La Roche, Inc., Nutley NJ*

DALE BLANKENSHIP • *University of Cincinnati Coll. of Med., Cincinnati, OH*

ALEX G. BONNER • *Sequemat/Genetic Design, Inc., Watertown, MA*

ALBERT BOOSMAN • *University of Washington, Seattle, WA*

WILLIAM A. BRADLEY • *Methodist Hospital Baylor Coll. of Med., Houston, TX*

W. F. BRANDT • *University of Cape Town, Rondebosch, South Africa*

H. BRYAN BREWER • *National Institutes of Health, Bethesda, MD*

CLARENCE A. BROOMFIELD • *Bel Air, MD*

BENJAMIN BURR • *Brookhaven National Laboratory, Upton, NY*

TIN CAO • *Southern Illinois University, Carbondale, IL*

J. P. CAPONY • *C.N.R.S., Montpellier, France*

MATS CARLQUIST • *Karolinska Institute, Stockholm, Sweden*
ALAN CARNE • *MRC, Laboratory of Molecular Biology, Cambridge, England*
STEVEN A. CARR • *Harvard Medical School, Boston, MA*
S. K. CHAN • *University of Kentucky, Lexington, KY*
JUI-YOA CHANG • *Ciba-Geigy, Basel, Switzerland*
ROBERT CHENG-CHI CHANG • *Veterans Administration Hospital, New Orleans, LA*
VINAY CHOWDHRY • *E. I. DuPont DeNemours and Company Wilmington, DE*
JOHN H. COLLINS • *University of Cincinnati Coll. of Med., Cincinnati, OH*
DORIS CORCORAN • *Uniformed Services Univ. of the Health Sci., Bethesda, MD*
JOHN W. CRABB • *University of Washington Med. Sch., Seattle, WA*
GARY DAVIS • *Yale University, New Haven, CT*
E. DHARM • *Hoffman-La Roche, Inc., Nutley, NJ*
MARY D. DIETLER • *USDA, Western Regional Res. Center, Albany, CA*
RUSSELL F. DOOLITTLE • *University of California-San Diego, La Jolla, CA*
LAWRENCE K. DUFFY • *University of Texas Medical Branch, Galveston, TX*
MARSHALL ELZINGA • *Brookhaven National Laboratory, Upton, NY*
AKE ENGSTROM • *Uppsala University, Uppsala, Sweden*
BRUCE W. ERICKSON • *Rockefeller University New York, NY*
LOWELL H. ERICSSON • *University of Washington, Seattle, WA*
VINCENT A. FISCHETTI • *Rockefeller University, New York, NY*
EVERETT FLANIGAN • *Revlon Health Care Group, Tuckahoe, NY*
IRWIN FLINK • *University of Arizona, Tucson, AZ*
GEOFFREY T. FLYNN • *Queen's University, Kingston Ontario, Canada*
ANGELO FONTANA • *Institute of Organic Chemistry University of Padova, Italy*
W. BARRY FOSTER • *Mayo Clinic, Rochester, MN*
AUDREE FOWLER • *Univ. of California School of Medicine, Los Angeles, CA*
JAY W. FOX • *University of Virginia Med. School, Charlottesville, VA*
GERHARD FRANK • *Inst. for Mol. Biology and Biophys. ETH, Zurich, Switzerland*

LEWIS FRIEDMAN • *Brookhaven National Laboratory, Upton, NY*
VALERIE S. FUJITA • *Northwestern University, Evanston, IL*
CURTIS S. FULLMER • *Cornell University, Ithaca, NY*
J. GAGNON • *University of Oxford, Oxford, England*
RACHID GHRIR • *Center for Molecular Genetics Gif-Sur-Yvette, France*
BRAD W. GIBSON • *Massachusetts Institute of Technology, Cambridge, MA*
R. W. GLANVILLE • *Max Planck Institute For Biochemistry, Munich, West Germany*
RAE GREENBERG • *USDA, Eastern Regional Res. Center, Philadelphia, PA*
KENNETH GRIST • *Brookhaven National Laboratory, Upton, NY*
WOLFGANG A. GUENZLER • *Gruenenthal Gmbh Center of Research, Aachen, West Germany*
DAVE HAWKE • *City of Hope Research Institute, Duarte, CA*
F. S. HEINEMANN, • *University of Connecticut, Farmington, CT*
ULF G. T. HELLMAN • *University of Uppsala, Uppsala, Sweden*
JOHN J. HEMPERLY • *Rockefeller University, New York, NY*
LOUIS HENDERSON • *Frederick Cancer Research Center, Frederick, MD*
AGNES HENSCHEN • *Max Planck Institute for Biochemistry, Munich, West Germany*
WALTER HERLIHY • *Repligen Corporation, Boston, MA*
MARK HERMODSON • *Purdue University, West Lafayette, IN*
JOHN R. E. HOOVER • *Smith Kline and French Laboratories, Philadelphia, PA*
MARCUS J. HORN • *Sequemat/Genetic Design, Inc., Watertown, MA*
WEI-YONG HUANG • *Veterans Administration Hospital, New Orleans, LA*
JOSEPH W. HUBER • *Burdick and Jackson Laboratories, Inc., Muskegon, MI*
M. HUNKAPILLAR • *California Institute of Technology, Pasadena, CA*
JEAN-MARIE IMHOFF • *Pasteur Institute, Paris, France*
WANDA M. JONES • *Rockefeller University, New York, NY*
HANS JORNVALL • *Karolinska Institute, Stockholm, Sweden*
YASHWANT D. KARKHANIS • *Merck Sharp and Dohme Research Laboratories, Rahway, NJ*
BORIVOJ KEIL • *Pasteur Institute, Paris, France*
WILLIAM R. KEM • *Univ. of Florida Coll. of Medicine, Gainesville, FL*

STEPHEN B. H. KENT • *Molecular Genetics Inc., Minnetonka, MN*
ANTHONY R. KERLAVAGE • *University of Pennsylvania, Philadelphia, PA*
DAVID G. KLAPPER • *Univ. of North Carolina School of Med., Chapel Hill, NC*
H. KRATZIN • *Max Planck Inst. for Exper. Medicine, Gottingen, W. Germany*
CHUN-YEN LAI • *Roche Institute of Molecular Biology, Nutley, NJ*
POR-HSIUNG LAI • *Sequemat/Genetic Design, Inc., Watertown, MA*
R. A. LAURSEN • *Boston University, Boston, MA*
E. LAZZARI • *University of Texas Dental Center, Houston, TX*
FLORENCE LEDERER • *Center for Molecular Genetics, Gif-Sur-Yvette, France*
MARJORIE B. LEES • *Eunice Kennedy Shriver Center, Waltham, MA*
ELLEN J. L. LEW • *USDA, Western Regional Res. Ctr., Berkeley, CA*
V. M. LIPKIN • *USSR Acad. of Sciences Shemyakin Inst., Moscow, USSR*
JAMES L. L'ITALIEN • *Yale University School of Medicine, New Haven, CT*
ROBERT LITWILLER • *Mayo Clinic, Rochester, MN*
RENNE CHEN LU • *Boston Biomedical Research Institute, Boston, MA*
WERNER MACHLEIDT • *Institute for Physiological Chemistry, Munich, West Germany*
NOBUYO MAEDA • *The University of Wisconsin, Madison, WI*
LEE MALOY • *NIAID—National Institutes of Health, Bethesda, MD*
B. N. MANJULA • *Rockefeller University, New York, NY*
P. MANJUNATH • *Clin. Res. Inst. of Montreal, Montreal, Quebec H2W 1R7, Canada*
MICHAEL N. MARGOLIES • *Massachusetts General Hospital, Boston, MA*
DANIEL R. MARSHAK • *Vanderbilt University, Nashville, TN*
J. C. MASON • *Queens University, Belfast BT9 7BL, Northern Ireland*
DAVID J. MC KEAN • *Mayo Clinic, Rochester, MN*
MARTIN MERRETT • *Wellcome Research Laboratories, Kent, England*
SHEENAH M. MISCHE • *Rockefeller University, New York, NY*
KUNIO S. MISONO • *Vanderbilt University Sch. of Medicine, Nashville, TN*
JOHN E. MOLE • *University of Alabama, Birmingham, AL*

PIERCARLO MONTECUCCHI • *Farmitalia Carlo ERBA, Milan, Italy*

HOWARD R. MORRIS • *Imperial College of Science and Technology, London, England*

I. V. NAZIMOV • *USSR Acad. of Sciences Shemyakin Institute, Moscow, USSR*

ROBERT NELSON • *University of Connecticut Health Center, Farmington, CT*

NGA YEN NGUYEN • *National Institutes of Health, Bethesda, MD*

RONALD L. NIECE • *University of Wisonsin Medical School, Madison, WI*

J. I. OHMS • *Beckman Instruments, Palo Alto, CA*

R. OLAFSON • *University of Victoria, Victoria, B.C. V8W 2Y2, Canada*

MARK OLSON • *University of Mississippi, Jackson, MS*

J. OZOLS • *University of Connecticut Health Ctr., Farmington, CT*

M. PANICO • *Imperial College of Science and Technology, London, England*

PAMELA S. PARKES • *Northwestern University, Evanston, IL*

ALLEN T. PHILLIPS • *Pennsylvania State University, University Park, PA*

DENNIS PISZKIEWICZ • *Duquesne University, Pittsburgh, PA*

A. R. PLACE • *University of Pennsylvania, Philadelphia, PA*

HERWIG PONSTINGL • *German Cancer Research Center, Heidelberg, W. Germany*

RUI-QING QIAN • *Max Planck Institute for Biochemistry, Munich, West Germany*

AUSTEN RIGGS • *University of Texas, Austin, TX*

BETTY JO H. ROBERTSON • *Plum Island Animal Disease Center, Greenport, NY*

ANTHONY C. ROCCO • *MCB-Harleco, Gibbstown, NJ*

STUART RUDIKOFF • *National Institutes of Health, Bethesda, MD*

M. R. SAIRAM • *Clinical Research Inst. of Montreal, Montreal, H2W 1R7 Canada*

JOHANN SALNIKOW • *Technical University, Berlin, West Germany*

MOHAMMAD SAMIULLAH • *Boston University, Boston, MA*

F. W. SCHIRMER • *Eli Lilly Research Labs., Indianapolis, IN*

DAVID H. SCHLESINGER • *New York University Sch. of Medicine, New York City, NY*

OYVIND SCHONBERGER • *Ludwig Mazimilians University, Munich, West Germany*

JYOTI SEN • *Columbia University, New York, NY*

PATRICIA J. SEPULVEDA • *Oklahoma Medical Research Foundation, Oklahoma City, OK*

RICHARD B. SETLOW • *Brookhaven National Laboratory, Upton, NY*

ELLIOTT N. SHAW • *Brookhaven National Laboratory, Upton, NY*

YASUTSUGU SHIMONISHI • *Osaka University, Osaka, Japan*

JACK SHIVELEY • *City of Hope Research Institute, Duarte, CA*

MARK R. SILVER • *E. M. Sciences, Gibbstown, NJ*

ROBERT D. SITRIN • *Smith Kline and French Laboratories, Philadelphia, PA*

ALAN J. SMITH • *University of California, Davis, CA*

JOHN A. SMITH • *Brigham and Women's Hosp. Harvard Med. Sch., Boston, MA*

DAVID W. SPEICHER • *Yale Medical School, New Haven, CT*

JOACHIM SPIESS • *The Salk Institute, San Diego, CA*

HOWARD M. STEINMAN • *Albert Einstein College of Medicine, Bronx, NY*

JAMES E. STRICKLER • *Yale University School of Medicine, New Haven, CT*

SUZANNE STRUSIAK • *Schering Corporation, Bloomfield, NJ*

F. WILLIAM STUDIER • *Brookhaven National Laboratory, Upton, NY*

N. TAKAHASHI • *Indiana University, Bloomington, IN*

Y. TAKAHASHI • *Indiana University, Bloomington, IN*

KOJI TAKIO • *University of Washington, Seattle, WA*

GEORGE E. TARR • *University of Michigan, Ann Arbor, MI*

PAUL TEMPST • *State University of Ghent, B-9000 Ghent, Belgium*

D. TETAERT • *Indiana University, Bloomington, IN*

SIU W. TONG • *Brookhaven National Laboratory, Upton, NY*

D. TRIPIER • *Hoechst AG, Frankfurt, Germany*

SALLY TWINING • *Medical College of Wisconsin, Milwaukee, WI*

JOZEF VAN BEEUMEN • *Laboratory of Microbiology, Ghent, Belgium*

MICHEL VAN DER REST • *Shriners Hospital, Montreal, Canada*

WILLIAM VENSEL • *Northwestern University, Evanston, IL*

ROBERT A. VIGNA • *Pierce Chemical Company, Rockford, IL*

WILLIAM VINE • *Rockefeller University, New York, NY*

H. VON BAHR-LINDSTROM • *Karolinska Institute, Stockholm, Sweden*

J. E. WALKER • *MRC, Laboratory of Molecular Biology, Cambridge, England*

K. A. WALSH • *University of Washington, Seattle, WA*

MARTIN WATTERSON • *Vanderbilt Medical School, Nashville, TN*

NANCY WEIGEL • *Baylor College of Medicine, Houston, TX*

SHARON L. WELDON • *University of California-San Diego, La Jolla, CA*

NORMAN WHITELEY • *Foster City, CA*

KENNETH R. WILLIAMS • *Yale University, New Haven, CT*

KEN WILSON • *University of Zurich, Zurich, Switzerland*

B. WITTMANN-LIEBOLD • *Max Planck Inst. for Molecular Genetics, Berlin, W. Germany*

JEANNE WYSOCKI • *Brookhaven National Laboratory, Upton, NY*

MAKOTO YAGUCHI • *National Research Council of Canada, Ottawa, Canada K1A OR6*

CHAO-YUH YANG • *Gottingen, West Germany*

PAU-MIAU YUAN • *City of Hope Research Institute, Duarte, CA*

IRVING ZABIN • *Univ. of California Sch. of Medicine, Los Angeles, CA*

CLYDE ZALUT • *Harvard Medical School, Boston, MA*

ALLEN F. ZIELNIK • *Dionex Corporation, Bensenville, IL*

ANITA ZOT • *University of Cincinnati Coll. of Med., Cincinnati, OH*

Conference Lecture

AN ANECDOTAL ACCOUNT OF THE HISTORY OF PEPTIDE STEPWISE

DEGRADATION PROCEDURES

Russell F. Doolittle

Dept. of Chemistry, Univ. California, San Diego

La Jolla, California U.S.A. 92093

Quite apart from the vast amount of biochemical data it has put at our disposal, the stepwise degradation of peptides and proteins, whereby one amino acid at a time is selectively removed from one end of a polypeptide chain, ranks among the most elegant of chemical manipulations. Thus, the simple alteration of two sets of solution conditions can result in the singular and progressive removal of monomeric units from the peptidyl polymer with remarkable ease and assurance. Although I myself have been only peripherally involved in the field, I have long been fascinated by the simple majesty of the chemistry, as well as by the non-chemical circumstances surrounding the development of the method.

In this short chapter I recount some of the historical developments that I have found most arresting. It is not meant to be a serious and comprehensive review, or even a balanced account. Rather, I have taken advantage of the organizers' invitation to make this a somewhat personalized rendering and have emphasized some of the human aspects that have contributed to the state of the art. Chance meetings, geography, timing, as well as personal style, personalities and ambitions, often play key roles in science. Here are some of those usually unrendered events as I recall or have reconstructed them on the basis of discussions with other investigators in the field.

During the summer of 1961 I was standing in line at a
large outdoor fish tank at the Marine Biological Laboratory
at Woods Hole, waiting my turn to get at a large dogfish in
order to obtain a blood sample. Also standing in line was
Leslie Smith, now at the Univ. of New Mexico, but at the
time hailing from Cambridge, England. Smith, of course,
had been a member of Sanger's team who had worked out the
sequences of a number of mammalian insulins during the
1950's, and he was now waiting to remove the pancreas from
the same dogfish that I was about to bleed. During the
small talk of the moment, I told him what I was up to, including the fact that I was planning to examine the fibrinopeptides of a number of vertebrate species, and also to
characterize the junctions split by thrombin in an effort
to explain the "species specificity" of interacting proteins. In particular I was trying to find why thrombin of
a given species clotted the fibrinogens of distantly related species much more slowly than it did the fibrinogen
from the same species. He allowed as that was interesting,
but, he remarked, weren't there some people in Sweden doing
much the same thing? He directed me to an article in a recent issue of Acta Chemica Scandinavia: "Studies on Fibrinopeptides from Different Species" by Birger Blombäck and
John Sjöquist (1960). The comparisons included the first
4-5 residues from the amino-termini of peptides from man,
pig and rabbit, these being compared with the complete bovine sequence that had recently been reported by workers at
the N.I.H., and also, the paper noted, independently by
them in Sweden. At least they hadn't been studying any
fish! They had determined their sequences by the "Edman
method", reference being made to some papers by Sjöquist
with regard to the methodology.

At that point I was just completing my Ph.D. dissertation on evolutionary aspects of blood clotting. Accordingly, I wrote to Blombäck to find if it would be possible to
come to Sweden in a postdoctoral capacity. I explained
that I had purified a large quantity of fibrinogen and fibrinopeptides from a primitive fish (the lamprey), and I was
anxious to determine the amino acid sequence of the latter.
In time I received a letter from Staffan Magnusson, also in
the same laboratory at Karolinska Institutet, explaining
that Blombäck was presently in Australia, but that Professor
Jorpes had said I was welcome in their department whenever
I wanted to come. Magnusson also noted that John Sjöquist,

the other half of the team, was newly arrived in Boston, where he was to spend a year or so with Vernon Ingram at M.I.T., and I should contact him also. Indeed, Sjöquist was invited by my mentor at Harvard, J. L. Oncley, to sit in on my final Ph.D. exam, and as a result I came to know him well.

And so, in September, 1962, aided by an N.I.H. fellowship, I began two years in Sweden, the first year and a half of which would be in Stockholm, and the last six months in Lund with Sjöquist. Upon my arrival in Stockholm I found the laboratory all abuzz about the "new Edman method" that the Blombäcks had learned in Australia. In fact, Birger and Margareta Blombäck had gone to Australia to install a new blood fractionation operation for the Commonwealth Serum Laboratories. But while they were there, they naturally looked in on Pehr Edman, who had emigrated to Australia from Sweden in 1957. Edman had taken his degree with Professor Jorpes at Karolinska Institutet back in 1945, and the Blombäcks were aware of his work. He told them that he had been improving his method very much, and they in turn suggested that they had some materials that were probably very appropriate for him to try his new method on: fibrinopeptides. In fact, Blombäck sent back to Sweden to have one of his technicians make the journey to Australia, and also to prepare a vast quantity of human fibrinopeptides.

Now, the fibrinopeptides come in two kinds, A and B, and in the case of human the A has 16 residues and the B 14 residues. They had quickly managed to obtain the sequence of the A peptide in a single stepwise degradation that went 15 cycles and left free arginine, the 16th residue, sitting alone in the reaction tube. It was an incredible feat, worthy of much more than the brief report they submitted to Nature. Unfortunately, the human fibrinopeptide B did not yield to the new method, as it had a recalcitrantly blocked amino-terminus.

The Blombäcks had returned from Australia only a week or two before my arrival. Other people working in Professor Jorpes's department at the time included Viktor Mutt, who was sequencing secretin, Staffan Magnusson, who was purifying prothrombin, and Agnes Henschen, who, as a student of Blombäck's, was already characterizing fibrinogen. All of them were anxious to learn the "new" method so they could apply it to their own work.

I had two assignments before we would actually begin
the sequencing of various fibrinopeptides. First, I was
asked to investigate the nature of the blocked amino-terminus in the bovine fibrinopeptide B. Workers at the N.I.H.
had reported that it was an acetylated threonine, but the
Swedish workers could find no evidence of acetyl release
upon mild hydrolysis, and Sjöquist was quite certain that
the single threonine was located four residues into the
chain. It was their thought that the block was most likely
due to a cyclized glutamic or glutamine (pyrrolidone ring).
Eventually I was able to demonstrate that they were correct
in that surmise, and I was actually able to open the ring
by mild alkaline hydrolysis with a yield sufficiently good
that we were able to initiate stepwise degradations. This
was subsequently put to good advantage in sequencing other
fibrinopeptides B that were similarly blocked. My second
assignment was to prepare vast quantities of fibrinopeptides
from sheep, goat, rabbit and pig. The list was later extended to include reindeer, the result of a mid-winter ex-
pedition that Blombäck and I took to the north of Sweden
where the Laplanders have an annual reindeer roundup.

Meanwhile, the Blombäcks were busy laying in reagents
and setting up the equipment that would be used in the new
procedures. By Christmastime all was ready, and on Dec.
28th we began the first run on sheep fibrinopeptide A,
which has 18 residues. We proceeded at about a step per day,
and, according to my notebook, by mid-January we had completed the 17th cycle. Amino acid analysis of the material
remaining in the reaction tube revealed mostly free arginine, the carboxy-terminal residue. This was certainly an
impressive demonstration of the power of the new method.
For my own part, having had very little prior sequencing experience, it was difficult for me to figure out just what
the big innovation was. Why had Sjöquist only been able to
determine the first 4-5 residues on comparable peptides?
Why didn't everyone just go 18 steps down an 18 residue
peptide?   What was the new chemical magic?

Actually, there were a number of factors, not all of
them having to do with the underlying chemistry. But two
things were clear in comparing our efforts with the earlier
work of others. First, we started with a very large amount
of material. In the case of sheep A (mol. wt. = 2,000),

we began the degradation with 20 mg of pure peptide, or
about 10 micromoles. Secondly, the protocol emphasized exceeding care in all operations in order to minimize losses.
Finally, in this particular case (sheep A), the peptide had
a "good" sequence for degradation: no serine or threonine
(or histidine) and the single arginine was at the carboxy-
terminus where it served as an anchor for holding the peptide behind in the aqueous phase during organic extractions.
Beyond that, there were indeed some novel features that I
will describe shortly.

As soon as the sheep A was completed, we went on to the
sheep B, and then the pig fibrinopeptides, and then several
others. The reindeer B had a blocked amino-terminus, but I
was able to open the pyrrolidone ring of a big enough fraction that we were able to initiate a stepwise degradation
and move in close to the first tryptic split. And by June
we had enough data to formulate the first fibrinopeptide
phylogeny.

At that point the Blombäcks went off to a blood clotting meeting at Glen Eagles, Scotland. They had been on
the organizing committee of the meeting, and they had thus
been able to arrange for Edman to come from Australia to
give a major address on his work. He stunned the audience,
they later told me, by reporting that he had constructed a
machine to do all these operations automatically; on its
maiden run they had managed to proceed 35 steps into a myoglobin sequence, all residues being successfully identified.
A tour de force by anybody's criteria.

Actually the Blombäcks had learned of the machine when
they were in Australia, and now they were readying to make
a return trip to learn more about it, as well as to look
after the blood fractionation operation they had set up.
They left in August, and with that I was free to begin the
sequence of the lamprey fibrinopeptides, which, after all,
was what had spurred my going to Sweden in the first place.
The lamprey fibrinopeptide B is large, by fibrinopeptide
standards, having 36 residues and, as I later found, a carbohydrate cluster. I actually undertook several degradations, but the main effort started with 20 mg, or about 3-4
micromoles. I was able to proceed 22 steps before it was
out of phase to a degree where I was unsure about the identification. I was also able to determine several residues
at the carboxy-terminus by traditional procedures. The

sequence was very interesting in that only a few recognizable features remained in this ancient molecule, including
a sulfated tyrosine and the characteristic carboxy-terminal
arginine. The rest was completely different from the mammalian fibrinopeptides. At any rate, I was happy enough
with the results and felt that I had accomplished what I
had set out to find. By the time Blombäck returned in December with more wondrous tales of the new machine, I was
packed and ready to move south to Lund where I was sched-
uled to work with John Sjöquist.

Sjöquist had been one of Edman's first students and
had contributed greatly to the early development of the
phenylisothiocyanate procedure. The lab he occupied had
been Edman's during the period 1948-1957, Edman having been
given a position at Lund after a postdoctoral stint in the
U.S.A. with Northrup and Kunitz, following the completion
of his dissertation in Stockholm. During my six-month stay
in Lund I learned very much about the subtleties of the sequencing art, and much more, also, about the network of personal events that gave rise to Edman's experiments and, in
addition, his emigration to Australia. Although Sjöquist
had temporarily turned his attention to the sequencing of
tRNA, he was still very interested in peptide sequencing.
Accordingly, we spent much time discussing the new innovations and their underlying rationale.

During this period I was also trying to arrange by
mail some sort of position back in the U.S.A. Through the
good offices of A. Baird Hastings, who had been the chairman of Biological Chemistry at Harvard when I began as a
graduate student, I was able to get an appointment as an
Assistant Research Biologist (in actuality, a senior postdoctoral position) in the laboratory of S. J. Singer at a
newly opened branch of the University of California, at La
Jolla. I arrived early in the summer and began working
with Singer on affinity-labeled antibodies. Although this
work was fascinating in its own right, like most young scientists I was impatient to be on my own and began efforts
to obtain an independent faculty position some place. In a
happy turn of events I was given a position in the Chemistry
Dept. at La Jolla, where I have remained ever since.

It was an ideal time to be at such a place. The masses
of undergraduate students that currently swarm over our
campus were yet to arrive, the primary educational program

at the time being completely research-oriented and devoted
to graduate students. While my laboratory was being out-
fitted, I immersed myself in the literature, and although
my major interest remained biochemical evolution, I was
also drawn by curiosity to the detailed history of events
leading to the "Edman method". The Dept. of Biology had
purchased the expensive three-volume set, "Chemistry of the
Amino Acids" by Greenstein and Winitz (1961), a remarkable
work that covers every conceivable aspect of amino acids
and peptides. In particular Volume II contains a comprehensive review of amino acid sequence procedures as reported
up to 1960. This review, combined with my experiences in
Sweden, is the basis for much of the history spelled out in
the following paragraphs, and should be consulted for many
of the primary references.

The Chemistry of Stepwise Degradations

All stepwise degradations follow the same general
principles. In essence, a coupling reagent is employed
that attaches to the α-amino group (I will not discuss carboxy-terminal procedures); the solution conditions for this
step are usually chosen such that both the coupling agent
and the peptide or protein are reasonably soluble. Typically a pyridine-water solution buffered in the region of
pH 9 is employed. It is naturally important that the reaction go as near to completion as possible, and for that
reason a great excess of coupler is usually provided. The
second step, then, is a laundering process to remove the
excess reagents and by-products, usually by extraction into
an appropriate organic solvent. Care must be taken not to
let the derivatized peptide get away in the process. After
suitable drying, the terminal residue is cleaved by an acid-
catalyzed rearrangement, following which the derivatized
amino acid is extracted away from the parent chain. The
removed residue must be identified, either directly or by
examination of an aliquot of the parent peptide to see what
has disappeared. For most of the remainder of this article
I emphasize three aspects of the process: (a) choice of
coupler, (b) cleavage conditions and (c) identification
schemes, for it is in these areas that most of the innovations have been claimed over the years. There are some
less obvious features, also, that can transform an ordinary
procedure into a sensational process, and I will comment on
some of these as I go along.

Table 1
Some Coupling Reagents Used in Stepwise Degradations

⬡–N=C=O	Abderhalden & Brockmann (1930)
S=C=S	Leonis (1948); Levy (1950)
⬡–N=C=S	Edman (1949)
$CH_3-CH_2-O-\overset{\overset{S}{\|\|}}{C}-S-CH_3$	Kenner & Khorana (1952)
$CH_3-\overset{\overset{O}{\|\|}}{C}-NH-\overset{\overset{S}{\|\|}}{C}-S-CH_3$	Elmore & Toseland (1954)
⬡–$\overset{\overset{S}{\|\|}}{C}-S-CH_2-COOH$	Barrett (1967)
$CH_3-\overset{\overset{S}{\|\|}}{C}-S-CH_2COOH$	Mross & Doolittle (1971)

The first legitimate stepwise degradation of a peptide was reported by Abderhalden and Brockmann in 1930, who successfully degraded the tripeptide alanyl-glycyl-leucine step-by-step. The coupling agent used in this pioneering effort was phenylisocyanate (Table 1). Three years earlier Bergmann and his co-workers (1927) had observed the labilizing effect of this agent when coupled to the α-amino-group of a peptide, but they had not appreciated the chemistry of the process or the potential it had for the progressive removal of subsequent residues. Thus, they had removed the "activated" residue with aqueous HCl, noting only that it could be "hydrolyzed" under conditions where the other peptide bonds remain intact. Abderhalden and

Brockmann (1930), on the other hand, recognized that hydrolysis was not involved, and as such they used methanolic HCl for the cleavage (Table 2). Moreover, their conditions were really quite mild, even by present standards, a temperature of 60-65^0 being used for half an hour. As for identifying the residues that were cleaved off, they crystallized them by the addition of water after having removed the methanolic HCl under reduced pressure. The method was plainly ahead of its time, if only because of the sophistication required to identify the crystalline derivatives removed at each step. But they knew exactly what they were doing, and they referred to the enterprise as, Stufenweiser Abbau von Polypeptidderivaten--stepwise degradation of derivatized polypeptides.

Peptide and protein chemistry was rejuvenated after the end of World War II, one of the great catalytic events being the introduction of paper chromatography, which allowed simple, if qualitative, amino acid analysis. The revival of endgroup labeling procedures by Sanger and his colleagues also contributed to an air of optimism that the detailed chemical structure of proteins might be determined. A number of laboratories were attempting to devise procedures for the removal of terminal residues sequentially, but it was Edman's brilliant choice of phenylisothiocyanate that was to shape the course of 30 years of unbelievably successful structure determination.

Phenylisothiocyanate had been widely available for over 60 years and had been used by Aschan for preparing amino acid derivatives as early as 1883. And yet it was overlooked as a degradative agent until Edman set about using it in 1948. It is a much superior reagent to the phenylisocyanate used by Abderhalden and Brockmann (1930), inasmuch as sulfur is a much better nucleophile under the circumstances than is oxygen. This is because the lesser overlap of p orbitals is a discouragement to double bond formation in the case of sulfur, and that attribute outweighs the greater electronegativity of the oxygen atom. The mechanism is strictly parallel in the two instances, contrary views not withstanding, cyclization and cleavage being achieved as a result of the attack of the sulfur or the oxygen on the carbonyl carbon of the peptide bond holding the terminal residue to the parent chain. The material released is in the form of a 5-membered thiazolinone (sulfur) or oxazolinone (oxygen). Plainly, water is not

Table 2

Some Reagents Used for Cleavage in Stepwise Degradations

Methanolic HCl	Abderhalden & Brockmann (1930)
Nitromethane/HCl	Edman (1950)
Dioxane/HCl	Fox et al. (1951)
Aqueous HCl	Fraenkel-Conrat & F.-C. (1951)
Aqueous citric acid	al. et Linderstrom-Lang (1952)
Acetic acid/HCl	Edman (1953)
Acetic acid/HCl vapors	Fraenkel-Conrat (1954)
Trifluoroacetic acid	Elmore & Toseland (1956)
Heptafluorobutyric	Edman & Begg (1967)
Pentafluoropropionic	Inglis (1976)

involved in the process. In both cases, however, the derivative rearranges in the presence of water into the ther-
modynamically more stable (thio)hydantoin (Fig. 1). The
reaction likely proceeds via the (thio)carbamyl form that
results from hydrolysis of the (thi)azolinone.

Originally it was not at all appreciated that hydantoin formation proceeded via of an intermediate, and no
conscious effort was made to separate the two reactions.
This is an important point, since the conditions favoring
one or the other of these reactions are different. When
I say "conscious effort", the implication is that unknow-
ingly the reactions were sometimes separated in time if not
in space. Thus, Abderhalden and Brockmann (1930) cleaved
the terminal residue with methanolic HCl, then took the
system to dryness and re-dissolved the residue in water, at
which point the conversion to the hydantoin likely took
place, leading to the crystallization of that derivative
and its separation from the parent peptide. Similarly,
Edman (1950) used HCl-saturated nitromethane to cleave his
coupled peptides, after which he evaporated the solvent and
dissolved the gemisch in water prior to extraction with
ethyl acetate. The conversion to the thiohydantoin doubt-
less took place in the aqueous part of the operation. As
we shall see, it was the separation of the two reactions in

space as well as time that became the key to successful extended degradations.

Edman reported his successful use of phenylisothiocyanate at the 1st International Congress of Biochemistry held at Cambridge, England in 1949, and the news quickly reached every peptide and protein laboratory in the world. As it happened, one of the Meccas of protein chemistry was the Carlsberg Laboratory in Copenhagen, a mere hour and a half away from the south of Sweden by ferry boat, and the connection between that lab and Lund was at once strengthened. Indeed, Linderstrom-Lang's group at Carlsberg had furnished Edman with some of the small peptides that he used in his experiments. The people at Carlsberg immediately began experiments with phenylisothiocyanate, and as experimentors are wont to do, they incorporated some changes of their own. The most controversial of these was the introduction of aqueous cleavage reagents (Table 2). They were driven to this innovation by the fact that the larger peptides and proteins they so eagerly wanted to study were not soluble in nitromethane. In fact, much to Edman's annoyance, their degradations with aqueous HCl went quite smoothly. Edman was annoyed because the chemistry of the cyclization reaction plainly did not involve water, and he realized that extended sequences could never be obtained unless every precaution was taken to prevent the random hydrolysis of peptide bonds in the parent material. Many workers shrugged off this concern and audaciously sequenced peptides and proteins without worrying about the mechanism. In at least one case, which I will describe shortly, this would result in a disaster.

It was clear from the start that the stepwise degradation could be combined with other approaches which made it unnecessary to identify the cleaved derivative directly. In 1954, for example, E. O. P. Thompson determined the amino terminus of "serum" (now plasma) albumin with the Sanger reagent (fluorodinitrobenzene). He then subjected the albumin to a degradative cycle with phenylisothiocyanate and performed a second endgroup determination with the fluorodinitrobenzene. By this route, which was the forerunner of the Dansyl-Edman that would be developed by Gray and Hartley in the 1960's, he was able to establish the beginning sequence Asp-Ala for the human protein and Asp-Thr for the bovine.

Alternatively, in the case of peptides one could perform a stepwise degradation and at each step remove an aliquot from the peptide material remaining and do a quantitative amino acid analysis to see what amino acid disappeared.
This "subtractive" procedure was used extensively by Moore
and Stein during the late 1950's on their way to determining the sequence of ribonuclease. Unfortunately they were
among those who weren't paying heed to Edman's concerns a-
bout mechanism and the potential problems of misapplied
cleavage systems.

In 1956 Edman published a paper showing that the material initially cleaved is in fact a thiazolinone and that
conversion to the thiohydantoin occurs subsequently and is
facilitated by the presence of water. But there was no hint
in this paper that it might be advantageous to separate the
two reactions physically. In fact, in a 1957 paper with
Heirwegh, Edman himself used extremely harsh aqueous cleavage conditions (1N HCl at 100^0 for an hour) in an endgroup
determination on pepsin, although it was clear that they
had no intention of undertaking an extended degradation with
these conditions.

At about this time Elmore and Toseland (1956) experi-
mented with trifluoroacetic acid (TFA) as a cleavage agent.
They were attracted to this reagent by a report that TFA
was an excellent solvent for proteins (Katz, 1954). Independently, as far as I can tell, Edman (1957) began using
TFA in his first experiments after moving to Melbourne, successfully determining the first four residues of pepsin.
The circumstances of the cleavage were exceptionally mild,
neat TFA being employed at 0^0 for only 5 minutes. No description was given of how the released derivative was extracted.

By this time, E. O. P. Thompson was also in Melbourne,
and it is he who apparently first suggested in print that
stepwise degradations would be better conducted if the conversion to the phenylthiohydantoin was conducted altogether
separately (Thompson, 1960). He also pointed out that for
those procedures that depend on following the disappearance
of an amino acid as opposed to direct identification of the
PTH-amino acid, i.e., Sanger-Edmans or the "subtractive"
method, harsh conditions are simply not required, since it
is immaterial whether or not the cleaved derivative is converted to another form. These remarks were contained in a

lengthy review of sequencing procedures, and by the time they appeared in print it was too late to help the Rockefeller team in their studies on bovine pancreatic ribonuclease. Thus, in 1960, Hirs, Moore and Stein reported the sequence of residues 11-18 to be:

Ser-Thr-Ser-Ser-Asn-His-Met-Glu.

As it happened, Gross and Witkop were in the process of devising a procedure for the chemical cleavage of proteins at methionine residues, and they were naturally drawn to ribonuclease as a methionine-containing protein whose sequence had been reported. Unfortunately, the composition of the fragments they obtained did not correspond to the reported sequence. Either an unusual rearrangement was occurring as a result of the cyanogen bromide treatment they were performing (Gross and Witkop, 1962), or the reported sequence was in error. It was the latter. In fact, the correct sequence, subsequently reported by Smyth, Stein and Moore (1962) was:

Gln-His-Met-Asp-Ser-Ser-Thr-Ser.

The question arises, how could such an extraordinarily different sequence have been determined? In the end it was the Edman degradation that took most of the blame, and in particular the harsh cleavage conditions used in the determination of the originally reported sequence.

What had occurred, of course, was that the glutamine at position 11 had cyclized during the course of peptide purification and could no longer be derivatized by either the Sanger reagent or phenylisothiocyanate. As a result, everything inflicted thereafter on this innocently blocked peptide was an artifact. The authors had performed a stepwise degradation using the subtractive method, and what they found was that after one cycle the amount of serine was down by about one residue, after two cycles the threonine was almost gone, after three cycles, the serine was further diminished, and so on. In fact, the serine and threonine were being progressively destroyed during their needlessly harsh cleavage conditions (glacial acetic saturated with HCl, 100°)! Beyond that they were caught in a tangle of unfortunate events. They had subjected the peptide to digestion with leucine aminopeptidase, but the enzyme obviously could not attack a blocked amino-terminus. Unfortunately the preparation they used contained small

amounts of carboxypeptidase, and the amino acids released
came from the wrong end of the sequence.

But then, in a classic non sequitur, the Rockefeller
group reported a set of experiments that showed that the
use of harsh cleavage conditions could result in the cyclization of glutamines, even though their particular glutamine had been cyclized long before the stepwise degradation
was ever undertaken! To make matters worse, it has to be
noted that Anfinsen and his co-workers had previously isolated the corresponding peptide under a different set of
conditions and found the amino-terminal to be either glutamic acid or glutamine; their isolation procedures had not
been conducive to pyrrolidone formation. But their result
had been ignored.

In 1960 Edman published an outline of the principles
involving the clear separation of cleavage and conversion
events. In an article in Annals of the New York Academy of
Science, he spelled out his three stage method: coupling,
cyclization (cleavage), and conversion. The key element
was to extract the thiazolinone away from the parent peptide after a mild anhydrous cleavage, and then convert it
into the thiohydantoin under more vigorous conditions in an
aqueous environment in a separate vessel. This was the
heart of the "new" method that the Blombäcks brought back
from Australia in 1962 and it reached a dramatic climax in
1967 when Edman and Begg described the spinning cup sequencer; by this time they had managed to proceed 60 steps into
a myoglobin sequence. The truly extended sequence procedure had arrived. Harsh cleavage conditions were now a
matter of history. Further improvements in the methodology
of stepwise degradations would have to come on other fronts.

Some Attempts at Alternatives

In the summer of 1966 George Mross came to work in my
laboratory as my first graduate student. Mross had been a
zoology major as an undergraduate, and he was attracted by
the prospect of formulating phylogenies based on amino acid
sequences. At the time I was busily collecting blood from
every creature the San Diego Zoo would let me near, particularly artiodactyls, the group that includes pig, deer,
sheep and cattle. The object was to derive as completely
as possible a step-by-step amino acid replacement scheme
for the diverse fibrinopeptides of a single group with a

well established fossil record. At the outset I explained
to George that the problem we were going to encounter was
that we would have relatively small amounts of peptide
(Persian Gazelle blood is not easy to come by), and as a result we could anticipate trouble with the Edman method.
More often than not we would get half-way down a sequence,
and the remainder would be washed out during the extraction
steps. We could also count on there being equivocal results
in the interpretation of chromatograms, the spots on thin
layers and papers occasionally going awry for unknown reasons. In an effort to bolster his chemical background, I
suggested that Mross attempt to solve these problems in two
ways. First, develop a solid phase procedure similar to
what workers down the hall were doing with Merrifield peptide synthesis. Secondly, come up with a stepwise degradation that regenerated the free amino acid so that we could
identify it quantitatively on the very reliable Moore and
Stein analyzer we had recently inherited. I turned him
loose on Greenstein and Winitz (1961) and went about other
matters myself, tackling especially the problem of finding
an enzyme that could unblock peptides or proteins with terminal pyrrolidone rings. A few months later George reported
to me that he had solved the solid phase procedure without
lifting a test tube: there it was, right in the November
20th issue of J. Amer. Chem. Soc., "A Solid-State Edman
Degradation" by Richard A. Laursen (1966). The good news
was that George could devote all his time now to a novel
degradation scheme.

At about the same time that Edman introduced phenylisothiocyanate as a degradation reagent, other investigators
were attempting to do the same thing with other reagents.
Levy (1950), following the lead of Leonis (1948), demonstrated that carbon disulfide could be coupled to the α-amino
groups of peptides and the terminal residue then removed
under acid conditions. In this case the reaction obviously
proceeded by way of a thiazolinone, since there was no opportunity to form the equivalent of a hydantoin. The derivative was readily hydrolyzed to the thiocarbamyl amino
acid upon exposure to water. Other workers explored equivalent reactions with dialkylxanthates (Table 1), and somewhat later dithioic acids were used in the same way. All of
these agents had one distinct advantage: the free amino
acids could be regenerated from the removed derivative under
very mild conditions. After considering the lot, Mross

synthesized several dithioic acids for use as coupling
agents; of these we mostly used a compound called thioacetylthioglycolic acid (TATG). For his dissertation, Mross explored and developed conditions for the effective use of
TATG in stepwise degradations, especially in conjunction
with peptides attached to solid supports (Fig 1).

Our first major success with TATG was achieved in
studies with another of my graduate students, Renne Chen.
For her thesis, Renné was isolating and characterizing those
parts of fibrin that become covalently cross-linked by factor XIII. She had laboriously isolated a set of cross-
linked peptides, as well as the corresponding set of non-
cross-linked peptides, all from the carboxy-terminus of the
γ-chain. On several occasions she had attempted to sequence
the major tryptic peptide by the Edman method, but each
time the degradation peetered out somewhere in the middle
of the sequence. After Mross had worked out conditions, we
were able to attach some of that precious material to polystyrene using a water-soluble carbodiimide, and using the
TATG procedure we were able to degrade the entire 15-resi-
due peptide right to the end on the first try, successfully
identifying every residue (Chen & Doolittle, 1971). It was
a very satisfying moment.

The method was not without its drawbacks, however.
For example, even though serine and threonine have always
been trouble for the Edman method, whether used in a direct
or indirect identification manner, the problem is exacerba-
ted in methods that depend on the regeneration of free amino acids. Tryptophan residues are also a problem, since
that residue is readily destroyed by even mild acid hydrolysis. But the principle drawback, and the most likely reason that this method has never caught on among other workers, is that the TATG reagent is not commercially available
and must be synthesized. Worse, it may be the foulest
smelling substance on earth.

Still, we ourselves have used the method to great advantage. We constructed a simple machine at very little
expense, and it functioned marvelously well. By this time
we were deeply engaged in determining the sequence of the
human fibrinogen molecule, and our home-made machine soon
became indispensable to that effort. We were able to proceed more than 40 steps on several large cyanogen bromide

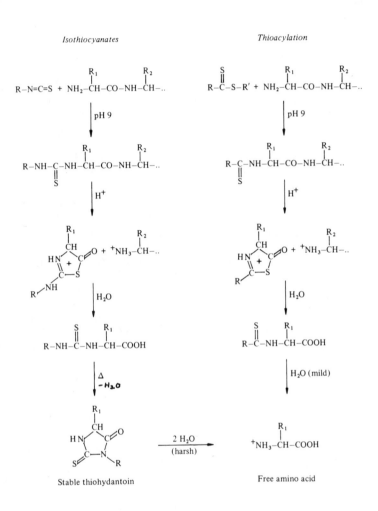

Fig. 1. Comparison of degradative reactions with phenylisothiocyanate and thioacetylthioglycolic acid (TATG). Note
that thioacetylation procedure does not include the equivalent of a hydantoin; regeneration to the free amino acid
being readily achieved instead.

fragments and some other large peptides. Indeed, the TATG
method was the mainstay of efforts that led to the completion of the sequence (Doolittle et al., 1979).

Recent Advances in Stepwise Degradation Procedures

The last decade has seen a number of technological advances in the stepwise degradation field, the most notewor-
thy result of which is the reduction in the amount of sample required. Abderhalden and Brockmann (1930) began their
degradation of Ala-Gly-Leu with 1-2 millimoles. In his
early studies Edman (1950) also used 1-2 millimoles, but by
1954 Fraenkel-Conrat had scaled the method down to where he
was using less than 0.5 micromoles of peptide, a remarkable
advance. By contrast, the extended degradations on the fibrinopeptides were most successful when 3-10 micromoles
were available. Edman and Begg (1967), using their automatic machine, had achieved those spectacular results on myoglobin with only 300 nanomoles. By the 1970's most workers
had pushed into the range of working between 10-100 nanomoles, whether employing traditional phenylisothiocyanate
methodology or more exotic procedures like the TATG method.
Most recently, there have been reports of studies in the 100
picomole-10 nanomole range (Hunkapiller & Hood, 1980;
Hewick et al., 1981). How have these improvements been made?

There are two major factors, neither of them having to
do with the underlying chemistry of the stepwise degradation. First, the engineering of the automatic devices has
been greatly improved, improving the overall efficiency of
the various operations. Secondly, and more importantly,
identification procedures for PTH-amino acids have been developed which are nothing less than remarkable. I have already commented on the inordinate difficulty of identifying
compounds by crystallization and elemental composition, as
was required in 1930. In 1950 Edman used paper chromatography, but only after back-hydrolyzing the PTH-derivatives
to free amino acids with $Ba(OH)_2$ (Table 3). This is a very
inefficient method. By 1953 Sjöquist, and independently
Landmann et al., had devised a scheme for the paper chromatography of the PTH-amino acids themselves. Sjöquist's
system could detect 50 nanomoles of PTH-amino acid, and it
was this fact, combined with an ingenious adaptation of the
degradation procedure, that allowed Fraenkel-Conrat to drop
three orders of magnitude with regard to the amount of

Table 3
Some Schemes Used for Identifying PTH-Amino Acids

"Back Hydrolysis" $\bar{c}$ Ba(OH)$_2$ Edman (1950)

Paper chromatography   Sjöquist (1953)
                                  Landmann et al. (1953)

Celite column   Sjöquist (1955)

Thin layer chromatography   Blombäck, Edman (1964)

Gas Liquid chromatography   Pisano et al. (1962)

Mass spectrometry   Hagenmeyer et al. (1970)

Reverse phase HPLC   Zimmermann et al. (1973)

---

material needed. It is interesting to note in passing that Fraenkel-Conrat was a postdoctoral fellow in Lang's Copenhagen laboratory in 1950 when Edman was getting under way, and that he was part of the original aqueous cleavage controversy.

The introduction of thin layer chromatography further lowered the amount of material required, and gas chromatography made available a quantitative method in the same sensitivity range (Table 3). But it was the development of reverse phase high performance chromatography (HPLC) in the 1970's that provided the biggest advances. PTH-amino acids can now be detected in the 1-10 picomole range on HPLC systems that can separate all of them in less than 10 minutes.

Of course the increase in the sensitivity of the detection devices meant that the degradative schemes themselves had to be conducted under conditions where background material and contaminants were proportionately reduced. The solvents employed had to be exceedingly pure, and the materials of the various reaction vessels and lines completely inert. Higher vacuum had to be used to remove last traces of volatile components. All of the latter have been dealt with effectively by the engineers of the field. The "clean machines" are impressive, indeed.

There have been some other useful innovations besides better detection devices, of course. Different coupling

reagents have been employed, for example, either for easier
detection or for providing better anchors in the aqueous
phase during the extractions. They have included radioactive, fluorescent and colored isothiocyanate reagents. A-
mong the most interesting has been diaminobenzilisothiocy-
anate (DABITC), a colored reagent introduced by Chang et al.
in 1976 (Table 4). The thiohydantoins produced with this
coupling agent are highly colored and can be identified on
polyamide sheets when applied in picomolar amounts. Interestingly, this was not the first color-yielding isothiocyanate to be used in stepwise degradations. Reith and Waldron, as long ago as 1954, synthesized a reagent that also
yielded brightly colored thiohydantoins (Table 4). Why was
that reagent never used by other investigators, whereas the
DABITC approach is one of the most popular in current use?
One answer may be that it was simply ahead of its time. But
another has to do with another of those simple innovations,
not unlike the separation of the cyclization and conversion
steps. Thus, these bulky reagents are slower to react than
phenylisothiocyanate, and it is difficult to get the reaction to approach completion. As a result, degradations get
out of phase after a few steps. Chang and co-workers (1978)
solved this problem and surmounted another obstacle simultaneously. First they coupled the peptide with only a modest excess of the colored coupling reagent. Then they added a great excess of phenylisothiocyanate in the conventional manner, thereby driving the reaction to completion.
At the same time, the excess reagent that is the source of
much background problem is colorless, whereas the released
amino acid derivatives to be identified are colored. This
is a genuine micro-method that can be executed by anyone.
No high technology is required.

The new sequencing machines are magnificently engi-
neered and miniaturized, but none of the principles employed
are new. In fact, Fraenkel-Conrat invented Gas-Liquid-Solid
Phase sequencing in 1954. He applied some peptide (less
than 0.5 micromoles) to a piece of filter paper (solid
phase), immersed it in appropriate coupling and washing solutions (liquid phase), and cleaved the terminal residues
by exposing the paper to HCl and acetic acid vapors in a
desiccator under reduced pressure (vapor phase). Using
this simple procedure, he was able to determine the first 14
residues of a peptide. This is not to minimize the marvel-
ous results of the new machines; just to keep matters in
perspective.

Table 4
Some Alternative Isothiocyanate Coupling Reagents

Structure	Reagent
⬡—N=C=S	Edman (1950)
CH_3, CH_3—N—(NO₂, NO₂)—N=C=S	Reith & Waldron (1954)
naphthyl—N=C=S	al. et Yamashina (1965)
CH_3-N=C=S	Laursen (1966)
SO_3^-—⬡—N=C=S	Birr et al. (1970)
CH_3, CH_3—N—⬡—N=N—⬡—N=C=S	Chang et al (1976)

In my view, then, the major advances in the development of stepwise degradations have fallen into two categories: the simple innovations that resulted from thoughtful analysis of the underlying chemistry, on the one hand, and those improvements that have resulted from increasingly precise engineering, on the other. The substitution of a sulfur atom for an oxygen, the conducting of all the operations on a piece of filter paper, the separation of the cyclization and conversion reactions in space and time, the attachment of

peptides to solid supports, the combination of an inefficient but colored coupling agent with an efficient but colorless one--these are the elegant contributions of the first
kind.

By chance the August 10th (1981) issue of the J. Biol.
Chem. arrived at our campus only a few days after the annual
announcement from the N.I.H. that institutional applications
were being solicited for large equipment items. It was not
surprising, then, that I was accosted by one of my colleagues
shortly thereafter.

"See here," he exclaimed, "we've got to put into the
N.I.H. and get one of these new Gas-Liquid-Solid Phase Sequencing machines."

"What on Earth for?" I asked, musing to myself about
Fraenkel-Conrat and his paper strips.

"You know that my research involves sorting out the
2,000 or so proteins that can occur in a eukaryotic cell,
and I can only resolve them on two-dimensional gels. Well,
these guys can actually remove one of those spots from a gel
and sequence it through 90 amino acids!"

That did seem impressive. "Here, let me see the article. Now wait a minute, the 90 steps is on myoglobin, which,
although remarkable, I'm sure didn't come off a gel. Here
in this table, let's find one that was eluted from a gel.
Melanoma tumor antigen: 13 steps. Not bad, I wonder if
they identified them all. What do you estimate that is in
dollars per residue?"

Plainly I was dragging my feet. The financial commit-
ment to a major new sequencing facility, even if the N.I.H.
came up with the money for such a machine, is not to be tak-
en lightly. My counsel to my colleague was, given the kinds
of needs he has, why not go with one of the elegant manual
procedures like DABITC? The cost investment is virtually
nil, and enough sequence data can certainly be obtained so
that suitable oligonucleotide probes can be synthesized and
the corresponding genes cloned and sequenced directly. Ma-
cines are not for everyone.

My colleague and others like him were not convinced. They want a machine. They pointed out that we are probably the only major university in California, perhaps in the country, that doesn't have a spinning-cup sequencer. Well, I wish them luck with their application, and I hope none of their spots have blocked amino-terminals. In fact, I hope they are all as well suited to automatic sequencing as is myoglobin. With that, I'd like to close with a quote from someone far afield from stepwise degradations:

"It seems to me that the endless multiplication of mechanical aids in fields which require judgement more than anything else is one of the chief dynamic forces behind Parkinson's Law."

E. F. Schumacher in Small is Beautiful

References

Abderhalden, E. & Brockmann, (1930) Biochem. Z., 225:386.

Barrett, G.C. (1967) Chem. Comm., 487.

Birr, C., Reitel, C. & Wieland, T. (1970) Ang. Chem. (Int), 9:731.

Blombäck, B. & Doolittle, R.F. (1963) Acta Chem. Scand., 17:1819.

Blombäck, B. & Sjöquist, J. (1960) Acta Chem. Scand., 14:493.

Blombäck, B., Blombäck, M., Edman, P. & Hessel, B. (1962) Nature, 193:1184.

Chang, J.Y., Creaser, E.H. & Bentley, K.W. (1976) Biochem. J., 163:607.

Chang, J.Y., Brauer, D. and Wittmann-Liebold, B. (1978) FEBS Letters, 93:205.

Chen, R. & Doolittle, R.F. (1971) Biochemistry, 10:4486.

Doolittle, L.R., Mross, G.A., Fothergill, L.A. & Doolittle, R.F. (1977) Anal. Biochem., 78:491.

Doolittle, R.F., Watt, K.W.K., Cottrell, B.A., Strong, D.D. & Riley, M. (1979) Nature, 280:464.

Edman, P. (1950) Acta Chem. Scand., 4:283.

Edman, P. (1956) Acta Chem. Scand., 10:761.

Edman, P. (1957) Proc. Roy. Aust. Inst., 24:434.

Edman, P. (1960) Ann. N.Y. Acad. Sci., 88:602.

Edman, P. & Begg, G. (1967) Europ. J. Biochem., 1:80.

Elmore, D.T. & Toseland, P.A. (1956) J. Chem. Soc., 188.

Fraenkel-Conrat, H. (1954) J. Am. Chem. Soc., 76:3606.

Greenstein, J.P. & Winitz, M. (1961) Chemistry of the Amino Acids, Vol. 2, pp. 1512-1687. John Wiley, New York.

Gross, E. & Witkop, B. (1962) J. Biol. Chem., 237:1856.

Hagenmeyer, H., Ebbighausen, W., Nicholson, G. & Vötsch, W. (1970) Z. Naturforsch., 25B, 681.

Heirwegh, K. & Edman, P. (1957) Biochim. Biophys. Acta, 24: 219.

Hewick, R.M., Hunkapiller, M.W., Hood, L.E. & Dreyer, W.J. (1981) J. Biol. Chem., 256:7990.

Hirs, C.H.W., Moore, S. & Stein, W.H. (1960) J. Biol. Chem. 235:633.

Hunkapiller, M.W. & Hood, L.E. (1980) Science, 207:523.

Inglis, A.S. (1976) J. Chromatogr., 123:482.

Katsuki, S., Scott, J.E. & Yamashina, I. (1965) Biochem. J., 97:25c.

Katz, J.J. (1954) Nature, 174:509.

Kenner, G.W. & Khorana, H.G. (1952) J. Chem. Soc., 2076.

Laursen, R.A. (1966) J. Am. Chem. Soc., 88:5344.

Landmann, W.A., Drake, M.P. & Dillaha, J. (1953) J. Am. Chem. Soc., 75:3638.

Leonis, J. (1948) C. R. Trav. Lab. Carlsberg, 26:315.

Levy, A.L. (1950) J. Chem. Soc., 404.

Mross, G.A. & Doolittle, R.F. (1971) Fed. Proc., 30:1241.

Pisano, J.J., Vanden Huevel, W.JA. & Hornig, E.C. (1962) Biochem. Biophys. Res. Comm., 7:82.

Reith, W.S. & Waldron, N.M. (1954) Biochem. J., 56:116.

Schumacher, E.F. (1973) Small Is Beautiful, Blond & Briggs, Ltd., London

Sjöquist, J. (1953) Acta Chem. Scand., 7:447.

Sjöquist, J. (1955) Biochim. Biophys. Acta, 16:283.

Smyth, D.G., Stein, W.H. & Moore, S. (1962) J. Biol. Chem., 237:1845.

Thompson, E.O.P. (1954) J. Biol. Chem., 208:565.

Thompson, E.O.P. (1960) Adv. Org. Chem., 1:149.

Zimmermann, C.L., Pisano, J.J. & Appella, E. (1973) Biochem. Biophys. Res. Comm., 55:1220.

Overview

AN EVALUATION OF THE CURRENT STATUS OF PROTEIN SEQUENCING

BRIGITTE WITTMANN-LIEBOLD

Max-Planck-Institut für Molekulare Genetik,

Abt. Wittmann,

D - 1000 Berlin 33 (Dahlem), Germany

In 1975, Richard Laursen organized the first of these conferences on Methods in Protein Sequence Analysis to enable scientists working actively in this field to exchange experience and recent results. Ever since the conferences have attracted protein chemists from all parts of the world. Again, at the present meeting newly developed and advanced methods and new sequences have been presented. The interest in highly sensitive sequencing techniques has increased and many newcomers to micro-sequencing came to this conference.

The success of the meeting was largely due to Marshall Elzinga's efforts. Although it was only decided last year to hold this conference in Brookhaven, excellent facilities were provided for this meeting, held at this most famous nuclear energy research center. The well preserved pleasant surroundings and the excursion to the nearby beach with the "Sunken Forest" provided an excellent atmosphere for the meeting, as had the fine scenery of Heidelberg with its castle, and the bright nights of Montpellier at the earlier meetings.

We are grateful to Lowell Ericsson, Agnes Henschen, Marcus Horn, Richard Laursen and John Walker for their contribution to the organisation of the meeting. The assistance of staff members of the Brookhaven National Laboratory, especially of Helen Kondratuk, is greatfully acknowledged.

We missed Erhard Gross at the meeting. He had been the spirit of the discussions in the preceding conferences. His brilliant contributions and his profound knowledge of peptide chemistry accelerated the exchange and contact between

*the participants. Those of us who had attented the workshop
of protein micro-sequencing held at Prague last year will
never forget his round table discussions. His early and
unexpected death overshadowed this conference considerably.*

INTRODUCTION

In the following the latest progress that has been
made in the methodology of protein and peptide sequence
analysis, as presented during this meeting, is summarized
briefly. The power of new approaches to micro-sequence analysis, the limitations and possible future developments
will be discussed.

Most progress has been made in the design of new micro-
sequencers, in a variety of applications of high pressure
liquid chromatography and in micro-sequence analysis by
mass spectrometry. Present strategies for sequence analysis
employing enzymes and improved chemical cleavages to fragment proteins have been discussed. Finally, the successful
combination of nucleotide and protein sequencing has considerably increased our knowledge of protein and peptide
structures during the past few years. These topics will be
touched upon in this chapter with respect to their value
for sequencing unknown polypeptides that are available in
limited amounts.

When trying to reach a conclusion about the state of
protein sequencing, as revealed at the present meeting,
one should bear in mind the wide variety of polypeptides
under investigation, differing from each other not only in
size (from peptides containing a few amino acid residues
to large proteins of more than thousand residues), but also
in their net charges, hydrophobicity and consequently their
solubility properties. Further, the occurrence of unusual
amino acids generated by post-translational modifications,
such as phosphorylation, acetylation, methylation, glycosylation and the attachment of covalently bound lipids, demand both intuition and flexibility of approach from the
researchers for sequence determinations. The wide variety
of protein structures makes amino acid sequence analysis
more difficult than the nucleotide sequencing of cDNA which
have a more uniform character.

In addition to their structural variability proteins
from biological sources can be available in quantities
ranging from micrograms to grams. This has led to a wide
range of sequencing methods. On the basis of certain re-

quirements of a project (and the attitude of the researcher)
protein chemists can be divided into groups (such as "con-
formists, reservists, nanoists, picoists, and femtoists")
who have good reasons for applying sequencing methods in a
wide range of sensitivity. To reach higher levels of sensitivity, requires the adaptation of a complete set of
methods at this range, e.g. change in the purification of
the polypeptides, new instrumentation, such as analyzers,
sequencers, or mass spectrometers, and the establishment of
more sensitive fragmentation and separation methods for all
types of peptides. Although this is in principle possible,
the adaptation to micro-analysis is expensive and only possible through a learning process enabling the interpretation of the results with the same safety. There is another
drawback to using more sensitive methods: more care is required to avoid background contaminations and a loss of
accuracy. Thus, the feeling of being bound to a certain sequence group, like being a member of an Indian cast, remains and hinders many protein chemists from stepping into
a project which demands another level of sensitivity.

However, the current methodology already provides
enough possibilities for sequencing proteins and peptides
purified in minute amounts from organs (or part of organs),
cell organelles, or tissue cultures. This will be discussed
at the end of this chapter in more detail. In the last
years micro-sequencing has surprisingly opened up a new di-
mension in the study of biological processes on a molecular
level.

METHODOLOGY

Strategies for Protein Sequencing

Various possible strategies for the structure determination of polypeptides were excellently summarized at the
conference (K. A. Walsh, B. Keil, M. Hermodson, and A. Fontana), and will not be repeated here. However, I would like
to discuss a few aspects important for both present and
future protein sequencing projects:

1. Limited proteolysis of native proteins with enzymes
such as trypsin, papain, or subtilisin which cleave only at
a few specific sites allows one to isolate a small number
of large fragments, rather than a large number of small ones.
This approach is usually preferred when sequencing large

proteins, as fragment separation is simplified and automatic Edman degradation techniques may be applied on the
fragments. This assumes that the peptides generated can be
easily separated without having problems with solubility
or low recoveries during purification. Unfortunately, this
is often not the case; many proteins from organelles, membranes or virus coats form rather hydrophobic peptides,
the larger of which are difficult to purify.

2. Restriction enzymes would be useful in protein sequence work which would enable to cut larger proteins at
restricted sites similar to those available for nucleotide
sequencing (K. A. Walsh, B. Keil). Such enzymes might become available in the future by introducing the appropriate
nucleotide sequence for the required peptide into the
structural gene by genetic engineering.

The possibility of cleaving proteins with less frequently applied enzymes, such as thrombin, collagenase, or
proteases from Myxobacter, Armillaria mellea, and a new
enzyme isolated from Achromobacter (cleaving at glycine or
alanine) has been discussed (B. Keil, H. Ponstingl).

3. Chemical cleavages, such as treatment with cyanogen
bromide (cleaving at Met-X peptide bonds), skatol derivatives and iodobenzoic acid and its dichloro-derivatives (at
Trp-X), hydroxylamine (at Asn-Gly), acid hydrolysis (at
Asp-Pro), or partial acid hydrolysis (at X-Asp-X), are
mainly used for generating large fragments. Such treatment
allows polypeptide chains to be fragmented with high speci-
fity at rarely occurring defined sites, and leave the N-termini of the polypeptide chains unblocked. The precautions
necessary to obtain high cleavage yields and to decrease
possible side reactions or side cleavages have been summarized (M. Hermodson, A. Fontana). N-bromo-succinimide, which
cleaves peptide bonds adjacent to tyrosine, cysteines,
methionines, and histidines is less specific and hence not
suitable as a means of primary fragmentation.

The drawbacks of any chemical reaction for cleaving
proteins on a micro-scale can be summarized as follows:

a) the generated fragment mixture is often rather
complex due to incomplete fragmentations. The complexity
increases considerably with increasing number of potential
cleavage sites;

b) the similarities in sequence of overlapping fragments generated (socalled "peptide families") make their
separation more complicated;

c) the fragments, mainly larger ones, are obtained in
variable yields, the hydrophobic of which may give low re-

coveries or are lost completely during purification;

d) the large excess of reagent necessitates its removal from the peptide mixture; further, byproducts of the reaction have to be removed;

e) the different solubilities of reagent and peptides have to be reconciled;

f) the traces of contaminants present in the reagents cause unwanted side reactions.

4. Minute protein amounts limit considerably the approaches which can be applied for sequence determination. In such cases the loss of any large or insoluble peptide during purification cannot be tolerated. With peptide mixtures containing hydrophobic peptides it is often more convenient to test several types of cleavages (e.g. after enzymatic or chemical cleavages) on slab gels, by thin-layer fingerprinting or by HPLC-separation with small quantities of proteins and to select the best ones to generate sets of small peptides, rather than a few large ones. By these means the recoveries can be increased and more suitable peptides for sequencing can be selected, e.g. hydrophobic sequence areas which cause difficulties in Edman degradation can be more safely sequenced in N-terminal areas of peptides where the effects of "overlapping" are less pronounced.

The latter strategy has been successfully applied to sequence ribosomal proteins, many of which contain difficult sequence regions (repetitive or hydrophobic sequences). It was found that many proteins were more quickly sequenced by isolating a large number of small sized peptides rather than the reverse (e.g. see sequence determination of proteins L20, L22, L23, L24 or S2, for references see ref. 1). However, for some ribosomal proteins the isolation of a few large peptides, e.g. for proteins S6 and L21, was the better approach. The largest ribosomal protein of E. coli, S1, which has 557 amino acid residues has recently been sequenced by isolating cyanogen bromide fragments as a first step and then to produce several sets of small peptides (2). Other examples of isolating small peptides are the structure determination of Sindbis and Semliki Forest virus (3). This was done because of solubility reasons and extensive repetitive sequence regions.

Another advantage of producing small peptides derives from the application of the newer isolation procedures, such as thin-layer techniques (described in ref. 4) and high pressure liquid chromatography (see below). Both methods allow complex mixtures of small peptides to be separated with high resolution but are at present less suitable for separating large fragments.

Combined DNA- and Protein Sequencing

Combined nucleotide and protein sequencing has recently become very popular. The modern technology of sequencing polynucleotides serves as a quick means to sequence the
structural genes of proteins once the cloning of the appropriate polynucleotide containing the expressed gene and its
location has been made. Accordingly, many sequences of polypeptides have been indirectly assigned by this technique.
In opposite to the relative tedious sequence analysis of
proteins by stepwise Edman degradation technique the nucleotide sequencing easily predicts the probable amino acid sequence by reading of the sequencing gels. In this way about
100 - 200 nucleotides per gel can be determined. Of course,
the main work lies in the cloning experiments rather than in
the final sequencing. But it often was thought that the
nucleotide sequencing might replace the work on the protein
level in all cases where the required part of the genome is
available.

However, it turned out that both types of sequencing
have kept their value. A combined approach of both, the
nucleotide and protein sequencing proved to be very powerful and is especially facilitated in cases where only limited protein amounts are available. By the combined approach
the search for peptides, their alignment and even their sequencing are easier. In several of the papers presented at
the conference (see for instance the chapters of J. Walker,
A. V. Fowler, and M. Kimura) the disadvantages of performing nucleotide sequencing alone were discussed and the main
points can be summarized as follows:

1. Without knowing the sequence of the expressed protein at the N- and C-terminal ends the correct finding of
the structural gene within the cDNA is not possible.

2. The reading frame has to be controlled by direct
amino acid sequencing.

3. Even in cases where the correct reading frame has
been ascertained errors of incorrect reading of one nucleotide can happen which results in the false prediction of
the corresponding amino acid.

4. At repetitive sequence areas not only sequencing on
the protein level is difficult, the same is true for mono-
tonous nucleotide sequences.

5. The occurrence of insertions within the nucleotide

sequence has to be controlled by direct protein sequencing.
   6. Post-translational modifications of the nascent
polypeptide chain occur frequently which necessitate a
proper amino acid sequence analysis. The nucleotide sequencing alone cannot give the nature of the mature protein.
   In recent years it could be demonstrated that these
post-translational modifications (e.g. phosphorylation,
methylation and acetylation) play an important role in the
biosynthesis of proteins. One example to underline this,
derives from sequencing the E. coli ribosomal protein S12
and its alterations in streptomycin resistant mutants.
There a case was found where one modified aspartic acid
residue of the wildtype has been altered to aspartic acid in
one of the mutants, and this change in the nature of one
amino acid residue (out of more than 7000 for the whole
ribosome) caused the difference in the phenotype. It is
interesting that this replacement is located in a sequence
region where also other alterations are clustered (5).
   In lectures and poster sessions many examples were presented where the combined approach of nucleotide and protein
analysis increased the speed of sequencing methods for both,
the structural gene and the mature protein considerably.
Examples were the sequence analysis of homologous protein
structures, of mutants, and of such polypeptides which were
difficult to obtain. The advances made recently in the separation of peptides by HPLC-techniques and in the minia-
turisation of sequencing methods as well as the application
of new techniques in the mass spectrometric analysis of
peptides now allows protein chemists to compete with nucleotide sequencing in speed and sensitivity. Where only preliminary results are necessary the application of nucleotide and mass spectrometric sequencing methods are very
useful (K. Biemann). Another reason for performing the combined approach of DNA-sequencing with the N-terminal sequence analysis of mixtures of proteins was to gain information on essential cysteine residues in biological systems
(K. Beyreuther).

HPLC-Separation of Peptides and Proteins

   Major progress has been made over the past few years in
separating complex peptide mixtures by HPLC-techniques, but
protein separations on HPLC-columns are still difficult.
Many examples of peptide separations using this technique
were presented during this meeting and demonstrate the high

capacity of HPLC-column chromatography (see for instance
the chapters of J. L'Italien, S. Fullmer, H. Kratzin,
H. Ponstingl, and K.J. Wilson).

Current experience allows us to conclude that reverse
phase separations, employing hydrocarbon supports, such as
C8 or C18 (particle sizes of 5 or 10 μm), at temperatures
of 30 to 50°C and flow rates of 0.5 to 1.5 ml/min, are most
suitable for peptides. Diverse aqueous buffers (made 5 - 50
mmol by the addition of salts) have been employed in linear
gradients with either acetonitrile, n-propanol, iso-pro-
panol, methanol or ethanol as organic modifiers. Although
phosphoric acid, perchloric acid and trialkylammonium salts
or phosphate buffers resolve the peptides very well, volatile buffers, such as 5 - 10 mM ammonium acetate or formate
of pH 4 or 5,5 mM ammonium bicarbonate, pH 7.5 or 0.05% tri-
fluoro acetic acid are superior for further sequencing purpose, especially when less than 20 nanomoles of peptide
material are injected for "preparative" runs. It was found
that propanol gradients were more suitable to the isolation of large peptides, and supports with increased bead
size were recommended (about 300 Å) (W. Machleidt).

Experience shows that one HPLC-chromatogram is normally
not sufficient to resolve all the peptides in a complex
mixture. H. v. Bahr-Lindstrom discussed the advantage of
using conventional separation techniques (e.g. Sephadex gel
chromatography) followed by HPLC-purification. This gives
optimal purification in the nanomol-range. However, if peptides are available in picomole amounts only, then HPLC-
techniques are well suitable although the repurification of
the peptides, if necessary, is difficult because of detection problems and the behaviour of hydrophobic peptides
which often stick to the glass walls. We employed a combination of HPLC-separation as the first step followed by
thin-layer fingerprinting to obtain unresolved peptides.
Thereby, we were able to purify peptides derived from several ribosomal proteins (e.g. proteins S2, L17, and L9;
to be published).

A typical HPLC-trace of the tryptic peptides from ribosomal protein S2 is given in Fig. 1. This protein with a
molecular mass of 27 000 consists of 30 tryptic peptides,
70% of which could be isolated from one HPLC-chromatogram
pure enough for direct analysis and micro-sequencing. The
remainder of the peptides were purified by thin-layer
fingerprinting. Tryptophan and tyrosine containing peptides
were found in the HPLC-trace by monitoring at 280 to 294 nm
(not shown), which is another advantage of peptide separa-

Fig. 1: <u>Separation of tryptic peptides of ribosomal protein S2 by reverse phase HPLC.</u> The protein (MG 26 613)
consists of 30 tryptic peptides (6), 70% of which could
be obtained in purified form by this procedure.

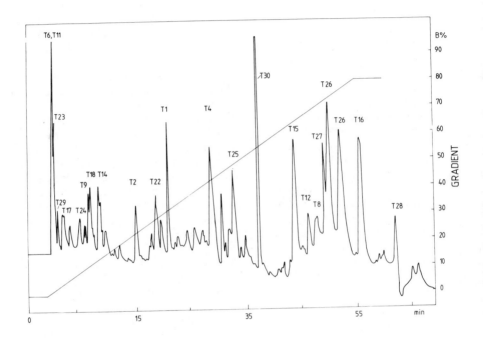

Column packing: Lichrosorb (Merck) RP-18; particle size:
5 µm; filled with Shandon filling apparatus; material: stainless steel (Knauer); length: 25 cm; int. diameter: 4.6 mm;
Sample Inject: 40 µl = 400 µg/15 nanomole, dissolved in buffer A + 2 µl HCOOH; Flow rate: 2 ml/min; Temperature: 40°C;
Pressure: 220 bar; Equipment: Dupont 850; UV detector 852
(variable); wave length 220 nm, range 0.16; sampler Wisp
710 A (Waters); Eluent Buffer A: 0.25 ml HCOOH (98%) + 0.4
ml NH4OH (25%)/ad 2 lt H2O, pH 4.2; buffer B: 80% methanol
(Uvasol, Merck) + 20% buffer A; gradient: 0%B to 50% B /
35 min, linear; 50% B to 80% B / 20 min, linear; 80% B to /
80% B /15 min; 80% B to 0% B / 2 min, linear; hold at 0% B /
15 min.

tion on HPLC.

However, the S2 example also demonstrates the limitations which are still implicit in HPLC-separations, namely losses and low yields of hydrophobic peptides. With hydrophilic peptides recovery on elution is about 80%. This decreases with increasing peptide length and hydrophobicity to 20 - 40%, especially for tryptophan and phenylalanine containing peptides which may not elute from the column at all. Peptides having covalently bound lipids are also trapped on the column (L. Henderson).

For proteins the situation is even worse. Similar sized and charged proteins cannot be yet satisfactorily separated on HPLC-columns. Therefore, main themes for discussions were how to chromatograph proteins in HPLC, how to get the trapped peptides back and how to know "what happens in hell" (McKean). Unfortunately, urea is not very suitable for HPLC-chromatography, as the pump sealings are affected, and addition of SDS to the buffers results in obvious loss of reproducibility of separation. However, bound peptides may be eluted at the end of the runs, by an additional elution with SDS. However, this raised the question of SDS removal before sequencing begins (especially in the picomole-range). Is gas-phase (M. Hunkapiller) or solid phase sequencing (J. Walker) a possible escape from the drawbacks of SDS on sequencing?

Other disadvantages of HPLC-separations of peptides are:

1. The peptide mixture must be applied in a concentrated form which is problematic for peptides which tend to precipitate.

2. It is difficult to predict the chromatographic behaviour of peptides of unknown amino acid composition, although this is - with limitations - possible once the amino acid analysis has been done (7). Therefore, any attempt to separate hard won peptide mixtures of unknown amino acid composition on HPLC is risky.

3. The supports do not allow the use of alkaline conditions for separations, a drawback for a peptide mixture only soluble under such conditions. It would be desirable to develop supports which withstand basic conditions; this may be feasible in the future.

4. Currently available detection systems impose limits depending on the detector and solvent system used; at least 100 - 500 picomoles of peptide are required for detection at 206 - 220 nm. With smaller quantities, a stream splitting device and detection with fluorescamine is necessary (8).

Alternatively, radio-labelled peptides have to be used. In
the future, more sensitive detectors may come onto the
market which could solve the detection problem without the
need to use radioactive peptides.

Several speakers mentioned the difficulties involved
when handling low picomole range amounts of protein or peptide (H. Hunkapiller, J. Spiess). The sample preparation
is the limiting step. Special care has to be taken to prevent the blocking of the N-terminal amino group during isolation and to avoid salt contamination, which would hinder
further sequence analysis. One proposal to overcome these
problems was to employ 0.01% mercaptoethanol in the aqueous
HPLC-buffer, to separate the peptides by HPLC after coupling them with DABITC to protect their N-termini and to
cleave the peptides with TFA only shortly before beginning
with micro-sequencing (J.-Y.Chang). Further, it becomes
necessary to purify the solvents for HPLC-separations to
the highest possible purity grade by combined treatment
with aluminia and quick redistillation under argon or
nitrogen.

HPLC-separations were applied to peptide mixtures obtained by enzymatic digest time-course experiments (R. Laursen, S. Fullmer), for the investigation of micro-hetero-
geneities in Hb-variants (K. R. Williams), for the study of
functionally important peptides of membranes (F. S. Heinemann), for the characterization of nucleotide binding sites
and antigenic determinants (A. R. Kerlavage) and for the
sequence determination of peptides such as costly peptide
hormones, e.g. ovine hypothalamic corticotropin releasing
factor (J. Spiess) which can only be isolated with considerable cost and effort.

Sequencing Techniques

The main micro-sequencing innovations were:
a) modifications of the Edman-type stepwise degradation method;
b) recent changes in sequencer design, and
c) new approaches to mass spectrometry for peptides.
The most common sequence analysis method is still the degradation with phenylisothiocyanate introduced by Edman
(9) performed in a variety of manual or automatic procedures.
The released PTH-amino acid derivatives have so far
been identified by:

    *   thin-layer techniques
    *   back hydrolysis and amino acid analysis
    *   gas chromatography
    *   mass spectrometry
    *   reverse phase HPLC-techniques

In recent years the most commonly applied methods for identification and quantitative determination of the derivatives have been HPLC-techniques. These methods imply isocratic (10-12) or gradient elution (13) on reverse phase columns. One example of PTH-amino acid separation is given in Fig. 2. Other examples are cited in a review (14) and in several of the preceding chapters. A different and sensitive approach for the identification of PTH-amino acids employs gas chromatography with glass capillaries in a splitless device (D. Tripier). Intrinsic labelling of a protein, prior to isolation, permits to use a very sensitive release of the amino acid derivatives. Alternatively, external labelling employing radioactive PITC has been applied. This technique has considerable drawbacks: The radioactive reagent cannot easily be checked for impurities and might give ambiguous results or leads to partial blocking of the N-terminal residue, if minute peptide amounts are sequenced.

The stepwise degradation technique is limited by several effects which decrease the yield of the released derivatives (see Table 1).

The reason for the occurrence of unspecific cleavage sites have been studied thoroughly by W. F. Brandt. This type of investigation is necessary if sequencing methods are to be improved. G. Frank and A. S. Bhown investigated the possibility of decreasing the presence of aldehydes or peroxides in the chemicals by the addition of aluminium

Table 1:   Limitations of stepwise degradation techniques

    *   Decrease in yield at each degradation cycle
    *   Increase of "overlapping"of the released amino acid
        derivatives
    *   Peptide losses
    *   Blockage of N-terminal amino acids by reaction with
        aldehydes or isocyanate
    *   Pyrrolidone carboxylic acid formation
    *   Transacetylation reactions
    *   Unspecific acid cleavage of labile peptide bonds

Fig. 2: Separation of PTH-amino acid derivatives by reverse phase HPLC.

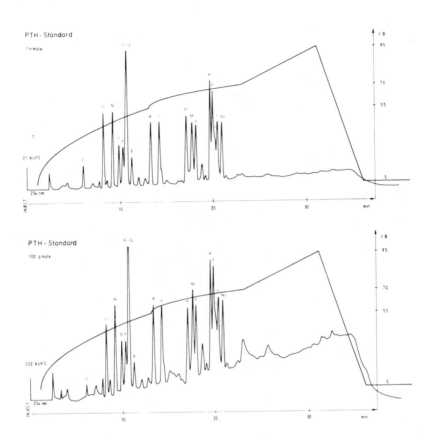

Column Packing: Hypersil MOS-C8 (Shandon), filled with
Shandon filling apparatus; material: stainless steel;
particle size: 5 μm; length: 25 cm; int. diameter: 4.6 mm;
Sample Inject: 1 nmol and 100 pmol of PTH-amino acid stan-
dard/20 μl CH_3OH; Flow Rate: 1.5 ml/min; Temperature: 40°C;
Pressure: 1500 psi; Equipment: Waters pumps 600A; system
controller 440; detector: 254 mm; sampler: WISP 710A;
Eluent buffer A: 10 mM NaOAc, pH 4.01; buffer B: CH_3OH/H_2O
(9:1); gradient: 5-55% B /12' exp. 5; 55-70% B / 10' exp.
5; 70-95% B / 8' exp. 6; 95 - 5% B / 5' exp. 6; hold at 5%
B / 10' exp. 6.

oxide, aminoethyl-cellulose or aminoethyl aminopropyl glass
beads to the solvents and quadrol buffer. A. S. Bhown proposed the use of fluorescamine and W. Machleidt that of
o-phthalaldehyde as an effective blocking reagent to reduce the amount of unreacted N-terminal residues at proline residues thereby eliminating the increasing "overlap" and background contamination. For the solid phase procedure Horn and Machleidt have used secondary and tertiary
amines to block unreacted acetic groups of the glass support. We have tested the latter proposal for liquid-se-
quencing and applied dilute ammonia (2.5%), triethylamine
in water (0.2%), or delivered quadrol at the end of each
degradation cycle, (in amounts that moisten the film) followed by the normal drying stage. This treatment slightly
increases the repetitive yield of the degradations. Besides
reaction with the acidic groups of the glass this increase
might be a result of the neutralisation of the remaining
acidic vapors in the cup compartment and teflon tubings.

   Manual methods: Usually, the manual sequencing methods
release the phenylthiohydantoin derivatives contaminated
with by-products of the reaction; therefore, the amino acid
sequence is derived indirectly by identifying the endgroup
of the residual peptide, e.g. by dansylation after degradation (method of Gray and Hartley, ref. 15), or the residual peptide is analysed by a subtractive method by amino
acid analysis (introduced by Schroeder or Konigsberg), or
more recently by mass determination (Y. Shimonishi). However, if precautions such as eliminating peroxides and aldehydes from the chemicals and exclusion of oxygen from the
reaction are taken, the direct identification of the released PTH-amino acid derivatives is possible (see for instance the chapter of G. E. Tarr).

   In the low nanomole to picomole range the PITC reagent
has been replaced by the more sensitive 4-N,N-dimethylamino-
azobenzene derivative in the DABITC-PITC double coupling
method (16). This technique has increased in sensitivity
through the identification of the released red-coloured
DABTH-amino acid derivatives on micro polyamide sheets (detection limit 20 picomole on 2.5 x 2.5 cm plates) and or by
HPLC. In the latter case the shift in absorbance of the derivatives towards the visible allows detection at 436 nm,
where the by-products of the degradation do not absorb so
strongly (J.Y. Chang). With this technique it is now possible to partially sequence all types of larger peptides
isolated for extended sequencer runs or functional studies,
which require a purity control or information on the N-ter-

minal sequence region (e.g. in fragment analysis of ribosomal proteins). In many instances microsequencing is pre-
fered to amino acid analysis in order to save material.

It is obvious from recent sequence studies that manual
sequencing is most popular and competes with the automated
techniques. The advantages are the simple, less expensive,
fast performance (many samples can be degraded simultaneously) and the easy control and reproducibility of the degradation, if the necessary precautions are taken (see
chapter of G. E. Tarr, and for ribosomal proteins that of
M. Kimura). In spite of this the manually performed degradations usually do not reach the high level of repetitive
yields realised in a good sequencer. The usual values reached by manual techniques are 90 - 93% and by solid-phase
sequencing about 93 - 95%, whereas in modified liquid-phase
machines 98% can be obtained. Small peptides (a maximum of
20 - 30 residues, depending on the individual sequence and
the ability of the person) or the N-terminal sequence region of longer peptides can be successfully sequenced
manually.

Although one might logically expect a further increase
in sensitivity by the application of fluorescing isothiocyanates, such as the proposed fluorescein derivative
(FITC), for the degradation (17) this has not yet found wide
application, obviously due to problems with the removal of
excess reagent and peptide losses during extraction. However, favoured by the quick development of sensitive fluori-
meters, this type of degradation may become of more interest
in the future for adaption to sequencing at the low picomole level.

Sequencers: Further progress has been made on automatic
techniques in the past two years. In addition to the already "classic" methods, the liquid-phase sequencer designed by Edman and Begg (18) and the solid-phase type introduced by Laursen (19), a combined liquid/gas-phase sequencer
employing a cartridge has been developed by Hewick and
Hunkapiller (20). This machine incorporates old ideas of
degrading peptides in a gas atmosphere (Sjöquist and Dreyer).
The different reaction compartments of the three sequencer
types are illustrated in Fig. 3. The indicated proportions
of the reaction chambers show the recent developments. The
dead-volumes of both the cartridge and the new columns
applied in solid-phase (see below) are much smaller than
the cup and its compartment in the liquid-phase machine.
The size of the reaction chamber has been reduced to the

Fig. 3: Different reaction chambers of the three devices available at present for automatic stepwise Edman degradation technique: a) liquid-phase film method (18); b) solid-phase (19) teflon column (21); and c) gas/liquid-phase cartridge version (20).

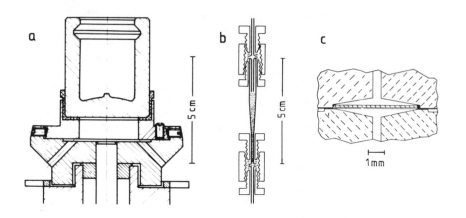

appropriate size for micro-sequencing. Although a miniaturization of the cup and its compartment is possible (A. Inglis, personal communication), this is limited by the space necessary for the drive magnet and the scoop-uptake.

However, not only have the dimensions of the reaction chamber to be kept as small as possible for micro-sequencing but also all other important parts of the sequencer, such as delivery valves, connecting lines, vacuum valves, etc, should have no or small dead-volumes. We have redesigned the commercially available liquid-phase sequencer (Fa. Beckman, Palo Alto) and incorporated newly constructed parts for micro-sequencing (reviewed in ref. 22 - 24), and further improvements, such as an automatic conversion device have been installed (25). A complete list of the modifica-

Table 2: <u>Improvements made to the liquid-phase sequencer</u>

Modifications made between 1973 and 1975:

* Reconstruction of the vacuum system
* Insertion of cooling traps (kept at liquid nitrogen temperature)
* Change of cup cover
* Newly constructed delivery and waste valves (see Fig. 4)
* Installation of automatic conversion device
* Permanent control of the performance of the sequencer (by monitoring the vacuum values in the cup compartment)

Recent modifications:

* Newly constructed reagent and solvent bottles
* Redesigned press valves
* Super-high speed for the drive (variable)
* Programme controlled by microprocessor
* Storage of programmes on cassettes
* On-line HPLC-identification

tions for liquid-phase sequencing so far made, are listed in Table 2. The advantages of these improvements are discussed in detail elsewhere (24,26).

As a consequence of the modifications, the repetitive yields as quantitised from test runs with ribonuclease (see Fig. 5) could be improved to 98 - 98.5%. To improve the degradation beyond this point would result in a considerable advantage for the degradation of long sequences whereas for small peptides a poorer yield can be tolerated. However, any further improvements require much more investment in the machine and more effort in purifying the polypeptide, the solvents, reagents and gases used. With conventional machines several compromises have to be made. Therefore, we have constructed our own machine which includes the improvements mentioned, together with others (see below).

In the cartridge version the important parts have all been miniaturized for sequencing picomoles of protein. The small porous sinter plate within the cartridge to which the sample is applied is comparable in surface dimensions to the surface area for forming the film in the cup version to obtain high repetitive yields. It is necessary to expose a large surface area of peptide to the reagent during de-

Fig. 4: Newly constructed dead-volume free delivery and
waste valves for liquid-phase micro-sequencing (24,26).
The valves are placed in series and connected by a central
zig-zag delivery line, emptied after the delivery of reagent or solvent by a purge of nitrogen which is delivered
through the last valve of the series.

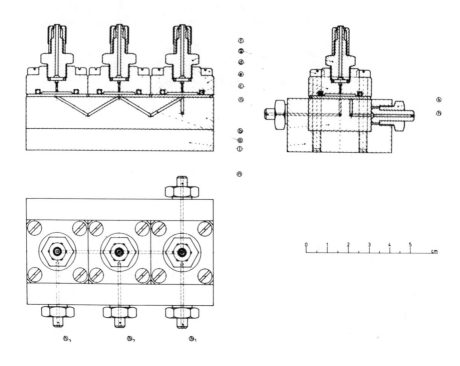

gradation. Thus the small cartridge is advantageous. In
addition, the machine is less complicated in construction,
should be easier to service and less costly to make than
any spinning cup machine. Difficulties with the new machine
may arise from the alternating gas and liquid phases during the reaction which could cause incomplete or less reproducible degradations. Other limitations derive from the
necessary use of polybrene, a synthetic carrier introduced for liquid phase sequencing (27,28). This carrier retains the polypeptide through its sponge-like properties.
Unfortunately, it also retains the by-products of the Edman
degradation. As a consequence, the identification of the

Fig. 5: <u>Test degradations of ribonuclease</u> (10 nanomoles) performed in a modified liquid-phase sequencer without any application of a carrier. Identification of the PTH-amino acid derivatives was made with sample amounts of each 1/5 of the released degradation cycles. Shown are the yields of PTH-alanine of step 4, 5, 6, 19 and 20 when alanine occurs within the sequence. Increased yields at step 19 and 20 are due to the modifications of the sequencer and extensive purification of reagents and solvents (26).

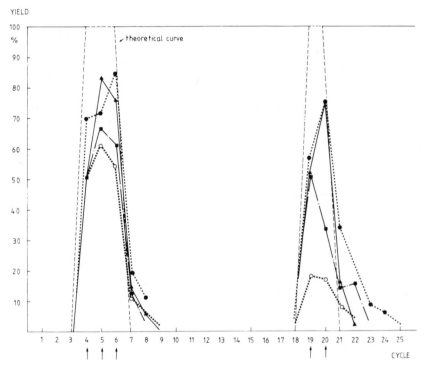

released PTH-amino acid derivatives shows more contaminants when polybrene is used. This is true although more stringent extractions are performed after the coupling stages than when polybrene is not used (e.g. 20 - 25 ml each of solvent S1 and S2 in the presence of carrier compared with 5 - 6 ml total amount of S1+S2 in the modified Beckman se-

quencer and 1 - 2 ml in the new Berlin sequencer without
polybrene). Examples of degradation cycle 1 to 29 of 10
nanomoles of ribonuclease are given in Fig. 6. This routinely can be done without any addition of a carrier in the
modified machine.

Usually liquid-phase sequencing is done with more than
20 nanomoles of substance in the presence of polybrene. In
this case an increase of contamination in the presence of
the carrier can be tolerated. With less amounts of protein
more of the degradation product has to be used for identification, and the contaminating by-products obscure the results, e.g. when the identification is made by HPLC-chromatography. This makes background subtractions and amplification of the signals necessary. These limitations with
polybrene were visible in the examples of degradation
cycles presented by J. Shiveley.Subtractions become inaccurate if partial cleavages of peptide bonds occur during
the acid cleavage stage of the degradation or when unusual
or modified amino acids are present within the sequence
(see for instance unusual residues in ribosomal proteins,
ref. 1). When sequencing less than 100 picomoles the ratio
of peptide to polybrene (6 mg are used) becomes so disad-
vantageous that the identification of the released amino
acid derivative is very unreliable. The present efforts
made with the cartridge version to sequence at the level of
a few picomoles (M. Hunkapillar) can be regarded as a pioneering project in protein sequencing.

New fields in protein chemistry would be opened if
micro-sequencing could be carried out with polypeptides
obtained from gels. This has been discussed during the
meeting, and the combination of solid-phase techniques with
elution of proteins from slab gels has been proposed
(J. Walker). For this purpose solid-phase techniques have
certain advantages, as proteins attached to a solid support
can be freed from salt traces or contamination with SDS.
This procedure might offer new possibilities due to the
recent miniaturization of the connecting lines and dead-

Legend to Fig. 6: *Micro-sequencing of ribonuclease (10 nanomoles) in an improved liquid-phase machine performed without the aid of any carrier. Sample amounts each 1/5 of the
total amounts released were injected for HPLC identification (as in Fig. 2). Details of the sequencer modifications,
the programme and purification of solvents are given in
ref. 26 (the dehydro-derivatives of PTH-Thr and -Ser
measured at 312 nm are not shown).*

Fig. 6 (1)

CYCLE 0
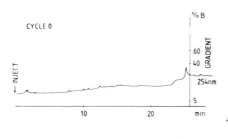

CYCLE 5
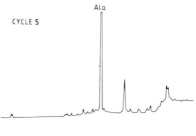
Ala

CYCLE 1

Lys

CYCLE 6
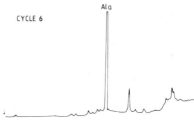
Ala

CYCLE 2
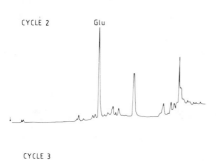
Glu

CYCLE 7

Lys

CYCLE 3
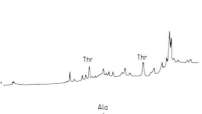
Thr   Thr

CYCLE 8
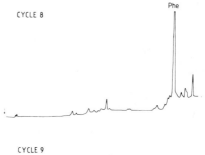
Phe

CYCLE 4
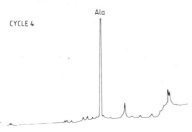
Ala

CYCLE 9
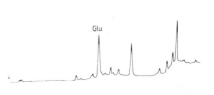
Glu

Fig. 6 (2)

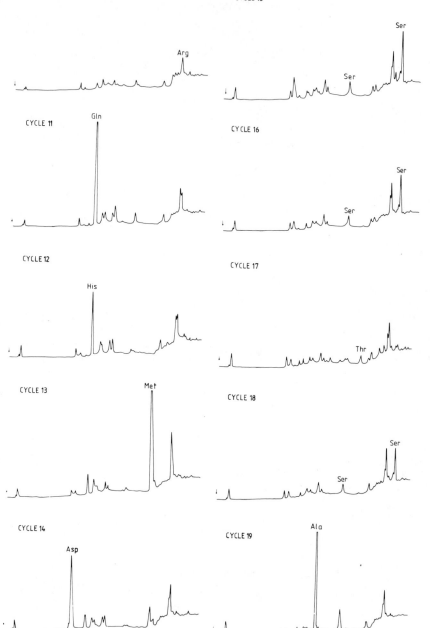

Fig. 6 (3)

CYCLE 20

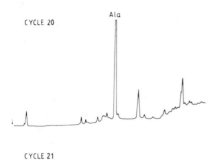

CYCLE 25

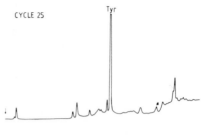

CYCLE 21

CYCLE 26

CYCLE 22

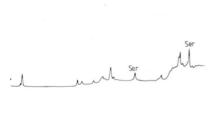

CYCLE 27

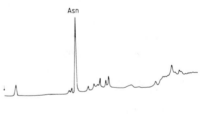

CYCLE 23

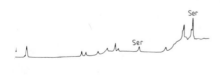

CYCLE 28

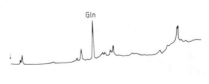

CYCLE 24

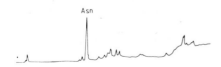

CYCLE 29

volumes (J. Walker) and the use of a small teflon tube for
the column of the solid-phase machine (J. Salnikow). Further, the attachment of polypeptides to glass support has
been studied in detail, and the application of the DABITC-
reagent to solid-phase micro-sequencing could be optimized
(21).

Our own efforts to construct sequencers aim at a machine in which the liquid-, solid- or gas-phase degradation can be performed. In this sequencer all the major
and expensive parts, such as miniaturized delivery valves
of the type as in Fig. 4, governing valves, the vacuum and
conversion device, entry and exit lines and the control
system are in common but depending on the type of degradation system required, the appropriate reaction vessel will
be inserted. The newly designed sequencer (see Fig. 7) has
all the improvements listed in Table 2 and 3 and additional
changes, as given in Table 3 are being made. The self-made
microprocessor allows one to write flexible programmes
suitable for special sequences. These can be stored on cassettes and printed for the protocols. Care has been taken
to provide the microprocessor with electronic parts of
high durability and to allow programming in Basic. An on-
line HPLC-detection system (realised for the first time in
a solid-phase machine by W. Machleidt) has been connected
with the flask of the conversion device. It takes aliquots
of the released degradation products (1/7 to 1/10 of total
amount) for direct identification by HPLC-chromatography
with isocratic elution of the derivatives similar to those
described (12) or by the use of a two-step gradient (to be
published). This on-line detection of the released amino
acid derivatives is a necessary step towards full automation and allows easy control of the sequencer.

Micro-sequencing, although already advanced, still has
its limitations. Improvements, in addition to those already
established, have to be made in both the technical and chemical aspects if reliable pico-sequencing is to become possible. In any case, a double check of the identification
of the released amino acid derivatives should be performed.
Alternatively, an independent approach, such as mass spectrometry or parallel DNA-sequencing has to be used to confirm the derived sequence. At present the degradation of
proteins with intrinsically labelled amino acids is the
most sensitive method, but it is of course limited to N-terminal sequencing studies. Complete structural determination with only a few picomoles is not yet possible. However,
one can be optimistic (at least "femtoists" can be opti-

Fig. 7

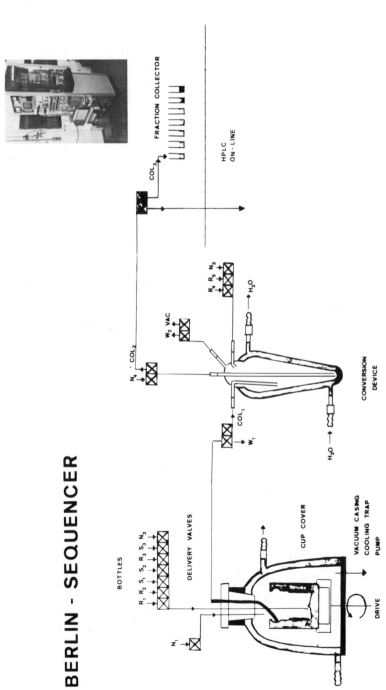

Table 3: <u>Design of the Berlin Sequencer</u>

Zig-zag valves   serve as:

* delivery valves for all reagents and bottles
  (in miniaturized version similar to Fig. 4)
* waste-valves
* pressurizing valves for all bottles
* buffer valves for HPLC (at low pressure side)
* vacuum valves

Microprocessor designed for:

* programming in Basic
* quick changes of programme steps (maximal 300
  per cycle)
* input commands via terminal
* monitoring of programme with screen
* storage of programmes on cassettes
* connection to printer for print-outs of programmes

On-line conversion and identification:

* detection of the released PTH-amino acid derivatives simultaneously with the degradation cycles

<u>Further modifications</u>:

* programmer control for:

  - increase in speed
  - choice of temperature
  - maintenance of sequencer
  - internal loops for particular cycles

* controlled dosage of reagents at super-high speed
* choice of different cup sizes
* cooling system for reagents and solvents
* redesigned vent system
* alternatively: liquid-, solid- or gas-phase
* glass capillary techniques

---

mistic) that such determinations may be possible in the
near future. It is an open question as to which of the
three automatic degradative principles will survive. It
has been suggested "atomic bombardment is killing us",
and the perspectives given by H. R. Morris in his talk
might have left this impression. However, it is likely

that the sequencers of the future will be of less complex
construction than any mass spectrometer and that almost
everybody who studies polypeptides will be able to do his
own sequencing. On the other hand there is no doubt that
in special cases of complicated structure mass spectrometry serves as a logical alternative for the sequencing
of peptides. Simplified sequencers can replace some of the
functions of amino acid analysers. With the same amount of
substance a sequencer can yield more information about the
specific nature of a polypeptide and its purity than can be
obtained from the amino acid composition.

Analyzer Equipment and HPLC-Systems

Most of the analyzers used at present, have been constructed for a range of 2 - 10 nanomoles per amino acid.
Recent peptide separation and micro-sequencing techniques
are highly sensitive and therefore require more sensitive
analyzers but without loss of accuracy (usually analysing
at a higher level of sensitivity is accompanied by a loss
in accuracy). Ninhydrin, o-phthalaldehyde (OPA) and fluorescamine are reagents for detection in the picomole range. OPA
is more suitable for staining amino acids as opposed to
fluorescamine which is better for peptide detection. OPA
reacts with ammonia to a lesser degree than fluorescamine.
However, the detection of proline is not yet satisfactory.
Hence, the best choice remains the ninhydrin reaction which
is sensitive enough for detection of picomole quantities
but suffers from pump fluctuations which cause increased
base-line noise in the conventional analyzers (equipped
with Milton-Roy type pumps).
Analyzers with the following technical characteristics
are required for analyses in the pico-range:
1. pulse-free pumps (HPLC-pumps) with chemical resistant sealings, easy to be removed;
2. buffer and reagent valves which are almost dead-
   volume-free;
3. buffer and reagent valves at the low pressure side;
4. optimum mixing and degassing in the heads of the
   pumps;
5. simple programme unit for governing the main functions on the analyzer, but separated from an independent integrating system;
6. stabilized reagent suitable to detect all types of
   amino acids;

7. suitable column support, based on either ion exchange or reverse phase resin, with a guaranteed
   quality (even from batch to batch), of low back-
   pressure value which is the basis for series of
   routine analyses;
8. automatic sampling and injection system for small
   volumes (less than 40 µl) to avoid contaminations
   from the buffer.

The main problems remaining when analysing small
amounts of peptide - even if the analyzer is of high performance - are contaminations, mainly by serine, glycine
and lysine, obtained at the isolation of the peptide and
preparation of the analysis. The ratio of sample volume to
peptide quantity becomes increasingly critical in the low
picomole range, and the elution of peptides from HPLC-columns, usually done in large volumes, becomes the limiting
step.
     At the conference, two main approaches for analysing
amino acids at the low picomole range were discussed:
     1. employing normal HPLC-reverse phase with derivatiz-
        ed amino acids, such as dansylated amino acids (29);
     2. combining ion exchange chromatography with HPLC-
        equipment.
The effort D. G. Klapper made, to construct his own HPLC-
analyzer is worth mentioning. At present, in many laboratories picomole amounts of peptides are used for sequencing
rather than for analysing, as a first approach.
     There are so many different HPLC-systems on the market
that judging which is best becomes impossible. Many of the
available systems are equipped with micro-processors which
are relatively easy to handle, but not durable enough. Often
they are sensitive to external electric interference or
overheating. Other problems are associated with changes in
the separation properties of the supports, due to quality
differences. The high back-pressure problem has been largely overcome: some suppliers now sell supports of 5 µm particle size which give a low back-pressure. Further, the different types of columns and some of their fittings do not
allow an easy change from one system to another or do not
enable an own refilling. More uniformly designed columns
would facilitate the application of HPLC-techniques.

RESULTS OF SEQUENCE ANALYSIS

At the conference the main emphasis was on advances in amino acid sequence analysis *techniques*. However, some *sequences* determined recently were covered by posters and mentioned in the lectures. Many contributions showed the importance of protein sequencing to understand biological processes at the molecular level, e.g. studies on:
* the tubulin family in order to explain the mitosis mechanism (H. Ponstingl);
* sequencing the fribrinogens to clarify blood coagulation (A. Henschen);
* determination of nucleotide binding sites and antigenic determinants for functional aspects (A. R. Kerlavage);
* investigation of membrane proteins (F. S. Heinemann, D. J. McKean) or nucleolar proteins (M. Olson) for permeability studies;
* analysing the antiphagocytic properties of the M-proteins (B. N. Manjula).

Micro-methods in protein chemistry, supported by DNA-sequencing, have allowed large proteins to be sequenced, e.g. beta-galactosidase related proteins (A. V. Fowler), E. coli DNA polymerase (V. M. Lipkin) and the largest E. coli ribosomal protein S1 (Fig. 8) which has 557 amino acid residues (2,30). Similarly, investigations of the functional domains and active sites of a number of proteins have become increasingly important (e.g. see studies by A. Carne, J. H. Collins, J. W. Crabb, H. Jörnvall, F. Lederer, A. T. Phillips, A. Place, and K. Takio). Further, the advanced techniques facilitate greatly the investigation of topics in medical research on a molecular level, such as immuno-defence, pathogenesis of cancer, nerve and brain activity. Micro-sequencing has permitted the studies on signal sequences and the different types of human interferon. Recent results were discussed by M. Hunkapiller (interferons), L.K. Duffy, J. Y. Chang and H. Kratzin (monoclonal antibodies), K. R. Williams (mutants of Hb-variants), M. B. Lees (active peptides from brain), H. R. Morris (antibodies and neuropeptides), K. J. Wilson (growth hormones), M. R. Sairam (ovine pituitary follitropin) and P. C. Montecucchi (active peptides extracted from amphibian skin). The complete sequence of the ovine hypothalamic corticotropin releasing factor was presented at the meeting (J. Spiess). Protein-chemical studies on biologically active peptides are the

Fig. 8: Complete amino acid and nucleotide sequence of protein S1 isolated from the small subunit of the E. coli ribosome. The differences in the protein and DNA structures are due to strain differences (for details see ref. 2 and 30). Protein S1 has 557 amino acid residues and a molecular mass of 61 159.

```
  1                               11                                       21
MET THR GLU SER PHE ALA GLN LEU PHE GLU GLU SER LEU LYS GLU ILE GLU THR ARG PRO GLY SER ILE VAL ARG GLY VAL VAL VAL ALA
ATG ACT GAA TCT TTT GCT CAA CTC TTT GAA GAG TCC TTA AAA GAA ATC GAA ACC CGC CCG GGT TCT ATC GTT CGT GGC GTT GTT GTT GCT

 31                               41                                       51
ILE ASP LYS ASP VAL VAL LEU VAL ASP ALA GLY LEU LYS SER GLU SER ALA ILE PRO ALA GLU GLN PHE LYS ASN ALA GLN GLY GLU LEU
ATC GAC AAA GAC GTA GTA CTG GTT GAC GCT GGT CTG AAA TCT GAG TCC GCC ATC CCG GCT GAG CAG TTC AAA AAC GCC CAG GGC GAG CTG

 61                               71                                       81
GLU ILE GLN VAL GLY ASP GLU VAL ASP VAL ALA LEU ASP ALA VAL GLU ASP GLY PHE GLY GLU THR LEU LEU SER ARG GLU LYS ALA LYS
GAA ATC CAG GTA GGT GAC GAA GTT GAC GTT GCT CTG GAC GCA GTA GAA GAC GGC TTC GGT GAA ACT CTG CTG TCC CGT GAG AAA GCT AAA

 91                              101                                      111
ARG HIS GLU ALA TRP ILE THR LEU GLU LYS ALA TYR GLU ASP ALA GLU THR VAL THR GLY VAL ILE ASN GLY LYS VAL LYS GLY GLY PHE
CGT CAC GAA GCC TGG ATC ACG CTG GAA AAA GCT TAC GAA GAT GCT GAA ACT GTT ACC GGT GTT ATC AAC GGC AAA GTT AAG GGC GGC TTC

121                              131                                      141
THR VAL GLU LEU ASP GLY ILE ARG ALA PHE LEU PRO GLY SER LEU VAL ASP VAL ARG PRO VAL ARG ASP THR LEU HIS LEU GLU GLY LYS
ACT GTT GAG CTG AAC GGT ATT CGT GCG TTC CTG CCA GGT TCT CTG GTA GAC GTT CGT CCG GTG CGT GAC ACT CTG CAC CTG GAA GGC AAA

151                              161                                      171
GLU LEU GLU PHE LYS VAL ILE LYS LEU ASP GLN LYS ARG ASN ASN VAL VAL VAL SER ARG ARG ALA VAL ILE GLU SER GLU ASN SER ALA
GAG CTT GAA TTT AAA GTA ATC AAG CTG GAC CAG AAG CGC AAC AAC GTT GTT GTT TCT CGT CGT GCC GTT ATC GAA TCC GAA AAC AGC GCA

181                              191                                      201
GLU ARG ASP GLN LEU LEU GLU ASN LEU GLN GLU GLY MET GLU VAL LYS GLY PHE VAL LYS ASN LEU THR ASP TYR GLY ALA PHE VAL ASP
   GAC       GAT CAG CTG CTG GAA AAC CTG CAG GAA GGC ATG GAA GTT AAA GGT ATC GTT AAG AAC CTC ACT GAC TAC GGT GCA TTC GTT GAT

211                              221                                      231
LEU GLY GLY GLY VAL ASP GLY LEU LEU HIS ILE THR ASP MET ALA TRP LYS ARG VAL LYS HIS PRO SER GLU ILE VAL ASN VAL GLY ASP GLU
CTG GGC GGC GTT GAC GGC CTG CTC CAC ATC ACT GAC ATG GCC TGG AAA CGC GTT AAG CAT CCG AGC GAA ATC GTC AAC CTG GGC GAC GAA

241                              251                                      261
ILE THR VAL LYS VAL LEU LYS PHE ASP ARG GLU ARG THR ARG VAL SER LEU GLY LEU LYS GLN LEU GLY GLU ASP PRO TRP VAL ALA ILE
ATC ACT GTT AAA GTG CTG AAG TTC GAC CGC GAA CGT ACC CGT GTA TCC CTG GGT CTG AAA CAG CTG GGC GAA GAT CCG TGG GTA GCT ATC

271                              281                                      291
ALA LYS ARG TYR PRO GLU GLY THR LYS LEU THR GLY ARG VAL THR ASN LEU THR ASP TYR GLY CYS PHE VAL GLU ILE GLU GLU GLY VAL
GCT AAA CGT TAT CCG GAA GGT ACC AAA CTG ACT GGT CGC GTG ACC AAC CTG ACC GAC TAC GGC TGC TTC GTT GAA ATC GAA GAA GGC GTT

301                              311                                      321
GLU GLY LEU VAL HIS VAL SER GLU MET ASP TRP THR ASN LYS ASN ILE HIS PRO SER LYS VAL VAL ASN VAL GLY ASP VAL VAL GLU VAL
GAA GGC CTG GTA CAC GTT TCC GAA ATG GAC TGG ACC AAC AAA AAC ATC CAC CCG TCC AAA GTT GTT AAC GTT GGC GAT GTA GTG GAA GTT

331                              341                                      351
MET VAL LEU ASP ILE ASP GLU GLU ARG ARG ARG ILE SER LEU GLY LEU LYS GLN CYS LYS ALA ASN PRO TRP GLN GLN PHE ALA GLU THR
ATG GTT CTG GAT ATC GAC GAA GAA CGT CGT CGT CGT ATC TCC CTG GGT CTG AAA CAG TGC AAA GCT AAC CCG TGG CAG CAG TTC GCG GAA ACT

361                              371                                      381
HIS ASN LYS GLY ASP ARG VAL GLU GLY LYS ILE LYS SER ILE THR ASP PHE GLY ILE PHE ILE GLY LEU ASP GLY GLY ILE ASP GLY LEU
CAC AAC AAG GGC GAC CGT GTT GAA GCT AAA ATC AAG TCT ATC ACT GAC TTC GGT ATC TTC ATC GGC TTG GAC GGC GGC ATC GAC GGC CTG

391                              401                                      411
VAL HIS LEU SER ASP ILE SER TRP ASN VAL ALA GLY GLU GLU ALA VAL ARG GLU TYR LYS LYS GLY ASP GLU ILE ALA ALA VAL VAL LEU
GTT CAC CTG TCT GAC ATC TCC TGG AAC GTT GCA GGC GAA GAA GCA GTT CGT GAA TAC AAA AAA GGC GAC GAA ATC GCT GCA GTT GTT CTG

421                              431                                      441
GLN VAL ASP ALA GLU ARG GLU ARG ILE SER LEU GLY VAL LYS GLN LEU ALA GLU ASP PRO PHE ASN ASN TRP VAL ALA LEU ASN LYS LYS
CAG GTT GAC GCA GAA CGT GAA CGT ATC TCC CTG GGC GTT AAA CAG CTC GCA GAA GAT CCG TTC AAC AAC TGG GTT GCT CTG AAC AAG AAA

451                              461                                      471
GLY ALA ILE VAL THR GLY LYS VAL THR ALA VAL ASP ALA LYS GLY ALA THR VAL GLU LEU ALA ASP GLY VAL GLU GLY TYR LEU ARG ALA
GGC GCT ATC GTA ACC GGT AAA GTA ACT GCA GTT GAC GCT AAA GGC GCA ACC GTA GAA CTG CCT GAC GGC GTT GAA GGT TAC CTG CGT GCT

481                              491                                      501
SER GLU ALA SER ARG ASP ARG VAL GLU ASP ALA THR LEU VAL LEU SER VAL GLY ASP GLU VAL GLU ALA LYS PHE THR GLY VAL ASP ARG
TCT GAA GCA TCC CGT GAC CGC GTT GAA GAC GCT ACC CTG GTT CTG AGC GTT GGC GAC GAA GTT GAA GCT AAA TTC ACC GGC GTT GAT CGT

511                              521                                      531
LYS ASN ARG ALA ILE SER LEU SER VAL ARG ALA LYS ASP GLU ALA ASP GLU LYS ASP ALA ILE ALA THR VAL ASN LYS GLN GLU ASP ALA
AAA AAC CGC GCA ATC AGC CTG TCT GTT CGT GCG AAA GAC GAA GCT GAC GAG AAA GAT GCA ATC CCA ACT GTT AAC AAA CAG GAA GAT GCA

541                              551        557
ASN PHE SER ASN ASN ALA MET ALA GLU ALA PHE LYS ALA ALA LYS GLY GLU
AAC TTC TCC AAC AAC GCA ATG GCT GAA GCT TTC AAA GCA GCT AAA GGC GAG TAA
```

basis of their chemical synthesis and further investigations of their biosynthesis and biological action.

These examples prove the value of micro-sequence analysis and emphasize the demand for further increases in sensitivity. Unfortunately, the difficulties involved can only be overcome by a considerable investment of energy and patience to develop sensitive isolation procedures, HPLC-systems, reliable sequencers, and new analyzers. Further, a great deal is needed to keep the machines in good performance and to make the repairs. Hopefully, more sequences of "energy releasing" factors will become available, and might support us in our work, such as:

$$\text{Asp-}{\text{Ile} \atop \text{Leu}}\text{-Ile-Thr-Tyr-}{\text{Ile} \atop \text{Leu}}\text{-Glx-Arg-Ser-Glu-Leu-Phe}$$

D   O   I   T   Y   O   U   R   S   E   L   F

Sequence Determination of Ribosomal Proteins

The E. coli ribosome has been extensively studied to establish its primary structure. This organelle consists of 53 proteins and three different RNA strands. The small protein amounts available forced us to adapt protein-chemical methods to a micro-scale. With a wide spectrum of micro-methods (as reviewed in ref. 4 and 24) we were able to fully determine the primary structures of all the proteins from the E. coli ribosome (Table 4). Therefore, we can now present a complete list of these protein sequences.

Since the primary structures of the three ribosomal RNA molecules are also known (reviewed in ref. 31), the E. coli ribosome is the first organelle whose components have been completely sequenced. As shown in Table 5, the chemical molecular weight of the 30S subunit amounts to 0.86×10^6, and that of the 50S subunit to 1.45×10^6. The 70S particle of the E. coli ribosome has a mass of 2.3×10^6 as derived from sequence analysis.

The E. coli ribosomal proteins differ in size, net charge, amino acid composition and sequence from each other. They do not show significant homologies in their primary structures. However, ribosomal proteins from various pro-karyotic species, e.g. E. coli and Bacillus subtilis, are related to each other (with 40 - 60% homology, depending on the particular protein pair). The same is true for ribosomal proteins isolated from various eukaryotes, e.g. yeast and rat. This finding leads to the interesting question of how ribosomes from different sources, whose structure has

Table 4: Number of amino acid residues and molecular weights
of ribosomal (and related) proteins from E. coli

Protein	Residues	Mol.wt.	Protein	Residues	Mol.wt.
S1	557	61 159	L1	233	24 599
S2	240	26 613	L2	271	29 602
S3	232	25 852	L3	209	22 258
S4	203	23 137	L4	201	22 087
S5	166	17 515	L5	178	20 171
S6	135	15 704	L6	176	18 832
S7K	177	19 732	L7 (2x)	120	12 220
(S7B	153	17 131)	L9	147	(16 500)
S8	129	13 996	L10	165	17 737
S9	128	14 569	L11	141	14 874
S10	103	11 736	L12(2x)	120	12 178
S11	128	13 728	L13	142	16 019
S12	123	13 606	L14	123	13 541
S13	117	12 968	L15	144	14 981
S14	97	11 063	L16	136	15 296
S15	87	10 001	L17	127	14 365
S16	82	9 191	L18	117	12 770
S17	83	9 573	L19	114	13 002
S18	74	8 896	L20	117	13 366
S19	91	10 299	L21	103	11 565
S20	86	9 553	L22	110	12 227
S21	70	8 369	L23	99	11 013
			L24	103	11 185
IF-1	71	8 119	L25	94	10 694
IF-3	181	20 695	L26 = S20		
EF-Tu	393	43 225	L27	84	8 993
NS1	90	9 226	L28	77	8 875
NS2	90	9 535	L29	63	7 274
			L30	58	6 411
			L31	62	6 971
			L32	56	6 315
			L33	54	6 255
			L34	46	5 381

Table 5   Components of the E. coli ribosome

	PROTEINS			RNA			PARTICLE	
	Number	AA	MM		NT	MW	Chem. MW	Phys. MW
30S	21	3108	$0,36 \times 10^6$	16S	1542	$0,50 \times 10^6$	$0,86 \times 10^6$	$0,9 - 1,0 \times 10^6$
50S	32	4232	$0,47 \times 10^6$	23S 5S	2904 120	$0,94 \times 10^6$ $0,04 \times 10^6$	$1,45 \times 10^6$	$1,6 - 1,8 \times 10^6$
70S	53	7340	$0,82 \times 10^6$		4566	$1,48 \times 10^6$	$2,30 \times 10^6$	$2,6 - 1,9 \times 10^6$

been changed during evolution, can perform the same functions in all organisms. More details on the chemical, physical and immunological properties and the evolution of
ribosomal proteins are described elsewhere (1,5,26,32,34,35).

Importance of Protein-Chemical Studies

As mentioned above, sequence analysis is a necessary
prerequisite to establishing the nature of the mature product. Any indirect approach, such as nucleotide sequencing,
can facilitate but not replace protein sequencing work. For
instance, direct protein-chemical studies have to be performed in order to assign modified residues and S-S-bridges,
which are essential for the conformation and function of the
polypeptide.

In the following, we show some examples of studies on
ribosomes, which were only possible after the primary sequences of the proteins had become available and after an
efficient isolation procedure for the peptides had been
established:

1. Amino acid replacements were localized in altered
ribosomal proteins isolated from approximately 40 antibio-
tic-resistant mutants. It was found that the replacements
are clustered in a few short regions of the protein chain
and that a replacement of one out of the 7340 amino acid
redidues in the ribosome suffices to confer the antibiotic
resistance (1).

2. Crosslinking between two proteins which are neigh-
bors within the ribosomal particle is an efficient method
for obtaining information about the spatial structure of
the ribosome. Knowledge of the primary structures of the
ribosomal proteins and the behavior of their peptides allowed the identification of the crosslink at the amino acid
level (ref. 33). Similarly, several protein-RNA crosslinks
in the ribosome were localized at the amino acid and nucleotide level (reviewed in ref. 34).

3. Limited proteolysis of ribosomal proteins often results in large peptide fragments which are isolated and used
for NMR- and CD-measurements. In this way, information about
the presence of structural domains in the studied protein
can be obtained (1).

4. Several ribosomal proteins have recently been crystallized (reviewed in ref. 35). Knowledge of the amino acid
sequence of the crystallized proteins is necessary for the
interpretation of the electron density maps at high resolu-

tion.

    5. The effect of chemical modification on initiation and elongation factors and on ribosomal proteins which specifically bind to ribosomal RNAs allows conclusions about the role of certain amino acids or protein regions in the function of these proteins (e.g. ref. 36).

    6. Comparison of the primary structures of ribosomal proteins isolated from a wide variety of organisms gives interesting information about the evolution of this particle (see above). Because ribosomes are the only organelle present in all organisms they are an ideal object for evolutionary studies.

    The studies mentioned above are examples to illustrate the great importance of protein-chemical investigations for a detailed insight into the structure and function of ribosomes and their proteins. The development and application of more sensitive and reliable methods for protein-chemical analysis will accelerate the progress in this field and will also influence other areas of biochemistry as it has done in the past.

REFERENCES

1. Wittmann, H.G., Littlechild, J., and Wittmann-Liebold, B.: In Ribosomes. Chambliss, G. et al. (eds.), University Park Press, Baltimore, pp. 51-88, 1980.

2. Schnier, J., Kimura, M., Foulaki, K., Subramanian, A.R., Isono, K., and Wittmann-Liebold, B. (1981) Proc. Natl. Acad. Sci. USA, in press.

3. Boege, U., Wengler, G., Wengler, G., and Wittmann-Liebold, B. (1981) Virology 113, 293-303.

4. Wittmann-Liebold, B. and Lehmann, A. (1980): in Methods in Peptide and Protein Sequence Analysis. Chr. Birr (ed.), Elsevier North-Holland Biomedical Press, Amsterdam, New York, Oxford, pp. 49-72.

5. Stöffler, G. and Wittmann, H.G. (1977): in Molecular Mechanisms of Protein Biosynthesis. Weissbach, H., and Pestka, S. (eds.), Academic Press, New York, pp. 117-202.

6. Wittmann-Liebold, B., and Bosserhoff, A. (1981) FEBS Lett. 129, 10-16.

7. Meek, J.L., and Rosetti, Z.L. (1981) J. Chrom. 211, 15-28.

8. Böhlen, P., Stein, S., Stone, J., and Udenfried, S. (1975) Anal. Biochem. 67, 438-445.

9. Edman, P. (1956) Acta Chem. Scand. 10, 761.
10. Frank, G., and Strubert, W. (1973) Chromatogr. 6, 522-524.
11. Beeumen, J. van, Damme, J. van, Tempst, P., and Ley, J. de (1977) see ref. 4, pp. 503-506.
12. Henschen-Edman, A., and Lottspeich, F. (1980), see ref. 4, pp. 105-113.
13. Zimmermann, C.L., Apella, E., and Pisano, J.J. (1977) Anal. Biochem. 77, 569-573.
14. Godtfredsen, S.E., and Oliver, W.A. (1980) Carlsberg Res. Commun. 45, 35-46.
15. Gray, W.R., and Hartley, B.S. (1963) Biochem. J. 89, 379-380.
16. Chang, J.Y., Brauer, D., and Wittmann-Liebold, B. (1978) FEBS Lett. 93, 205-214.
17. Muramoto, K., Meguro, H., Tuzimura, K., Sakurai, A., and Tamura, S. (1976) Agr. Biol. Chem. 40, 2503-2504.
18. Edman, P., and Begg, G. (1967) Eur. J. Biochem. 1, 80-91.
19. Laursen, R.A. (1971) Eur. J. Biochem. 20, 89-102.
20. Hewick, R.M., Hunkapiller, M.W., Hood, L.E., and Dreyer, W.J. (1981) J. Biol. Chem. 15, 7990-8005.
21. Salnikow, J., Lehmann, A., and Wittmann-Liebold, B. (1981) Anal. Biochem., in press.
22. Wittmann-Liebold, B. (1973) Hoppe-Seyler's Z. Physiol. Chem. 354, 1415-1431.
23. Wittmann-Liebold, B., Geissler, A.W., Marzinzig, E. (1975) J. Supramol. Struct. 3, 426-447.
24. Wittmann-Liebold, B.: in Polypeptide Hormones (1980) R.F.J. Beers and E.G. Bassett (eds.) Raven Press, New York, pp. 87-120.
25. Wittmann-Liebold, B., Graffunder, H., and Kohls, H. (1976) Anal. Biochem. 75, 621-633.
26. Wittmann-Liebold, B. (1981): in Chemical Synthesis and Sequencing of Peptides and Proteins. T. Liu, A. Schechter, R. Heinrikson, and P. Condliffe (eds.), Elsevier North Holland Biomedical Press, Amsterdam, New York, Oxford, pp. 75-110.
27. Tarr, G.E., Beecher, J.F., Bell, M., and McKean, D.J. (1978) Anal. Biochem. 84, 622-627.
28. Klapper, D.G., Wilde, C.E., and Capra, J.D. (1978) Anal. Biochem. 85, 126-131.
29. Spiess, J., Rivier, J., Rivier, C., and Vale, W. (1981) Proc. Natl. Acad. Sci. USA 78, 6517-6521.
30. Kimura, M., Foulaki, K., Subramanian, A.R., and Wittmann-Liebold, B. (1982) Eur. J. Biochem., in press.

31. Noller, H.F. (1980), see ref. 1, pp. 3-22.
32. Wittmann-Liebold, B. (1980) in: RNA Polymerase, tRNA and Ribosomes. S. Osawa, H. Ozeki, H. Uchida, and T. Yura (eds.), University of Tokyo Press, Tokyo, pp. 639-654.
33. Allen, G., Capasso, R., and Gualerzi, C. (1979) J. Biol. Chem. 254, 9800-9806.
34. Wittmann, H.G., Dijk, J., and Brimacombe, R. (1981) in: Current Topics in Cellular Regulation 18, 487-504.
35. Wittmann, H.G. (1982) Ann. Rev. Biochem. 51, in press.
36. Ohsawa, H., and Gualerzi, C. (1981) J. Biol. Chem. 256, 4905-4912.

Strategy

STRATEGIC APPROACHES TO SEQUENCE ANALYSIS

KENNETH A. WALSH

Department of Biochemistry, SJ-70

University of Washington, Seattle, WA 98195

It has now been nearly 30 years since Sanger and his colleagues first showed that a protein (insulin) has a unique amino acid sequence[1]. The tools they developed to generate fragments, to isolate small peptides, to determine amino acid compositions, and to establish end groups defined the frontier of protein chemistry at that time, but have been replaced over the years by procedures with greater efficiency, speed and sensitivity. Moore, Stein and their colleagues[2,3] introduced more quantitative procedures. Edman replaced end group analysis with the stepwise degradation procedure bearing his name[4] and led us towards automated instruments which allow us to perceive answers before we forget the questions. It is becoming routine that the sequence of a purified protein is determined as a normal part of a complete program to study the structural basis of its function and control.

By 1976, about 80,000 residues had been placed[5]. That figure doubled in the next three years[6] and continues to grow at an ever accelerating pace as new strategies evolve. During the last decade the present series of symposia, has played a valued role in catalyzing and dispersing knowledge of recent innovations in technology and strategy. Hopefully, this conference will continue the tradition of leading us towards greater capability with large proteins, greater sensitivity with small quantities, and greater efficiency by improved strategies.

Not long age, it was a major achievement to determine the sequence of 200 residues in a single chain. Even today

less than 5% of completed analyses describe proteins larger
than 500 residues. In part this is simply related to the
size distribution of proteins, but it also reflects a
reluctance to commit the time and resources required to
analyze a 600-1500 residue chain, no matter how significant the task. In each case of such a large protein,
the investigators were faced with problems which required
development of unconventional strategies (e.g. with phos-
phorylase[7], β-galactosidase[8], RNA polymerase[9] and collagen
[10]).
    Fortunately, the potential for creation of new methodo-
logies exists. Two alternate procedures of growing significance are found in the areas of DNA sequencing, and of
mass spectrometry, both of which are discussed in this
symposium. Although DNA sequencing appears to be a wave
of the future, it is important to remember that it describes only the instructions for assembling the nascent
polypeptide chain, not the covalent nature of the mature
protein after proteolytic editing and covalent conjugation.
It seems to be the general rule rather than the rare
exception that proteins are acylated, methylated, glycosylated, phosphorylated, crosslinked, trimmed by proteolysis
or conjugated to prosthetic groups during or following
translation[11]. Since it is the mature protein which
ultimately excercises function in vivo, it is important to
identify these modulations which are not evident in the
genome or even in its cDNA counterpart. Those with experience in both protein and DNA sequencing maintain that
ideal results are obtained by a blend of the two approaches.
The nature of potential errors in each is quite different.
The protein chemistry can supply the covalent modulations,
ensure the exclusion of intron misinformation and preclude
frameshift errors. The nucleic acid chemistry can link
together segments of sequence or composition and minimize
the need for overlapping peptides. In an ideal world,
these two approaches would be carried out together, and in
tandem with xray crystallography so that a complete three-
dimensional model would result. However, practical considerations minimize the possibility that all three approaches
will be accomplished simultaneously in one laboratory.
    The second alternative is found in new applications of
mass spectrometry to sequence analyses[12-15]. Although
this method has long been the procedure of choice for
identifying anomalous or blocked residues, the difficulties
of derivatization and volatilization, and the considerable

expense have limited its widespread application. However,
recent applications include the complete mass spectrometric
analysis of the 162-residue protein dihydrofolate reductase
12 and the development of fast atom bombardment systems[13],
[14]. It may well turn out that combinations of mass spectrometric systems and DNA sequencing will ultimately largely
supplant our current chemical strategies. Time will tell
whether improvements in the accessibility of DNA and of
mass spectrometric instrumentation will lead us away from
our preoccupation with the Edman degradation.

Meanwhile, most approaches are based on selective fragmentation, difficult purifications and endless degradations.
The last few years have seen improvements in each of these
areas. New chemical or enzymatic cleavage procedures (e.g.
at Glu, Lys or Trp) have been summarized[6]; highly selective
cleavage at Asn-Gly[16] or Asp-Pro[17] are being more widely
applied; limited proteolysis is more frequently used to
partition large proteins into more tractable fragments[7].
Perhaps the most significant impact on efficiency has been
the introduction of HPLC techniques for peptide purification, as demonstrated by the range of contributions in this
volume. In our own laboratory, we are finding that a long-
standing philosophy of generating a smaller number of
larger fragments is reversing as it becomes clear that complex mixtures of smaller peptides are readily resolved by
HPLC[18]. The rate-limiting step in sequence analysis may no
longer be the purification hurdle.

In the area of Edman degradations, a major improvement in
the automated methodology has resulted from the introduction of the polycationic carrier Polybrene [19], [20] which
facilitates peptide retention regardless of size. It even
is found in the new gas-phase instrument of Hewick et al[21]
which replaces the spinning cup with a miniaturized flow-
through chamber containing sample imbedded in a matrix of
polybrene. This instrument is the latest development in a
series of mechanical improvements which began with the
solid phase instrument[22], and was followed by major modifications of the spinning cup instrument[23-25]. Unfortunately there is a significant gap between development of
these changes and their commercial availability.

It is clear that micro-refinements of current techniques
based on the Edman degradation broaden the spectrum of
proteins accessible to sequence analysis. Early methods
required μmols (mgs) of proteins; most now analyze peptides
in the 2 - 50 nmol range with PITC, DABITC or Dansyl

chloride. Methods with radiolabelled phenylisothiocyanate
entered the subnanomol range, and methods using intrinsically radiolabelled proteins are satisfactory in the sub-
picomol range[6]. Until recently it appeared that subnanomol analysis would require radiolabelling approaches,
but HPLC introduction and instrumental refinements in
Berlin[23] and more recently in Pasadena[21,24,25], are leading
to analyses at the 10 picomol level. This opens avenues
of analysis to trace proteins of biological interest as well
as facilitating strategies of analysis of large proteins,
where quantity of starting material becomes a limiting
factor.

When beginning a sequence analysis, there is a tendency
to simply fragment the protein with trypsin or CNBr and
grind out a mass of data without any general plan in mind.
Particularly with larger proteins, preliminary experiments
should first be explored to define simple methods to
divide the protein into smaller segments, for example by
limited proteolysis or by cleavage at Asn-Gly or Asp-Pro
bonds. Promising approaches can be recognized on SDS-gels
and by sequenator analysis of unfractionated mixtures[7].
It is much easier to analyze four 200-residue segments, and
link them together, than to analyze one 800-residue protein.

Other tactics are useful to extend information retrieval
from a large peptide. If an extended sequenator analysis
places a unique cleavage locus near its end, it is efficient to begin again with the blocked (e.g. succinylated)
peptide, cleave it at that locus and repeat the sequenator
analysis without the need for any separation of fragments
[26]. Effective variations of this technique were done
directly in the spinning cup with CNBr[27] or by acid cleavage at an Asp-Pro band (Walsh, unpublished).

Ideally, Edman degradations should be interpretable for
100 cycles or more, but there are few reports of such
achievements. The acid lability at hydroxyamino acid
residues is an intrinsic problem[28], and further ineffic-
iencies result from mechanical problems and impurity of
chemicals. Frank[29] has reported that novel scavenging
techniques increase yields and extend analyses. Others
have recently described a method to block the background
peptides with fluorescamine, which can be timed to leave
the major peptide with an amino-terminal proline residue
accessible to continued degradation[30].

The final product of a successful strategy can be a
hollow victory. Since predictive methods for three-

dimensional structures have not become useful beyond the
level of secondary structure, meaningful understanding of
protein function relies on the combination of xray data
and sequence data. Without the xray data, we can only
identify families of related proteins and locate affinity
labels or sites of conjugation in the linear sequence.
Our growing awareness of the domain substructure of proteins lends confidence that segments of sequence information which dictate domains will continue to be identified
with functional substructures. As DNA recombination
events become better understood, we may recognize patterns
of evolutionary assembly of these functioning substructures of complex proteins. However, it remains true that
our ability to read and interpret our sequential data lags
behind our ability to generate new data to interpret.

REFERENCES

1. Sanger, F., and Tuppy, H. (1951) Biochem. J., 49, 463.

2. Spackman, D.H., Stein, W.H., and Morre, S. (1958) Anal.
    Chem., 30, 1190.

3. Hirs, C.H.W., Moore, S., and Stein, W.H. (1960) J. Biol.
    Chem., 235, 633.

4. Edman, P. (1956) Acta Chem. Scand. 10, 761.

5. Edman, P. (1977) Carlsberg Res. Commun. 42, 1.

6. Walsh, K.A., Ericsson, L.H., Parmelee, D.C., and Titani,
    K. (1981) Ann. Rev. Biochem. 50, 261.

7. Walsh, K.A., Ericsson, L.H., and Titani, K. (1978). in
    Versatility of Proteins, Li, C.H., ed., Academic
    Press p. 39.

8. Fowler, A.V., and Zabin, I. (1978) J. Biol. Chem. 253,
    5521.

9. Ovchinnikov, Y.A., Monastyrskaya, G.S., Gubanov, V.V.,
    Guryev, S.O., Chertov, O.Y., Modyanov, N.N., Grinkevich,
    V.A., Makarova, I.A., Marchenko, T.V., Polovnikova, I.
    N., Lipkin, V.M. and Sverdlov, E.D. (1981) Eur. J.
    Biochem. 116, 621.

10. Fietzek, P.P., Allmann, H., Rauterberg, J., Henkel,
    W., Wachter, E., and Kuhn, K. (1979) Hoppe-Seyler's
    Z. Physiol. Chem. 360, 809.

11. Uy, R., and Wold, F. (1977) Science 198, 890.

12. Morris, H.R. (1979) Phil. Trans. R. Soc. Lond. A. 293,
    39.

13. Morris, H.R., Panico, M., Barber, M., Bordoli, R.S.,
    Sedgwick, R.D. and Tyler, A. (1981) Biochem. Biophys.
    Res. Commun. 101, 623.

14. Williams, D.H., Bojesen, G., Auffret, A.D., and Taylor,
    C.E. (1981) FEBS Lett. 128, 37.

15. Nebelin, E. (1980) in Methods in Peptide and Protein
    Sequence Analysis, Birr, C., ed., Elsevier/North
    Holland Press p. 173.

16. Bornstein, P., and Balian, G. (1977) Methods Enzymol.
    47, 132.

17. Landon, M. (1977) Methods Enzymol. 47, 145.

18. Bloxham, D.P ., Parmelee, D.C., Kumar, S., Wade, R.D.
    Ericsson, L.H., Neurath, H., Walsh, K.A., and Titani,
    K. (1981) Proc. Natl. Acad. Sci. U.S.A. 78, in press.

19. Tarr, G.E., Beecher, J.F., Bell, M., and McKean, D.J.
    (1978) Anal. Biochem. 84, 622.

20. Klapper, D.G., Wilde, C.E. III, and Capra, J.D. (1978)
    Anal. Biochem. 85, 126.

21. Hewick, R.M., Hunkapiller, M.W., Hood, L.E., and
    Dreyer, W.J. (1981) J. Biol. Chem. 256, 7990.

22. Laursen, R.A., (1971) Eur. J. Biochem. 20, 89.

23. Wittman-Liebold, B., (1980) in Polypeptide Hormones,
    Beers, R. F. Jr., Bassett, G., eds., Raven Press, p.
    87.

24. Hunkapiller, M.W., and Hood, L.E. (1978) Biochemistry 17, 2124.

25. Hunkapiller, M.W., and Hood, L.E. (1980) Science 207, 523.

26. Koide, A., Titani, K., Ericsson, L.H., Kumar, S., Neurath, H., and Walsh, K.A. (1978) Biochemistry 17, 5657.

27. Boosman, A. (1980) in Methods in Peptide and Protein Sequence Analysis, Birr, C., Ed., Elsevier/North Holland Press, p. 513.

28. Brandt, W.F., Henschen, A., and von Holt, C. (1980) Hoppe-Seyler's Z. Physiol. Chem. 361, 943.

29. Frank, G. (1979) Hoppe-Seyler's Z. Physiol. Chem. 360, 997.

30. Bhown, A.S., Bennett, J.C., Morgan, P.H. and Mole, J. E. (1981) Anal. Biochem. 112, 158.

Improvements in Instrumentation and the Chemistry of Isothiocyanate Degradation

A NEW PROTEIN MICROSEQUENATOR USING GAS PHASE EDMAN REAGENTS

Michael W. Hunkapiller, Rodney M. Hewick,
William J. Dreyer, and Leroy E. Hood

Division of Biology
California Institute of Technology
Pasadena, California 91125

Since its introduction in 1967, the use of automated Edman degradation in the spinning cup sequenator has been the most widely used method for determining the primary structure of polypeptides (1). For many years, the major limitations of the technique were the large amounts of sample required (10 to several hundred nmol) and difficulty in sequencing short peptides. These problems were mainly due to sample loss during the repeated extractions with organic solvents used to remove nonvolatile reagents and by-products from the sample film. Early attempts to minimize sample loss included the use of more volatile coupling buffers (2, 3), reduction in the volume of extraction solvents (4, 5), and the addition of various nonprotein carriers to the cup to add mass and stability to the sample film (3). However, the major breakthrough came with the introduction of a polymeric quaternary ammonium salt, Polybrene, to spinning cup sequencing technology (6). This substance effectively anchors small quantitites of both proteins and peptides in the cup and allows sequencing of even short hydrophobic peptides to completion. By combining the use of Polybrene with more extensive purification of reagents and solvents, improvements in spinning cup sequenator design, and analysis of > PhNCS-derivatives by reverse phase HPLC, we have previously obtained extended NH_2-terminal sequence information on subnanomole quantities of a variety of peptides and proteins (7, 8).

The solid phase sequencing system of Laursen (9), wherein the protein or peptide is covalently attached to a derivatized glass or polystyrene support, was developed in parallel with the spinning cup system and provides another answer to the problem of extractive sample loss. However, the major limitations of this technique are that it is rarely possible to achieve quantitative attachment of sample to the support phase, another set of reagents and solvents is required for covalent sample attachment, and gaps appear in the amino acid sequence where attachment has occurred.

GAS PHASE SEQUENATOR DESIGN AND OPERATION

We now describe a new type of sequenator which combines the simplicity of sample application of the spinning cup sequenator with the miniaturization and mechanical simplicity of the solid phase sequenator. These features are made possible by the use of gas phase rather than liquid phase reagents at two critical points in the Edman degradation, a technique introduced for manual Edman degradation (10) and later attempted in automated sequencing (11, 12). In our system, the polypeptide is embedded in a matrix of Polybrene dried onto a porous glass fiber disc located in a small cartridge-style reaction cell. The sample, though not covalently attached to the support, is essentially immobile throughout the degradative cycle, since the only liquid streams it is exposed to are relatively apolar extraction solvents.

The gas phase sequenator has certain features in common with our liquid phase spinning cup sequenator described previously (7, 8) and some critical differences, as shown in the schematic (Fig. 1).

The similarities include the following: the use of argon, delivered through a series of filters, to pressurize the reagent and solvent reservoirs; similar argon pressure-regulation manifold, reservoir pressurizing valves, and reservoir-venting manifold; similar reagent/solvent reservoirs, some of which are modified to deliver vapor as opposed to liquid; the use of zero-holdup, pneumatically actuated diaphragm valves for delivery of reagents, solvents, and argon; automated conversion of anilinothiazolinones to > PhNCS derivatives (7, 13); and control of valve function by means of a solid state programmer.

The major differences are summarized as follows: the replacement of the cup and drive unit by a cartridge assembly which houses a miniature glass reaction chamber (internal volume, $\sim$0.050 ml) and the presentation of the polypeptide sample in this chamber

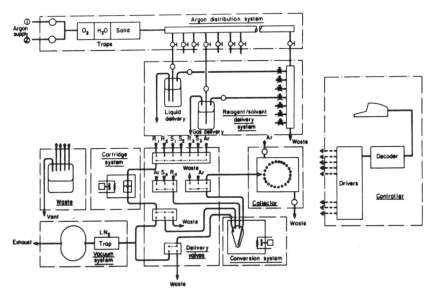

Fig. 1. Schematic diagram of gas-liquid solid phase sequenator.

on a porous glass fiber disc. The cartridge is mounted in an insulated oven equipped with heating and cooling elements for programmed temperature control.

The pneumatically actuated diaphragm valves have been miniaturized to reduce their internal volume to one-tenth of that in the spinning cup system. These valves also control the vacuum to the cartridge and the > PhNCS conversion flask. The large electromagnetic vacuum valves that are used in the spinning cup instrument are not required. The reservoirs and various delivery valve blocks are mounted in the vertical plane directly above and below the reaction cartridge as appropriate to minimize the length of connecting tubing. The automatic conversion flask and fraction collector are miniaturized, and the vacuum pump and liquid nitrogen trap also are correspondingly smaller.

The reaction cartridge is constructed from two pieces of Pyrex glass rod with finely ground vacuum-flat surfaces at both ends. Each piece of glass is ultrasonically machined so that it has a 0.020-inch diameter central capillary flared out to give a conical recess at one end. The recessed ends of the two pieces of glass are clamped together in a metal cylinder to form a central chamber. The conical cavity in the upper piece of glass has a

small recess sufficient to hold the glass fiber disc (GF/C Glass
Microfibre Filter, Whatman Ltd., England) on which the protein
or peptide is supported. Clamped between the two pieces of glass
is a fibrous porous Teflon disc (Zitex filter membrane, extra-coarse
grade, Chemplast Inc., Wayne, NJ). This Teflon disc physically
supports the glass fiber disc and is crushed between the abutting
glass surfaces to give a vacuum seal. Enlarged detail of this arrangement is shown in Fig. 2.

The reagents and solvents used in the gas phase sequenator
are listed in Table I along with their approximate per cycle consumption. Great emphasis has been placed on the purity of reagents
and solvents used for the Edman degradation and this purity is
just as essential for the gas phase as for other sequenators.

The sequenator program is generally similar to that described
(7) for our modified spinning cup sequenator. However, since

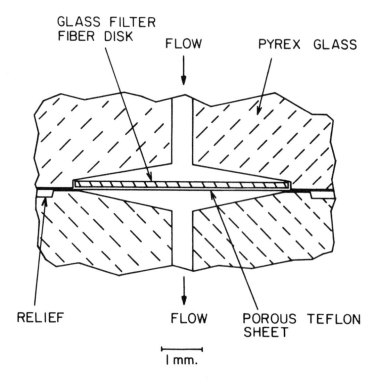

Fig. 2. Enlarged detail of reaction chamber.

Table I. Reagents Used in Gas Phase Sequenator

Reagent/solvent	Volume per cycle
	ml
R1: 15% phenylisothiocyanate in n-heptane	0.05
R2: 25% trimethylamine in water	5 cc/min argon flow
R3: Trifluoroacetic acid, 0.01% dithio-threitol	5 cc/min argon flow
R4: 25% trifluoroacetic acid in water, 0.01% dithiothreitol	0.05
S1: Benzene	1.1
S2: Ethylacetate, 0.05% acetic acid, 0.002% dithiothreitol	2.7
S3: 1-Chlorobutane, 0.001% dithiothreitol	1.2
S4: Acetonitrile, 0.001% dithiothreitol	0.3

the present instrument uses a miniaturized cartridge system and gas phase reagents, there are several features worth noting. 1) Only sufficient phenylisothiocyanate solution to completely wet the glass fiber disc is delivered ($\sim$20 µl). The heptane is removed by briefly flushing the cartridge with argon. 2) The coupling stage is effected by slowly bleeding trimethylamine/water vapor through the cartridge to waste. Likewise, cleavage is effected by slowly bleeding trifluoroacetic acid vapor through the cartridge to waste. 3) Argon is used to pressurize the cartridge after a vacuum step or to flush the delivery valve block, lines, and cartridge of liquids or other gases. 4) The miniaturized conversion flask has two argon supplies: line flush, which flushes the flask delivery valve block and lines and pressurizes the flask; and flask argon, which delivers argon into the bottom of the flask to aid drying and agitation of fluids. 5) The conversion flask, in addition to the vapor-vent valve, has a liquid waste outlet (see Fig. 1), so that the flask can be thoroughly rinsed with solvent during each cycle. 6) To confine the sample to the lower tip of the miniaturized flask and also perform efficient extraction, the amino acid anilinothiazolinone is extracted from the glass fiber filter with several small volumes of chlorobutane. Continuous drying of the flask contents is necessary and is effected by flushing the flask with argon during these steps in order to minimize the time between extraction from the disc and addition of reagent 4 to the flask.

In the present program, the cartridge is maintained at 42°C.
The basic program incorporates a single coupling stage with two
deliveries of phenylisothiocyanate and two complete cleavage
stages. The amino acid anilinothiazolinone released by the shorter
first cleavage stage is extracted and directed to the conversion
flask, and the derivative extracted after the second cleavage
is directed to waste. This procedure reduces exposure of the
more labile amino acid derivatives to strong anhydrous acid, since
most of the derivative is released during the first cleavage, yet
maintains an adequate overall cleavage time. In most cases, a
sequence run is initiated with a double coupling. This is performed
by completing the first degradative cycle with the reagent 3 valve
switched off.

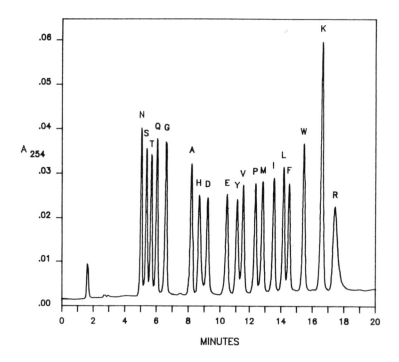

Fig. 3. Separation of >PhNCS amino acids by HPLC on an IBM
Cyano column. Column conditions: 32°C; aqueous phase - 0.015 M
sodium acetate buffer, pH 5.8; hydropholic phase - MeCN/MeOH
(4:1); flow rate - 1 ml/min. Sample was 10 μl of MeCN containing
1 nmol of each >phNCS amino acid.

Sequenator fractions are analyzed by reverse-phase HPLC using IBM Cyano columns with an adaptation of the procedure described previously (14). The resolution of >PhNCS amino acids obtainable with this system is shown in Fig. 3.

RESULTS

Sperm whale apomyoglobin was sequenced at several levels of sample loading in order to test the efficiency of the gas phase sequenator in performing the Edman degradation. The first 90 amino acid residues were identified from a single sequenator run with a 10-nmol load of protein (Fig. 4). The yields of individual >PhNCS-derivatives are similar to those reported previously (7); those of authentic >PhNCS serine and >PhNCS threonine, two of the most troublesome, are 15 and 30%, respectively. The reproducibility of the background of >PhNCS-derivatives (Fig. 4) afforded by the automatic conversion system allows accurate residue assignment even late in the run when the signal-to-background ratio is low. Based upon quantitation of the amount of protein

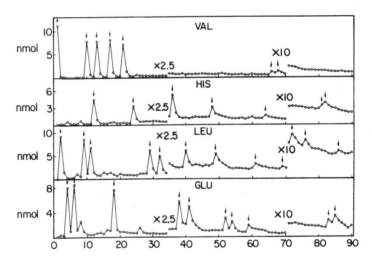

Fig. 4. Yields of >PhNCS valine, histidine, leucine, and glutamic acid derivatives from an NH_2-terminal amino acid sequence analysis of 10 nmol of sperm whale apomyoglobin. Aliquots (40%) from each cycle were analyzed by HPLC. HPLC peak heights were converted to nanomoles for each derivative using values from a standard mixture of >PhNCS-derivatives, and the yields were normalized to 100% injection.

loaded on the cartridge disc by amino acid analysis of a companion aliquot, the yields of >PhNCS valine and >PhNCS leucine at cycles 1 and 2, respectively, were in excess of 90%.

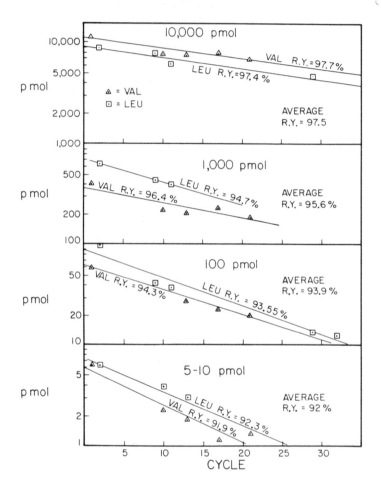

Fig. 5. Sequenator repetitive cycle yield with various amounts of sperm whale apomyoglobin. Semilogarithmic plots of >PhNCS valine ( ) yields (cycles 1, 10, 13, 17, and 21) and >PhNCS leucine ( ) yields (cycles 2, 9, 11, 29, and 32) versus sequenator cycle are shown for analysis of 10 nmol, 1 nmol, 100 pmol, and 10 pmol of protein. The repetitive yield (R.Y.) at each cycle was calculated from the slopes of the linear-least-squares-fitted straight lines for the plots of >PhNCS yields.

The repetitive cycle yield for runs with myoglobin loads ranging from 10 nmol to 5 pmol varied from 98% (10 nmol) to 96% (500 pmol) to 94% (50 pmol) to 92% (5 pmol) (Fig. 5). The drop in repetitive yield as the sample load is decreased most likely reflects the effect of sample washout during the solvent extractions and/or trace levels of oxidants in the sequenator system. Trace levels of oxidants are suspected, because only in the lowest sample load (5 pmol) was identification of serine and tryptophan residues, both of which are very sensitive to oxidative destruction, impossible. Even with these effects, partial sequence data to residue 22 was obtained on the 5-pmol myoglobin run (Fig. 6).

Human angiotensin II was used to test the ability of the new sequenator to handle short, relatively hydrophobic peptides. The yields of > PhNCS-derivatives for runs with angiotensin loads ranging from 5 nmol to 50 pmol are shown in Fig. 7. The complete sequence of the octapeptide could be determined using 5 nmol and 500 pmol of sample, and all but the last two residues (proline and phenylalanine) could be determined using 50 pmol. The HPLC

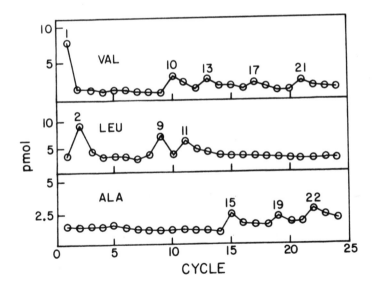

Fig. 6. Yields of > PhNCS valine, leucine, and alanine derivatives from an NH_2-terminal amino acid sequence analysis of 10 pmol of sperm whale apomyoglobin. Yields were calculated as described in the legend to Fig. 4.

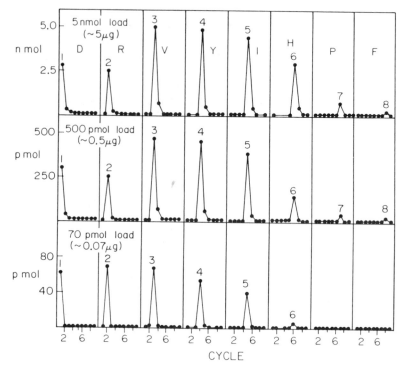

Fig. 7. Yields of > PhNCS-derivatives from NH$_2$-terminal amino
acid sequence analyses of 5 nmol, 500 pmol, and 70 pmol angiotensin II (sequence: H-Asp-Arg-Val-Tyr-Ile-His-Pro-Phe-OH).
Yields were calculated as described in the legend to Fig. 4.

traces from a 50-pmol angiotensin run are shown in Fig. 8. Very
little extraneous 254 nm absorbing material appears in the chromatograms. Those artifacts that are present, mainly a small peak
eluting just after > PhNCS methionine derivative, do not seriously
interfere with identification. Interference with identification
of low levels of > PhNCS aspartic acid and > PhNCS glutamic acid
derivative, which can be obscured by the reduced and oxidized
forms of dithiothreitol, is avoided by methylating the acidic side
chains with a 1 N methanolic HCl solution prior to HPLC analysis
(15).

In order for the new gas phase sequenator to be a general
purpose instrument, it was designed to handle as wide a variety
of polypeptides as possible. Proteins can be analyzed using less
than 10 pmol (0.2 µg) of sample; peptides can be sequenced with

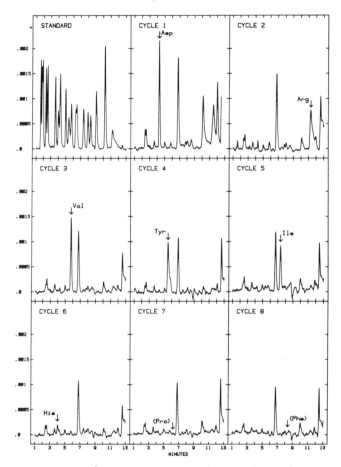

Fig. 8. HPLC traces from an NH_2-terminal amino acid sequence analysis of 50 pmol of human angiotensin II. The order of elution of > PhNCS-derivatives (12.5 pmol of each) in the standard mixture (upper left corner) is Asn, Ser, Thr, Gln, Gly, Ala, His, AspOMe, GluOMe, Tyr, Val, Pro, Met, Ile, Leu, Phe, Trp, Lys, and Arg. Ten-µl aliquots, representing 40% of each sample, were injected. The positions of the > PhNCS-derivatives assigned in the traces for cycles 1 through 6 are indicated by the arrows and the three-letter amino acid designations. The positions of the expected > PhNCS-derivatives for cycles 7 and 8 are indicated in a similar manner, although they are not assignable in this experiment.

less than 100 pmol (0.1 µg) of sample. Both hydrophobic peptides and proteins, including integral membrane proteins, can be sequenced.

Proteins purified by SDS-polyacrylamide gel electrophoresis and by electrofocusing in polyacrylamide gels can be analyzed after the proteins are electrophoretically eluted from the gels. Polypeptides that contain proline residues followed by amino acid residues with bulky side chains also pose no serious problems if the cleavage time is doubled to minimize overlap due to incomplete reaction (16).

CHARACTERISTICS OF GAS PHASE SEQUENATOR

An important characteristic of the reaction cartridge built around the sample support disc is the ease with which the sample containment area can be miniaturized. This, along with the simple flow-through nature of the cartridge assembly, allows the reagent and solvent consumption to be reduced to one-tenth or less of that used in commercial instruments. This has several benefits worth noting. The first is a significant reduction in operating costs. A second is the increased practicality of providing the required amounts of ultrapure reagents and solvents, an important consideration since many of the commercially available chemicals require additional purification to provide the desired level of purity. Yet a third advantage is the increase in speed with which samples can be sequenced. This results from the decreased time required for mass transfer in the miniaturized system and from the very rapid changeover from one sample to another. Cycle time is only 45-55 min, and sample reloading (including cartridge cleanup and Polybrene precycling) is only 3-4 h. A final benefit and the one that is perhaps most important, is that lower reagent and solvent usage per cycle results in a reduced accumulation of impurities accompanying the > PhNCS-derivative samples that are analyzed by HPLC. The low HPLC backgrounds shown in Fig. 8 attest to the low background level of the sequenator, and this miniaturization of artifacts is essential to sequencing at ultra-micro levels.

The efficiency with which the new sequenator performs the Edman chemistry, as judged by its repetitive cycle yield, is at least as good as the best available spinning cup sequenators and better than the Laursen-type solid phase instruments. Average repetitive yields of 98% are obtainable with as little as 10 nmol of protein, and 92% cycle yields can be obtained with only 5 pmol of protein (Fig. 5). Sequencing efficiency also is high for peptide analysis, with complete sequencing of small to medium peptides possible with subnanomole quantities.

Several factors contribute to this high efficiency. One is the thoroughness with which the sample cartridge can be cleansed with the extracting solvents. Another is the protection of the sample from removal from the reaction chamber. Other factors include the thoroughness with which the entire reaction system can be protected from leaks and cross-contamination, purged with argon, and evacuated. Also, the coupling base used in this system, trimethylamine, is very easily purified by distillation, whereas Quadrol and other buffers used in classical sequenators are difficult to purify and are prone to contain impurities, such as aldehydes, which block the Edman chemistry. Further, since the Edman reactions can proceed to completion rapidly at 42°C (rather than the usual 55°C), the acid-catalyzed splitting of the polypeptide chain that generates background > PhNCS-derivative signals and reduces repetitive yield is lessened (1, 17). The miniaturization even increases the efficiency of the analysis of the > PhNCS-derivative fractions. The > PhNCS-derivatives generated in the automatic conversion flask are transferred directly into the 300-µl conical tubes used in the autosampler for the HPLC system. This saves time, avoids sample loss, reduces chances of sample contamination, and improves quantitation.

Note: Portions of this paper have appeared in J. Biol. Chem. **256**, 7990-7997 (1981) by the present authors.

REFERENCES

1. Edman, P., and Begg, G. (1967) Eur. J. Biochem. **1**, 80-91.
2. Hermodson, M. A., Ericsson, L. H., Titani, K., Neurath, H., and Walsh, K. A. (1972) Biochemistry **11**, 4493-4502.
3. Niall, H. D., Jacobs, J. W., van Rietschoten, J., and Tregear, G. W. (1974) FEBS Lett. **41**, 62-64.
4. Crewther, W. G., and Inglis, A. S. (1975) Anal. Biochem. **68**, 572-585.
5. Brauer, A. W., Margolies, M. N., and Haber, E. (1975) Biochemistry **14**, 3029-3035.
6. Tarr, G. E. Beecher, J. F., Bell, M., and McKean, D. J. (1978) Anal. Biochem. **84**, 622-627.
7. Hunkapiller, M. W., and Hood, L. E. (1978) Biochemistry **17**, 2124-2133.
8. Hunkapiller, M. W., and Hood, L. E. (1980) Science **207**, 523-525.
9. Laursen, R. A. (1971) Eur. J. Biochem. **20**, 89-102.
10. Schroeder, W. A. (1967) Methods Enzymol. **11**, 445-461.
11. Waterfield, M. D., Lovins, R. E., Richards, F. F., Salomone, R., Smith, G. P., and Haber, E. (1968) Fed. Proc. **27**, 455.

12. Dreyer, W. J. (December 27, 1977) U.S. Patent No. 4,065,412.
13. Wittmann-Liebold, B., Graffunder, H., and Kohls, H. (1976)
     Anal. Biochem. **75**, 621-633.
14. Johnson, N. D., Hunkapiller, M. W., and Hood, L. E. (1979)
     Anal. Biochem. **100**, 335-338.
15. Tarr, G. E. (1975) Anal. Biochem. **63**, 361-370.
16. Brandt, W. F., Edman, P., Henschen, A., and von Holt, C. (1976)
     Hoppe-Seyler's Z. Physiol. Chem. 357, 1505-1508.
17. Brandt, W. E., Henschen, A., and von Holt, C. (1980) Hoppe-
     Seyler's Z. Physiol. Chem. **361**, 943-952.

TOWARDS LONGER DEGRADATIONS ON A SEQUENATOR

Gerhard FRANK

Institut für Molekularbiologie und Biophysik

ETH-Hönggerberg, 8093 Zurich   SWITZERLAND

INTRODUCTION

Despite new and fascinating micro sequencing techniques successful sequencing nowadays is not necessarily dependent upon very specialized equipment. It has to be realized that the great majority of laboratories are still working with commercially available sequenators,for example the Beckman sequencers. Although there are limitations they are still fit for many tasks, provided that sufficient amounts of protein are available, or that it can easily be obtained in radioactive form. With these prerequisites met such instruments can provide very long degradations which have a substantial effect on the efficiency of a sequence determination. The good efficiency (the number of residues finally placed in sequence divided by the number of fragments isolated to prove that sequence, see Walsh et al. (1)) reduces both the amount of protein and the labour which have to be employed in the production of fragments. Thus, with sufficient amounts of protein available, sequence determination with standard equipment can be quite economical.

PROFITABLE MEASURES TO
IMPROVE SEQUENATOR PERFORMANCE

To avoid the inborn difficulties with volatile buffers

91

we used 0,25N quadrol as the only buffer for all experiments.
In order to achieve optimum performance of a sequenator several requirements have to be met, the most important:an adequate program and sufficiently pure chemicals. For the
spinning cup type of instrument an optimum contact between
polypeptide film and the reagents is also required.

                              The Program

        A very accurate adjustment of the reagent levels -
with the exception of PITC - is imperative and cannot be
completely substituted by an appropriate repeated change
of the cup rotation speed. The latter, as proposed by
Hunkerpiller and Hood (2), may be helpful but alone is not
the ultimate solution. Unfortunately the most widely used
system, the delivery through a constant nitrogen pressure,
causes difficulties in this respect: Differences in the pressure on the reservoir bottles can occur due to inaccurate
adjustment after topping up or because of corrosion of either
the manual or the solenoid vent valves. Furthermore partially clogged delivery valves and a faulty programmer (infrequent errors which occur in the last decade for example
with the Beckman sequencer) have been the cause of rather
"hard-to-detect" troubles. The main problem with these
effects is that they never cause a complete breakdown but
rather reduce the ultimate performance. As a result the chemicals are quite often erroneously suspected.

        Closely linked to the program is a well controlled re-
action temperature for the cleavage. Due to the preceding
drying steps and as a consequence of the rather high heat
of evaporation of the solvent, the protein film has been
cooled to rather low temperatures before the cleavage reaction has to take place. Unfortunately the temperature
cannot be measured with desirable accuracy since the ther-
mometer is located at a position where poor heat conduction
from the film is expected. Thus optimum conditions can only
be achieved by trial and error. This becomes even more difficult since the temperature drop depends on the amount of
solvent that remains in the cup and hence factors like cup
speed (50 or 60 cycles power supply) efficiency of the scoop,
thickness of the film etc. start to play a role. Even the

obvious cure, to allow a sufficient time for reequilibration, is not free of danger since the rather unstable intermediate of the reaction is then exposed for a longer time to the heat and the gases in the bell jar.

Incomplete contact between film and reagent caused by the film peeling from the wall of the cup can also be due to an inadequate program. This does not refer to the obvious situation where the film comes completely off the cup and is no longer moistened by the reagents but to cases where the film is only partially detached and may well appear completely moistened. In such instances the "only" effects are a more or less reduced repetitive yield and increased overlaps. The most frequent cause of such trouble is too vigorous drying before the addition of reagent. Another cause can be an oil film on the cup which may be attributed to back diffusion of oil from the pumps. This can be reduced considerably if any fine vacuum step is avoided and replaced e.g. by the "high vacuum plus nitrogen flush" condition of the Beckman sequencer. Obviously cool traps in the vacuum lines are another very good, but expensive, measure against this complication.

The Chemicals

In contrast to the widely accepted demands that chemicals for sequenator procedures have to be extremely pure and thus have to be used shortly after purification, our demands are less stringent. We mainly want the chemicals to be free of impurities which actually interfere with the Edman reactions and we thus used the principle of last minute purification (3) to scavange just the dangerous part of the impurities. Since this approach appeared to be rather successful, particularly in case of the quadrol buffer, we have attempted to make even better use of the principle. A main problem was that the aminoethylcellulose scavanger showed a tendency to age and take on a slimy appearance. We thus replaced it by β-N-aminoethyl-(3-aminopropyl) silica which we produced from LiChroprep Si 100 particle size 40-63 μm (from Merck, Darmstadt, FRG) and N-(β-aminoethyl)-γ-aminopropyl-trimethoxysilane (from Pierce, Rockford, Ill.).

The reaction was carried out in acetone as described by
Bridgen (4). This new material has the advantage that it
exhibits a higher capacity than aminoethylcellulose (Fig.1).
It does not alter its consistency with time and although
silica is unstable in alkaline solutions the pH 9.0 in the
mainly organic solvent of the quadrol buffer does not
appear to have any ill effect on the material. Long term
experiments are in progress and our present impression is
that the results will be quite satisfactory. A particular
virtue of the aminoethyl-aminopropyl cellulose is that it
can be titrated. This will enable us to determine the time
course of the reaction with quadrol buffer impurities and
thus calculate the adequate time for replacement. We expect this to be far more than a year since our buffer solutions are kept over scavanger while in stock. (Due to our
initiative the quadrol buffer delivered by Fluka, Buchs,
Switzerland, does already contain a scavanger which is particularly convenient in this respect)

Since aminoethyl-aminopropyl silica is available in
large amounts and at reasonable prices it presents itself
as a suitable material for the reaction with PITC thus pro-

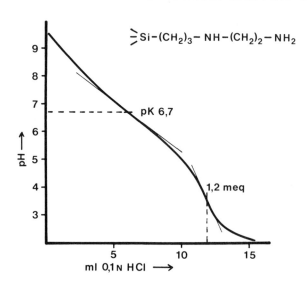

Fig.1) Titration curve of 1 g of β-N-aminoethyl-(3-amino-
    propyl)-silica.

ducing a scavanger for all impurities interfering with the phenylthioureau derivative resulting from the coupling reaction. In our experiments the capacity of the material was reduced to approximately 0,5 meq/g showing that not all amino groups were accessible for the PITC. Nevertheless the product should be fit for the purpose but it is probably rather unstable as can be deduced from a typical smell which has to be attributed to some degradation product. In spite of this complication the material has been added to the ethylacetate but no effect, neither favourable nor adverse, could be observed.

Future experiments are planned to understand and probably control the degradation and the addition of the material to the quadrol buffer as a second scavanger is also considered.

RESULTS

The performance

Using 0,25N quadrol buffer with either peptides and proteins and provided that all the above mentioned prerequisites can be attained, we can determine well over 80 cycles for most proteins(assuming the background is not excessively high as in proteins of over 200 residues or, under particular circumstances, even in smaller proteins). In the presence of polybrene we normally determine peptides of up to 50 residues completely (with the occasional exception of the last 2 or 3 residues depending on the particular amino acids at those positions e.g. serine or threonine) and larger peptides can be sequenced for 70 to 80 cycles depending on the actual size and properties.

Micro sequence determination with radioactively labelled proteins has so far only been attempted in one case (5). In spite of great differences in the radioactivity of the different amino acids and a slightly heterogenous starting material there was a clear result for 30 residues. After these 30 cycles the experiment was stopped since the result thus obtained, was more than enough to prove the assumed relationship to an already known DNA sequence. Furthermore the evaluation of additional steps would have been quite time

consuming since the HPLC separations of the PTH amino acid
had to be fractionated before the products could be counted.
From the data of the first 30 cycles it can be concluded
that about 50 cycles would most likely have been accessible
for a sure determination particularly with a more homogenous starting material.

In most cases where these standards could not be met it
could be proved that the starting material was not a pure
polypeptide. This happens particularly with large peptides
from specific cleavages and is a problem which deserves
further attention.

The actual basis for such rather long runs (both micro
and normal ones) is the fact that the values of the background amino acids, although very high, are extremely constant. Unfortunately this favourable fact is quite often
blurred by the less constant yield of the extraction from
the sequenator cup. Thus the high constancy of the background will only become apparent when there is a good internal standard within the cup. Presently we do not know a
suitable material but in one instance we accidently found
a favourable condition. In the β-chain of C-Phycocyanin (6)
there was no valin between residues 51 and 127. Thus we
could use the valine background as an internal standard for
the calculation of residues 79 up to 100 of this run. Fig.2
shows a plot of the yields of isoleucine and alanine versus
the number of degradation cycle, based on the assumption
that the valine background was 30 units all the time. As
can be seen the scatter is less than 10% of the level of
the background. It would probably be even less for alanine
if the valine values would not be slightly influenced by
isoleucine. In our HPLC separation valine follows isoleucine and the separation is a nearly baseline one. However
minute tailing of isoleucine could cause the valine values
to rise a little. In our way of calculation the consequence
of the elevated standard valine would give too low a value
for all the other amino acids. It is thus probably no accident that the alanine background comes out particularly low
just at those cycles where the isoleucine - and subsequently
the valine - is particularly high.

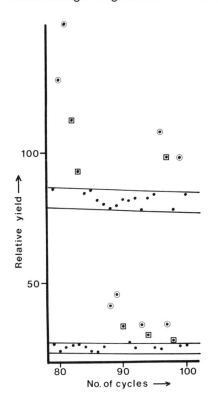

Fig.2) Yields for alanine (top) and isoleucine (bottom) of the last 22 cycles of a sequenator run. ⊙ significant for the amino acid, ▣ overlap, • background. ‥‥ range within which the background scatters.

The strategy

Based on the presently available performance we are using a very simple strategy for the determination of complete sequences: After the aminoterminal sequence of the whole chain has been determined to an extent as great as possible we look for a specific cleavage site close to it's end. After cleavage at that site the isolation of the overlapping peptide is quite often simplified because of the knowledge about the composition of some other fragments which can be derived from the already known part of the sequence. The overlapping peptide is then identified beyond doubt by its amino terminal composition and is also sequenced as far as possible. Then the most suitable cleavage site for the next cleavage is determined etc.

The best result we have so far obtained by this method
was to get a 161 residues sequence of a 162 residues protein
subunit with just 3 sequenator runs. The last 3 residues
could then be determined with carboxypeptidase thus establishing both the carboxyterminal serine and the overlap
into the sequenator results.

CONCLUSIONS

Long sequenator runs apparently are never the result
of a single adaptation, adjustment or other improvement.
They are always the consequence of a large number of measures which all tend to look rather simple and unimportant.
Many of them are hard to achieve since the instruments which
are mainly used for the purpose are not optimally constructed. Logically one is tempted to alter parts of the
instruments in order to improve the realization of known
advantages. This is absolutely right and meritorious but it
must not lead to the wrong conclusion that the particular
alteration which has been executed is an indispensible prerequisite for a successful long run. The fact that we could
do long runs without any alteration simply proves the contrary.

Long runs obtained with a commercially available instrument apparently are a very economical way of carrying
out sequence determinations. However the rather high amounts
of protein needed can compensate the advantageous effect.
The more difficult it is to obtain the proteins to be sequenced the more true this becomes. Thus the usefulness
of our approach varies greatly from protein to protein but
if it can be applied it has another great advantage: Due to
the very simple and straightforward strategy which can be
followed, the amount of work which goes into a sequence determination is small and cannot be compared with the great
efforts which are presently made in the field of microsequencing.

Finally it should be stated that all the measures explained above could yield a multiple benefit if a workable
way to reduce the background would be found. It is therefore

one of my main expectations and a great hope for this
meeting that it might bring us closer to the solution of
this problem.

REFERENCES

1. Walsh, K.A. et al. Ann. Rev. Biochem. 1981 50:216-84

2. Hunkerpiller, M.W. and Hood, L.E. 1978.
   Biochemistry 17: 2124-33

3. Frank, G. 1979. Hoppe-Seyler's Z. Physiol. Chem.
   360:997-99

4. Bridgen, J. 1975. FEBS Lett. 50:159-62

5. Schwyzer, M. et al. 1980. J. Biol. Chem. 255:5627-34

6. Frank, G. et al. 1978. Hoppe-Seyler's Z. Physiol. Chem.
   359-1491-1507.

THE NATURE OF NON-SPECIFIC PEPTIDE BOND CLEAVAGE DURING

THE ISOTHIOCYANATE DEGRADATION OF PROTEINS

WOLF F. BRANDT, AGNES HENSCHEN* AND CLAUS VON HOLT
Department of Biochemistry, Chromatin Research Unit, Council
for Scientific and Industrial Research, University of Cape
Town, Private Bag, Rondebosch. 7700 Republic of South Africa.

INTRODUCTION

Only in favourable cases do the number of amino acids that
can be positioned in a single run during the isothiocyanate
degradation exceed sixty residues. The premature termination of the degradation is caused by an accumulation of
small losses in yield of only a few percent in each cycle,
and a gradual appearance in the chromatograms of an increasing general background of all Pth-amino acids.

The relative amounts of the background Pth-amino acids
generally reflects the amino acid composition of the protein or peptide degraded. Depending on the protein between 20-50% of all amino acids present have been converted
into the Pth derivatives after 40 cycles. In addition an
unknown amount of the fragmented protein will have been
lost in the form of soluble peptides.

Thus the reduction of background Pth-amino acids is a prerequisite for more extended isothiocyanate degradation.
This prompted us to investigate the origin and nature of
peptide bond cleavage during the isothiocyanate degradation.

MATERIAL AND METHODS

Proteins and Peptides.   Histones H4 and H3 (1) as well as
sperm whale myoglobin, (Miles) converted to apomyoglobin
(2), served as model proteins.   The pentapeptide Phe-Asp-
Ala-Ser-Val was the generous gift of Professor E. Wünsch,
Martinsreid.

* Max-Planck Institut für Biochemie, Martinsried bei München

Peptide and Protein Modification. The pentapeptide Phe-
Asp-Ala-Ser-Val (1mg) was esterified in 1 ml freshly prepared methanolic HCl (5% HCl w/v) at 80°C for 2 minutes.
Proteins were glycinated at all free carboxyl groups at pH
4.7 in   1 M glycine methylester using a water soluble car-
bodimide as condensing agent (3).   In some experiments the
glycine derivative was replaced by taurine ($NH_2CH_2CH_2SO_3^-$).
Proteins (2-4mg) were selectively O-acetylated by 1 ml acetyl chloride in TFA (5% w/v) at room temperature for
15 minutes (4).

Acid Incubation.   Between 1-4mg of freeze dried protein or
peptide were incubated in 1 ml of acid.   Both HFBA AND TFA
were purified and dried as described by Edman (2).   The
amount of H_2O in the acid was measured on a Beckman DK. 2
spectrophotometer at 1 900 mµ (5).

N-terminal Group Determination.   The dansylating procedure
was essentially that described by Gray (6).   Dansyl
amino acids on the polyamide plates were quantitated by
fluorescent quenching using a 520 nm secondary filter.

Sequenator Analysis.   Sequencing reagents were purified as
described by Edman (2).   The instrument used was a 890[B]
sequencer (Beckman Instruments).   The programme is similar
to the Beckman D x 10;   details together with the modifications used have been described previously by us (7).

Pth-amino acids were identified and quantitated by high
pressure liquid chromatography(HLPC) using an acetate buffer
and a methanol gradient (7).

RESULTS AND DISCUSSION
The Nature of the Problem.   In order to investigate the
stability of peptide bonds under commonly used sequencing
conditions, histone H4, myoglobin, insulin and various peptides were exposed for prolonged periods at 50-55°C to
various acids and buffers including those used in sequence
analysis.   Histone H4 was chosen as a model compound in the
first instance, because of its blocked N-terminus.

The exposure of histone H4 to HFBA results in a twenty fold
increase in newly generated N-terminal amino acids compared
to the untreated protein (Table 1A).   No relationship
appears to exist between the yield of the various Pth-amino
acids and the amino acid composition of the protein.   This
suggests that the peptide bonds are not cleaved randomly in
HFBA.

In order to pinpoint the cleaving sites, 4 mg of histone

H4 was exposed to HFBA at 55°C for 30 hours and subsequently subjected to 5 automatic isothiocyanate degradations.

The quantitation of the Pth-amino acids, together with the knowledge of the primary structure of the protein, allows one to deduce the cleavage sites and their approximate yields (Table 1B) (8).

It appears that mainly two types of fragmentation occur to any appreciable extent namely a fragmentation at Asp residues and the N → O shift at the hydroxyl amino acids. Very similar results have been obtained with apomyoglobin. Here results however are more difficult to quantitate due to the presence of 6 Asp and 11 hydroxy amino acid residues (Table 3).

The position and the extent of cleavages at Asp and Ser can be more reliably measured on a small synthetic peptide Phe-Asp-Ala-Ser-Val. After exposing this peptide to HFBA, new N-terminal amino acids are liberated as a function of time (Fig. 1). The appearance of Ala as an N-terminal amino acid indicates a cleavage on the C-terminal side of Asp. The presence of Ser shows the extent of the N → O shift. The appearance of free Val is rather unexpected. However, on considering the structure of the peptide, it becomes evident that the C-terminal carboxyl group and the β-carboxyl group in Asp, resemble one another closely in their relative positions to the peptide bond. The carboxyl group could, in both cases, form an anhydride in the anhydrous HFBA, resulting in a peptide bond cleavage i.e.

To test the involvement of the carboxyl group in this cleavage, the pentapeptide was esterified in methanolic-HCl prior to the esposure to HFBA. This modification results in a drastic reduction of the cleavage at Asp and the C-terminal Val (Fig. 1B).

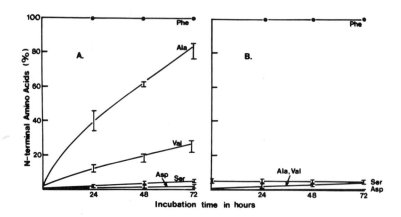

Figure 1. Effect of heptafluorobutyric acid on the formation
of N-terminal amino acids in the peptide Phe-Asp-Ala-Ser-Val.
N-terminal amino acids (determined by dansylation) liberated
as a function of incubation time in HFBA at 55°C acid before
(A) and after (B) the carboxyl groups had been esterified with
methanolic-HCl.

The only very small amount of free amino acids (Ala-5%, Val-
4% other below 1%) after 70 hours HFBA incubation indicates
that cleavage occurs only on the C-terminal site of Asp. The
partial cleavage of the C-terminal peptide bond in the HFBA
appears to be a general phenomenon and can be shown to occur
in other proteins and peptides.

Quantitative Implications
The most prominent fragmentation of peptide bonds occurs at
the C-terminal site of aspartic acid residues (Table 1B).
Typically, 32 to 38% of all three aspartyl peptide bonds present in histone H4 have been cleaved in the acid under the
conditions used.

Fragmentation at the hydroxyamino acids serine and threonine
varies from 3 to 19% and is most likely due to an N → O
shift which is known to occur in strong acids (9).   Steric
reasons may be responsible for the varying extent of the
shift.   Under the conditions used cleavage at other residues appear to be at much lower levels.   For example,

TABLE 1   FRAGMENTATION OF HISTONE H4 BY HFBA

A.

Pth-derivative	Yield in (mol/100 mol protein)		
	Untreated control[a]	HFBA treated (30h at 55°C)	On complete hydrolysis
Asp	0	5	300
Asn	0	27	200
Glu	0	3	400
Thr	0	0	600
Gly	2	7	1700
Gln	0	0	200
Ala	1	38	800
Ser	1	4	
Tyr/Thr	0	4	400
His	0	0	200
Thr	0	1	600
Pro/Thr	0	6	100
Met	0	0	100
Val	1	30	800
Arg	1		1400
Lys	3[b]	4[b]	1100
Phe			200
Ile			600
Leu	1	2	800
Total	10	135	10200

B.

Cleavage sites at residue R number	Approx. cleavage yield (%)
Asp	
24 Asp-Asn	32
68 Asp-Ala	38
85 Asp-Val	30
N → O Ser	
1 Ac-Ser	19
47 Ile-Ser	7
N → O Thr	
29 Ile-Thr	14
53 Glu-Thr	17
70 Val-Thr	9
79 Lys-Thr	2
81 Val-Thr	10
96 Arg-Thr	2
Glu	
52 Glu-Glu	2
63 Glu-Asn	0
74 Glu-His	1
Gly-Gly	1

A. Liberated N-terminal amino acids were measured as Pth derivatives after one isothiocyanate degradation cycle. [a]Histone H4 possess a blocked terminal acetyl serine residue. [b]Pth-Lys, Pth-Phe and Pth-Ile have been determined together and expressed as Pth-Lys equivalents.

B. The degree of cleavage in HFBA has been calculated from the recoveries of Pth-amino acid during the first five degradation cycles, and the primary structure of H4 (1).

only 2 nanomoles of Leu per 100 nmoles of protein (Table 1A)
have been exposed as new N-termini.   If this were the result of random peptide bond cleavage only, 0.17% of any one
peptide bond involving Leu have been fragmented during the
30 hours acid exposure (8 Leu residues are present per molecule).   Similar low yields in the first degradation cycle
have been obtained for Glu, Gln, Tyr, His, Arg, Lys, Phe and
Ile.

We estimate that a protein would have been exposed for a
total of about 10 hours to HFBA during 60 automatic spinning
cup degradation cycles using the double acid cleavage program, taking into account the delivery time, the reaction
time, the following vacuum steps and to a lesser extent the
exposure due to residual acid present during the extraction
step.   Therefore the general background generated by random fragmentation should be at the .0015% level per peptide
bond per step.   New premature sequential degradation will be
initiated at a rate of 0.20% at each aspartyl peptide bond
per degradation cycle.   The same applies to Ser, Thr, these
reactions will thus lead to a decrease in yield accompanied
by an increase of background Pth-residues.   From these results, we concluded that what appears as random and non-
specific background of Pth-amino acids during sequencing is
mainly caused by the accumulation of fractional but preferential cleavage at Asp, Ser and Thr residues in each degradation cycle rather than by non-specific hydrolysis or acido-
lysis.   We calculated that such losses, taking the position
of the labile residues into account, would lead in myoglobin
to a drop in yield of above 40% at sequencing cycle 61.   In
addition there will be losses due to the extraction of soluble peptides.   A measure of the loss can be determined directly by subjecting PTC-^{14}C myoglobin to automated degradation.
By measuring the radioactivity in all the extracts it becomes
apparent that for example during cycle 10 alone 0.7% of the
label is lost from the cup.   Similarily 6% of the amino acid
by weight can be recovered from the collected waste from the
first 10 steps.   The amino acid composition of the lost material differs from that of myoglobin.

The Effect of Water on Peptide Bond Stability
The HFBA prepared by the recommended method (2) contains between 0.02 - 0.05% H_2O.   Increasing the H_2O content in the
HFBA to 1% has no effect on the amount and nature of the cleavages.   At a water concentration of 8.4% (1 mole H_2O/mole
HFBA) a decrease in the total number of peptide bonds cleaved
can be observed (Table 2).   This is due to a decrease in the

TABLE 2.   EFFECT OF H_2O ON PEPTIDE BOND STABILITY IN HFBA

Pth-amino acid	Yield in moles/100 moles protein			
	Histone H4 degradation cycle 1		Myoglobin degradation cycle 1	
	Dry HFBA	HFBA-8.4% H_2O	Dry HFBA	HFBA-8.4% H_2O
Asp	5	5	6	7
Asn	27	6	2	0
Glu	3	2	3	7
Gly	7	20	7	15
Gln	0	1	0	0
Ala	38	12	50	35
Ser	4	7	16	16
Tyr/Thr	4	7	14	14
His	0	1	0	0
Pro/Thr	7	6	2	14
Met	0	0	0	0
Val	30	8	100	100
Arg	4	13	34	10
Lys }				
Phe }	4	10	49	30
Ile }				
Leu	2	4	51	22
Total	135	100	227	170

Histone H4 was incubated for 30h and myoglobin for 40 h at 55°C in the acid.

specific cleavages especially at Asp residues and only a slight increase in other cleavages.

The water concentration in the HFBA in the spinning cup during a typical degradation was found to be 0,5%. This was measured using tritiated H_2O in the quadrol buffer (no corrections for exchangable protons were made). We concluded from these findings that hydrolysis does not play a major role in the formation of the general background.

The Effect of Temperature, Concentration and Nature of the Acid on Peptide Bond Stability. The cleavage rate in HFBA at aspartic-and hydroxy amino acids is strongly dependent on the temperature. Virtually no fragmentation can be observed after a 24 hr incubation at room temperature while at 80°C no uncleaved protein remains as determined by polyacrylamide electrophoresis. Formic acid, TFA and HFBA all cause a similar degree of fragmentation of peptide bonds while

acetic is inert.   Diluting the HFBA in chlorobutane (20%
v/w) decreases specific cleavages dramatically as does the
reduction of temperature.   The effect of these two changes
have not yet been tested on automated isothiocyanate degradation.

Modification of Asp, Ser and Thr.   In order to prevent the
fragmentation at aspartic acid residues the carboxyl residues
of proteins were converted to the amides with glycine methyl
ester or taurine ($NH_2CH_2CH_2SO_3^-$).   Such modified myoglobin
shows a substantial reduction in N-terminal amino acid li-
beration due to cleavages at Asp after HFBA incubation (Table
3).

TABLE 3. THE EFFECT OF SIDE GROUP MODIFICATION ON PEPTIDE
BOND STABILITY

Phenylthio-hydantoin derivative	Amounts of phenylthiohydantoin derivatives (nmol) Degradation cycle no.					
	1			2		
	Unmodified	Glycinated	Glycinated and acetylated	Unmodified	Glycinated	Glycinated and acetylated
Asp	6	0[a]	0[a]	2	1[a]	1[a]
Asn	2	1	0	5	1	1
Glu	3	1[a]	0[a]	19[c]	27[a]	7[a]
Gly	7	8	7	16	6	4
Gln	0	0	0	22	3	3
Ala	50[b]	13	10	52[c]	33	10
Ser	16	20	0	2	1	1
Thr/Tyr	14	7	0	2	1	1
His	0	0	0	7[c]	18	3
Pro/Thr	2	6	0	2	1	0
Met	0	0	0	0	0	0
Val	100[b]	100	100	37[c]	20	6
Arg	34[b]	3	1	22	3	1
Lys	4	9[d]	7[d]	30[c]	9[d]	8[d]
Phe	30[b]			14		
Ile	60[b]			0		
Leu	51[b]	13	13	93[c]	87	120

[a] Quantitated as methyl esters

[b] Cleavage at Asp residues in myoglobin would yield Val, Ile,
Arg, Leu, Phe, Ala, Ile in the first degradation cycle.

[c] Cleavages due to N → O shift yield Ser and Thr in the first
cycle followed in cycle 2 by Glu,His,Glu,His and Glu for
seryl cleavages and Leu,Glu,Val,Ala and Lys for threonyl
cleavages. [d] Pth-Lys, Pth-Phe, Pth-Ile co-eluted and are
expressed as Pth-Lys equivalents.

A further reduction can be achieved by reversing the N → O shift by pre-incubating the protein in the sequencing buffer prior to the addition of the PITC (10). However O-acetylation by acetylchloride in TFA was found to be more effective. This modification of the hydroxyamino-acid residues reduces newly formed terminal residues after exposure to HFBA from 300 nmoles in the unmodified to 50 nmoles in the glycinated and acetylated protein (Table 3).

The prevention of the N → O shift by acetylation is however unsatisfactory. Although a six fold decrease in background was observed when modified myoglobin was sequenced the yield decreases more rapidly when compared to the unmodified protein. This has been pin-pointed to the destruction of Trp under the O-acetylation conditions. In addition N-terminal O-acetyl Thr rearranges to blocked N-acetyl Thr derivative. It has nevertheless been establsihed that specific cleavages at Asp and Ser, Thr are the major causes for the drop in yield and increasing background during isothiocyanate degradation. Though the Asp cleavage can well be controlled by suitable substitution a satisfactory substituent for hydroxyamino acids preferably of the ether type linkage has not yet been found.

REFERENCES

1.   von Holt, C., Strickland, W.N., Brandt, W.F. and Strickland, M.S. (1979) FEBS Lett. 100, 201-218.

2.   Edman, P. and Henschen, A. (1975) in Protein Sequence Determination (Needleman, S.B., ed.) 2nd edn., pp. 232-279, Springer-Verlag, Berlin.

3.   Gibson, D. and Anderson, P.J. (1972) Biochem. Biophys. Res. Commun. 49, 453-459.

4.   Previero, A., Barry, L.G. and Coletti-Previero, M.A. (1972) Biochem. Biophys. Acta 263, 7-13.

5.   Begg, G.S., Pepper, D.S., Chesterman, C.N. and Morgan, F.J. (1978) Biochemistry 17, 1739-1744.

6.   Gray, W.R. (1972) Methods Enzymol. 25, 121-143.

7.   Strickland, M., Strickland, W.N., Brandt, W.F., von Holt, C., Wittman-Liebold, B. and Lehman, A. (1978)

Eur. J. Biochem. $\underline{89}$, 443-452.

8.   Brandt, W.F., Henschen, A. and von Holt, C. (1980)
      Hoppe-Seyler's Z. Physiol. Chem. $\underline{361}$, 943-952.

9.   Iwai, K. and Ando, T. (1967)   Methods Enzymol. $\underline{11}$,
      263-282.

10.   Thomsen, J., Bucher, D., Brunfeldt, K., Nexø, E. and
      Oleson, H. (1976)   Eur. J. Biochem. $\underline{69}$, 87-96.

Microsequencing

ISOLATION AND SEQUENCE DETERMINATION OF POLYPEPTIDES AT THE PICOMOLE LEVEL

J.Y.CHANG, R.KNECHT and D.G.BRAUN

Pharmaceuticals Research Laboratories

CIBA-GEIGY Ltd. 4002 Basle, Switzerland

Increase of sensitivity for protein sequencing depends on the development of micro-methods for amino acid analysis, peptide isolation and sequence determination. While sensitive automatic methods for both amino acid analysis (1,2) and sequence determination (3) are available, there is no straight forward method for peptide isolation. Resolution of a complex peptide mixture usually requires the successive use of combined chromatographic methods (4). The procedure is time-consuming and costly with regard to material since peptide loss occurs with each purification step. Therefore, an ideal peptide isolation method should be not only sensitive but also alone powerful enough to completely resolve peptide mixtures. We have developed new techniques to meet these requirements by precolumn derivatization of amino acids and peptide with chromophoric reagents. The methods used and the results obtained are summarized and discussed in this paper.

Amino Acid Analysis in the Picomole Range by Precolumn Derivatization with Dimethylaminoazobenzene Sulfonyl Chloride (DABS-Cl)(5,6)

Amino acids labelled with DABS-Cl (DABS-amino acids) can be separated on RP-HPLC and detected in the visible region of the light spectrum (436 nm). The derivatization is quantitative and reproducible and, unlike OPA and thio-

113

hydantoin derivatives, DABS-amino acids are both photo and
chemically stable at room temperature. The sensitivity
limit of detecting DABS-amino acids, dependent upon the
quality of the detector, can be lower than 1 pmole. Fig. 1
shows the separation of 5 pmoles each of DABS-amino acid
on a Zonbax-ODS column using the acetonitrile/acetate
buffer (pH 4.1) system. The sensitivity and accuracy of
this method were demonstrated by the composition analysis
of 7.5 ng of melittin and 30 ng of ribonuclease acid
hydrolysates (Fig. 2). The quantitative results show that
the DABS-Cl method is as reliable as a conventional amino
acid analyzer (Biotronik) (data not shown). The combination of this sensitive amino acid analysis technique with
the peptide cleavage by carboxypeptidase has made possible
the determination of the carboxyl terminal sequence of
polypeptides at the picomole level.

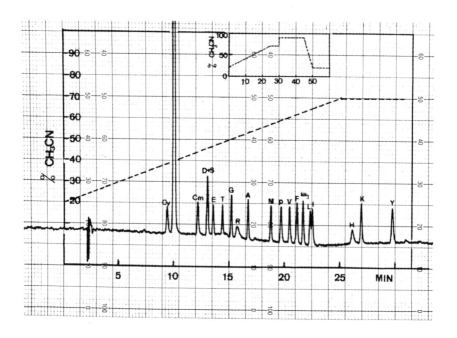

Fig. 1. Separation of 18 DABS-amino acids (symbolized by
one letter abbreviations of their corresponding amino
acid). Solvent A is 0.045 M acetate (pH = 4.1). Solvent B
is acetonitrile. Gradient as indicated. Flow rate is 1.2
ml/min. Column temp. is 22°C. (Reproduced from Ref. 5).

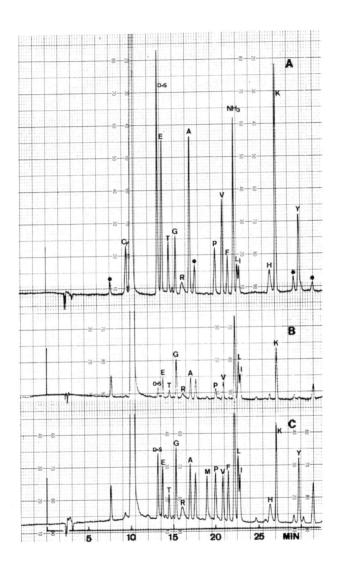

Fig. 2. Original chromatogram of amino acid analysis by the DABS-Cl method. (A) 33 ng of oxidized ribonuclease. (B) 7 ng of melittin. (C) 10 pmole of Hamilton standard.

Isolation of Peptides in the Picomole Range by Precolumn
Derivatization with Dimethylaminoazobenzene Isothiocyanate
(DABITC) (7)

Peptides precolumn derivatized with DABITC (DABTC-
peptides) can be separated by RP-HPLC and detected in the
visible region (436 nm). As little as 1 ng (2 pmole) of a
DABTC-pentapeptide were identified against a stable baseline. DABTC-peptides can be subsequently recovered from
the column and their N-terminal amino acid, amino acid
composition and sequences may be analyzed at picomole
level.

In column chromatography of polypeptides at the
picomole level this precolumn derivatization method offers
distinct advantages over those conventional methods in
which detection is based on U.V. absorption (8,9) or
post-column reaction with fluorescent reagents (10,11):
(1) Detection in the visible region completely avoids
interference by U.V. absorbing contaminants or baseline
rise due to gradient change; (2) Precolumn derivatization
limits the source of contaminations to only the sample and
coupling buffer.

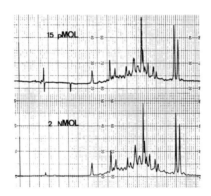

Fig. 3. Analytical run (15 pmole, top panel) and preparative run (2 nmole, bottom panel) of the DABTC-peptides
derived from the trypsin cleavage of light chain 7S34.1
at Arg residues (lysine blocked with methyl isoeyanate).
Solvent A is phosphate (pH 7.2). Solvent B is acetonitrile. Gradient is 15% B - 70% B in 30 min. Flow rate
1 ml/min.

Fig. 3 depicts the separation of DABTC-peptides from a tryptic digest of a light chain from monoclonal hybridoma-derived antibody 7S34.1. The loading capacity of each run varied from 5 pmole (analytical run) to 2 nmole (preparative run). Purity of every peak was analyzed by direct 1-step acid degradation of their N-terraini from a 200-400 pmole chromatogram. DABTC-peptide were collected in the amount of 0.5 to 3 nmole for sequence determination.

Sequence Determination of DABTC-peptide by the Automatic Sequenator and Manual DABITC-PITC Method at Picomole Level

A commercial sequenator (Beckman 890C) without extensive modification (liquid nitrogen trap connected Alcatel ..... pumps) and extra-purified reagent can be used to determine at 1-3 nmole of DABITC-peptide up to 20 to 40 residues with clear and confident results. The background resulted from sequencing of peptide is minimal. The PTH-amino acids were analyzed by HPLC on a Zorbax-CN column (Fig. 4). Alternatively, the manual DABITC-PITC method (12) can be used, and resultant DABTH-amino acids may be quantitatively analyzed by HPLC on a Zerbax-OPS.

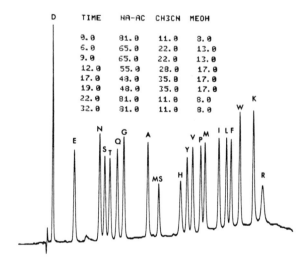

Fig. 4. Separation of PTH-amino acid (250 pmole each) on a Zorbox-CN column. Flow rate is 1 ml/min. Column temperature is 33°C. Gradient is as indicated.

Conclusion

We have demonstrated new methods for amino acid and peptide analysis. The high sensitivity was achieved by precolumn derivatization of amino acids and peptides with chromophoric reagents followed by their separation on HPLC. The rationale for this approach is based on the least background (and therefore higher sensitivity) and require simpler instrumentation in comparison to the post-column detection system.

References

1.   Spackman,D.H., Moore,S. & Stein,W.H. (1958) Anal.Chem. 30, 1190-1205
2.   Hare,P.E. (1977) Method in Enzymology 47, 3-18.
3.   Edman,P. & Begg,G. (1967) Eur.J.Biochem. 1, 80-91.
4.   Allan,G. (1981) Sequencing of Proteins and Peptides North-Holland
5.   Chang,J.Y., Knecht,R. & Braun,D.G. (1981) Biochem.J. in press.
6.   Chang,J.Y., Martin,P., Bernasconi,R. & Braun,D.G. (1981) FEBS Lett., in press.
7.   Chang,J.Y. (1981) Biochem.J., in press.
8.   Fullmer,C.S. & Wasserman,R.H. (1979) J.Biol.Chem. 254, 7208-7212.
9.   Hancock,W.S., Bishop,C.A., Prestidge,R.L. & Hearse, M.T. (1978) Anal.Biochem. 89, 203-212.
10.   Rubinstein,M., Chen-Kiang,S., Stein,S. & Udenfriend,S. (1979) Anal.Biochem. 52, 177-182.
11.   Lai,C.Y. (1977) Method in Enzymology 47, 236-246.
12.   Chang,J.Y. (1981) Biochem.J. , in press.

METHODOLOGIES IN THE MICROSEQUENCING OF PROTEINS AND PEPTIDES

Ajit S. Bhown, John E. Mole and
J. Claude Bennett
Division of Clinical Immunology and Rheumatology
Department of Medicine and Department of Microbiology,
University of Alabama in Birmingham
Birmingham, Alabama   35294

In the second edition of "The Proteins" Caufield and Anfinsen commented that "the rather tedious business of amino acid sequence determination will become increasingly automated over the coming years, and such efforts may eventually become the province of well-trained machine operators." This prediction has become nearly true due to the availability of a highly successful automated sequenator for stepwise degradation and computerized instrumentation for quantitative identification of released amino acid derivatives. Commenting on this prediction, Niall (1) stated"...the present author does not believe, or at least would prefer not to admit that protein sequence determination has yet become a mere technical operation requiring a minimum of intellectual input...." and correctly foresaw the need for continued intellectual involvement in the field of protein sequence analysis. This realization was conceived only when the scientific community began seeking at the molecular level answers to questions such as tissue graft rejections; cell-cell interactions; immunological phenomenons, and other intriguing biological functions. Since these seem to be mediated by cellular membrane and/or plasma proteins which are available in extremely minute quantities, a quest began for methods to determine their structural details. This resulted in a number of modifications which have led to the decreased requirement for sample quantities. There is probably no single reason for the decreased

119

METHODOLOGIES IN THE MICROSEQUENCING

requirement of protein and peptide load, rather it is due
to a combination of improvements achieved in different
aspects of protein sequencing. In order to effectively
achieve the microsequencing at subnanomole to picomole
levels we have developed simple adaptations and modified
certain existing techniques which (2-4) essentially
covers different aspects of automated protein sequence
analysis. In the next few minutes I will present these
modifications and finally would like to show quantitative
data on proteins and peptides obtained as a result of
incorporating those changes over a period of 2-3 years.

Instrumentation

Although different components of the sequenator have
varying effects on sequencing efficiency, we have

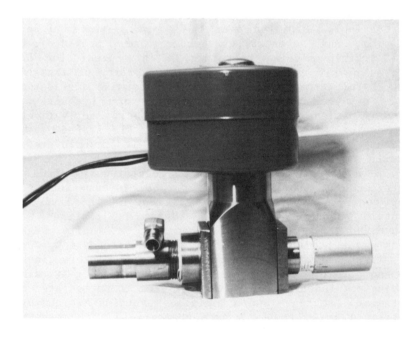

Figure 1.   Stainless steel solenoid valve.

METHODOLOGIES IN THE MICROSEQUENCING

concentrated our efforts on improving the high vacuum
system on the cup with minimum instrument plumbing (2).
To achieve this the air cylinder/bimb a valve, which
controls the high vacuum to the cup has been replaced
with a solenoid valve (Figure 1), which is actuated by a
relay powered with an independent 120 v line.   The
solenoid valve is installed between cold trap and cup
after the low and restricted vacuum lines (Figure 2).
The schematic representation is shown in (Figure 3).

Figure 2.   Configuration of solenoid valve in the Beckman
sequencer with cold trap.

METHODOLOGIES IN THE MICROSEQUENCING

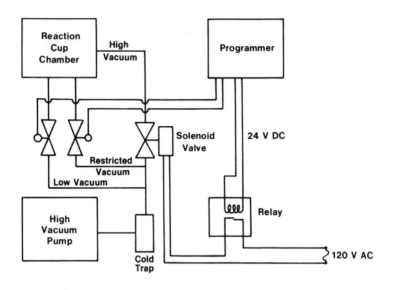

Figure 3. Schematic representation of arrangment of solenoid valve and relay.

Chemicals

Although the purity of all the chemicals used in automated sequenator is critical quadrol is more prone to impurities because it suffers from autoxidation causing reappearance of impurities, e.g. aldehydes. Wittmann Liebold (5) and Hunkapiller and Hood (6) have suggested prepurification of solvents and reagents, while Frank (7) has recommended the use of aminoethyl cellulose to absorb the amino group reactive impurities in quadrol. In order to avoid this autocontamination of quadrol we have successfully employed aminoethyl aminopropyl glass (AEAP) beads to absorb out all amino group reactive impurities as and when they reappear (3). Figure 4 shows the schematic arrangement of in situ quadrol purification.

METHODOLOGIES IN THE MICROSEQUENCING

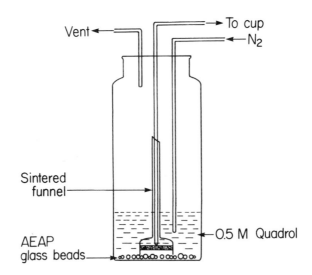

Figure 4. Quadrol reservoir (R_2) with sintered funnel and AEAP Glass beads. (with permission from Analytical Biochemistry).

Identification

Another aspect of achieving high sensitivity in microsequencing involves the method of identification of phenylthiohydantoin (PTH) derivatives of amino acids. High pressure liquid chromatography (HPLC) is probably the most commonly employed method for PTH amino acid identification using reverse phase columns and a gradient of polar to non-polar solvents. The most commonly used as non-polar solvents. The major problems with this system in the past have been i) base line separation and ii) base line drift after a few sample injections. We have found (3) that by employing a short (0.46 x 150 cm) 5 μm ultrasphere ODS column and Omnisolv methanol manufactured by MCB Manufacturing chemicals a base line separation with practically no rise in base line even after 25-30 sample injections is possible (Figure 5).

METHODOLOGIES IN THE MICROSEQUENCING

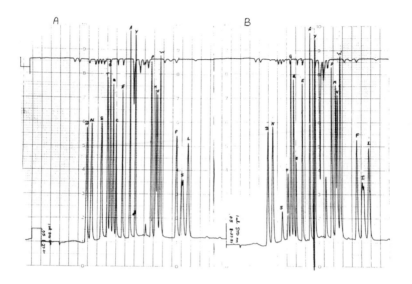

Figure 5.   Separation of standard PTH amino acid mixture;
A) before   analyzing   any   unknown   sample;   B) after
analyzing 25 samples from sequencer.

Amino   acid   sequence   analysis   both   manual   and
automated   is   always   associated   with   a   non-specific
peptide bond cleavage of the molecule due to repeated
exposure to low pH during degradation cycle.   This
generates additional amino termini responsible for an
increase in background as the sequence analysis proceeds.
Consequently,   interpretation   of   sequence   data   to
establish an extended amino acid sequence becomes
impossible.   In order to suppress this rise in background
we have successfully employed fluorescamine (Fluram) to
chemically block the newly generated amino termini
without effecting the sequencing of the parent molecule.
Although the experimental details have been published
(4), in brief, when ever a proline residue is expected in
the sequence, the sample is reacted with Fluram in the
cup at ambient temperature and then the regular program
is   initiated.   The results of the sequence analysis
without (Fig. 6A) and with (Fig. 6B) fluram blocking

METHODOLOGIES IN THE MICROSEQUENCING

clearly indicates that fluorescamine can be very effective in blocking the newly generated amino termini. This simple addition to the growing list of strategies developed to aid amino acid sequence analysis seems very effective and promixing the use of fluorscene may permit sequences determination of up to 100 or even more residues in "multiple sequencer runs" (MSR), as explained in Figure 7.

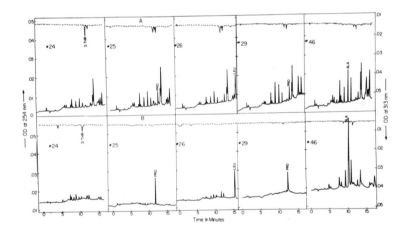

Figure 6. Chromatograms of sequence analyses of serum amyloid P-component: A) without fluram blocking; B) with fluram blocking at cycles 12, 25 and 29. (with permission from Analytical Biochemistry).

METHODOLOGIES IN THE MICROSEQUENCING

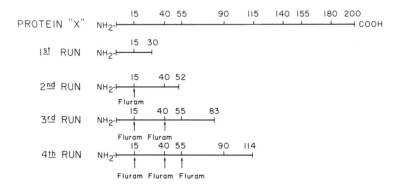

Figure 7.   MSR strategy using fluram as an effective
suppresser   of   background.   (With   permission   from
Analytical Biochemistry).

Employing these techniques extended (35-45 residues)
amino acid sequence information on proteins have been
obtained while small peptides have been sequenced up to
their carboxy termini at subnanomole level (Table I).
Figue 8 shows yields of PTH derivatives of the first 45
amino terminal amino acids of 5 nanomoles of neutrophil
elastase and Figure 9 shows the regression curves of
different proteins and peptides sequenced at subnanomole
and picomole levels.

METHODOLOGIES IN THE MICROSEQUENCING

TABLE I

	SAMPLE	AMOUNT	RESIDUES SEQUENCED
A.	PROTEINS		
	1. ELASTASE	500 PM	30
	2. FACTOR D	8 NM	57
	3. P COMPONENT	10 NM	55
	4. H-2K^K	5 NM	30
B.	PEPTIDES		
	1. CYANOGEN BROMIDE FRAGMENTS OF:		
	A. FACTOR D		
	CNBr-2	10 NM	51
	CNBr-3	15 NM	15 (c)
	B. FACTOR B		
	CNBr-20K	1 NM	15
	CNBr-15K	3 NM	48
	C. P27		
	CNBr-12-3A	2 NM	17 (c)
	CNBr-12-3B	2 NM	13 (c)
	2. MILD ACID CATALYZED CLEAVAGE:		
	A. FACTOR B		
	I) 8 K	5 NM	40
	B. REV P30		
	I) REV-1-A-4	1 NM	30
	3. TRYPTIC PEPTIDES:		
	A) PRC-P27		
	I) TP-6	5 NM	6 (c)
	II) TP-14	1 NM	27 (c)

(C)-Sequence completed upto carboxyl end.

METHODOLOGIES IN THE MICROSEQUENCING

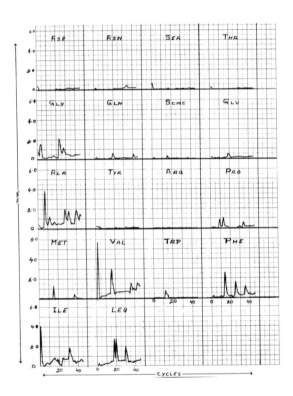

Figure 8.   HPLC tracings of sequence analyses of human neutrophil elastase (6nm).

METHODOLOGIES IN THE MICROSEQUENCING

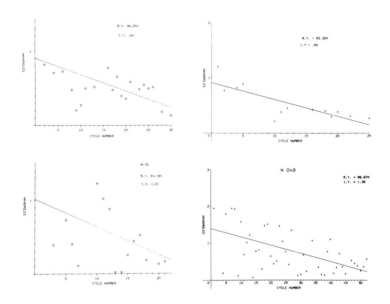

Figure 9.   Regression curves of sequence analysis of
proteins and peptides at subnanomole to picomole levels.

METHODOLOGIES IN THE MICROSEQUENCING

REFERENCES

1.   Niall, H.   (1977)   The Proteins, ed. H. Neurath and
      R.L. Hill.   Academic Press, New York, Vol. III p.
      181.

2.   Bhown, A.S.   Cornelius, T.W., Mole, J.E., Lynn,
      J.D., Tidwell, W.A., and Bennett, J.C.   (1980)
      Anal. Biochem. 102:35.

3.   Bhown, A.S., Mole, J.E. and Bennett, J.C.   (1981)
      Anal. Biochem. 110:355.

4.   Bhown, A.S., Bennett, J.C., Morgan, P.H. and Mole,
      J.E.   (1981)   Anal. Biochem. 112:158.

5.   Wittmann-Liebold, B.   (1980)   Polypeptide Horneoues
      ed. R.F. Beers, Jr. and E.G. Bessett, Raven Press,
      N.Y. p. 87.

6.   Hunkapiller, M.W. and Hood, L.   (1978)   Biochem.
      17:2124.

7.   Frank, G.   (1979)   Hope-Seyler's Z. Physiol. Chem.
      380:997.

MICRO-SEQUENCE ANALYSIS OF OVINE HYPOTHALAMIC CORTICOTROPIN RELEASING FACTOR

Joachim Spiess, Jean Rivier, Catherine Rivier, and Wylie Vale
The Salk Institute, Peptide Biology Laboratory

P. O. Box 85800, San Diego, CA 92138

INTRODUCTION

In 1955, first direct evidence was provided for a hypothalamic corticotropin-releasing factor (CRF) that stimulated corticotropin secretion from the pituitary (1,2). Since then several groups attempted to purify and characterize this factor (see 3 for review).

Starting material of the CRF purification performed in this laboratory originated from side fractions of the purification (4) of luteinizing hormone releasing factor (LRF) from 490,000 ovine hypothalami carried out in the Laboratories for Neuroendocrinology of the Salk Institute.

The complete characterization of ovine hypothalamic CRF was only possible after refinement of the initial purification (5), application of reverse-phase HPLC to the final purification (6) and establishment of improved spinning cup methodology for amino acid sequence analysis (7).

While most of the experimental data of the sequence analysis of CRF have been described elsewhere (7), this presentation is focused on details of the spinning cup methodology.

EXPERIMENTAL PROCEDURES

Edman degradation was performed with a Wittmann-Liebold (8,9) modified Beckman 890C spinning cup sequencer as described earlier (10). The modification included replacement of the delivery valves by pneumatically driven

131

diaphragm valves, replacement of the vacuum system by
Alcatel 2030C pumps protected by cooling traps (automatically filled with liquid nitrogen) and introduction of an
automatic converting flask. Vacuum block, delivery valves,
and converting flask were manufactured in the Max-Planck-
Institute for Molecular Genetics (Berlin). The sequencer
program was established on the basis of programs earlier designed for the modified spinning cup sequencer (9,11).
Polybrene (4 to 6 mg) purified by adsorption and ion exchange chromatography was subjected to approximately 10 to
12 sequencer cycles prior to the peptide application. The
conditions of sequence analysis are summarized in Table I.

Table I.   Conditions of Sequence Analysis

Reaction	Reagent or Solvent	Volume applied	Time (Temp)
Coupling	5% (v/v) phenylisothiocyanate in heptane (R1)	0.18 ml[a]	29 min (42°C)
	0.33M Quadrol trifluoro-acetate in 43% (v/v) n-propanol pH 8.9 (R2)	0.47 ml[b]	
Extraction after coupling	a) Benzene (S1) b) 0.1% (v/v) acetic acid in ethylacetate (S2)	8.5 ml 6.5 ml	
Cleavage	Heptafluorobutyric acid (R3)	0.21 ml	400 sec (42°C)
Extraction of ATZ amino acids	15% (v/v) acetonitrile in butylchloride, 0.3 mM dithioerythritol (S3)	4.2 ml	
Conversion	25% (v/v) trifluoroacetic acid (R4)	1.2 ml	55 min (52°C)
Extraction of PTH amino acids	25% (v/v) acetonitrile in methanol (S4)	5.0 ml[c]	

[a] R2 was applied after R1 delivery and partial removal of
heptane.

[b] Total volume, applied in two portions (0.25 ml, 0.22 ml).

[c] Total volume, applied in three equal portions.

Shortening of the cleavage time by 100 sec generated significant increase in carry-over of PTH-amino acids. For the
removal of N-terminal proline residues, two cleavages, each
extended to 500 sec and followed by extraction with butylchloride (S3), were performed. Neither with CRF nor with
other polypeptides of CRF size did we obtain any evidence
for unspecific cleavage of peptide bonds by heptafluorobutyric acid under the conditions of this sequencer program.
Although no special effort was undertaken to protect N-
terminal glutamine residues during the cleavage reaction of
the Edman degradation, significant cyclization of glutamine
was not observed.

The addition of acetonitrile to butylchloride (S3) decreased the carry-over of polar PTH-amino acids by allowing
for more efficient extraction of polar amino acid anilinothiazolinones (ATZ) from the cup.

PTH-amino acids were determined by reverse-phase HPLC
in a Hewlett-Packard 1084B liquid chromotograph equipped
with an auto sampler and a programmable variable wavelength
detector. DuPont Zorbax ODS columns (0.46 x 25.0 cm; particle diameter: 5 to 6 μm) were eluted with mixtures of
0.010 to 0.016M phosphoric acid (adjusted to pH 5.7 to 5.8
with triethylamine) (A) and acetonitrile containing 3 to 5%
tetrahydrofurane (THF) (B). THF was added as modifier to
improve the resolution of PTH-phenylalanine and PTH-ε-N-
phenylthiocarbamyl lysine. Serine residues were identified
as PTH-serine derivative eluting before PTH-alanine as described earlier (10,11). This serine derivative was apparently a reaction product with dithioerythritol. If dithioerythritol was replaced by dithiothreitol in S3, a serine
product was formed which was eluted slightly after the PTH-
serine dithioerythritol product. Determination of threonine
residues as PTH-dehydrothreonine required a (programmed)
wavelength change from 266 to 318 nm. The elution pattern
of PTH-arginine and PTH-histidine varied slightly from
column to column. Resulting separation problems could be
solved by adjustment of the ionic strength of buffer A in
the usual fashion. PTH-norleucine, added to the tubes of
the sequencer collector before collection of the PTH-amino
acids, served as internal standard. All common PTH-amino
acids were resolved by reverse-phase HPLC within 18 to 22
min (Fig. 1).

RESULTS

CRF fractions extensively purified by reverse-phase

HPLC and hydrolyzed with 4M methane sulfonic acid or constant
boiling HCl were subjected to amino acid analysis. Under
the assumption that CRF contained 7 Glx residues, the
following amino acid composition was suggested for CRF: 4
Asx, 2 Thr, 3 Ser, 7 Glx, 2 Pro, 4 Ala, 1 Val, 1 Met, 2 Ile,
8 Leu, 1 Phe, 2 Lys, 2 His and 2 Arg. The methionine residue was determined as methionine sulfoxide. On the basis
of these data, a minimal molecular weight of 4,666 was calculated. More details of the amino acid analysis of CRF
are described elsewhere (7).

Direct micro-sequence analysis of natural CRF with a
modified spinning cup sequencer was accomplished with a
total of 44 µg of peptide (corresponding to approximately
10 nmol). With 2.9 µg of peptide (0.6 nmol) applied to the
cup, the N-terminal 27 residues could be determined. Residues 1 to 39 of CRF were identified with 16.8 µg of peptide
(3.5 nmol) subjected to sequence analysis in the spinning
cup. In the first sequencer cycle of this experiment, the
peptide was coupled to 1.2 µmol of 3-sulfo-phenylisothio-
cyanate (3-SPITC) (12) to facilitate the binding of the

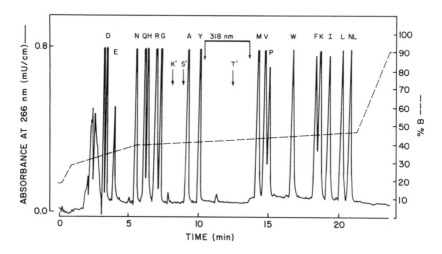

Fig. 1. Reverse-phase HPLC of PTH-amino acids (100 pmol/
PTH-amino acid). The retention times of the characteristic
PTH-derivatives of serine (S'), threonine (T') and ε-N-
acetyl lysine (K') are indicated. Temperature: 50°C; flow
rate: 1.0 ml/min. The percentage of solution B in the
eluent mixture is given by the dashed line.

peptide to the positively charged Polybrene. The reaction
of the peptide with 3-SPITC was followed by a coupling to
PITC under the normal conditions of the sequencer program.
The N-terminal serine residue and the lysine residues had
not completely reacted with 3-SPITC as indicated by the appearance of the usual PTH-derivatives of serine and ε-N-
phenylthiocarbamyl lysine (Table II). The repetitive yield
based on the yields of PTH-leucine, was approximately 95%.

The sequencing data (Table II) accounted for 39 of the
41 amino acids determined by amino acid analysis. It was
accordingly suggested that the C-terminal structure of CRF
was not recognized. Therefore, CRF (2 nmol) was acetylated,
selectively cleaved at its arginine residues with trypsin
and applied to the cup. Simultaneous Edman degradation of
two fragments, CRF (17-35) and CRF (36-41), was observed.
Residue 40 of CRF was tentatively identified as isoleucine.
No evidence for further C-terminal residues was found.

On the basis of digestion experiments with carboxypeptidase Y (CPY) in conjunction with C-terminal tritiation

Table II. Sequence Analysis of Ovine Hypothalamic CRF

Cycle	PTH-amino acid	Yield (pmol)	Cycle	PTH-amino acid	Yield (pmol)
1	Ser	195	21	Met	843
2	Gln	1652	22	Thr	111
3	Glu	1735	23	Lys	189
4	Pro	1110	24	Ala	779
5	Pro	1316	25	Asp	791
6	Ile	1435	26	Gln	480
7	Ser	218	27	Leu	540
8	Leu	1671	28	Ala	554
9	Asp	1743	29	Gln	418
10	Leu	1544	30	Gln	460
11	Thr	217	31	Ala	379
12	Phe	1401	32	His	281
13	His	1126	33	Ser	85
14	Leu	1318	34	Asn	415
15	Leu	1484	35	Arg	133
16	Arg	961	36	Lys	42
17	Glu	851	37	Leu	225
18	Val	951	38	Leu	261
19	Leu	984	39	Asp	86
20	Glu	619			

experiments (for details see 7), it was suggested that -Ile-
Ala-NH$_2$ represented the C-terminal sequence of CRF.

This interpretation was confirmed in digestion experiments with thermolysin. The C-terminal dipeptide F-Ile-
Ala-NH$_2$ was cleaved from CRF with thermolysin, dansylated,
purified from other dansylated products (by extraction from
an alkaline medium into chloroform) and characterized by
reverse-phase HPLC (Fig. 2).

The purity of CRF was estimated to be 85 to 90% on the
basis of CPY experiments, amino acid composition and sequence analysis data.

It was concluded from the analytical data, that CRF
had the following primary structure: H-Ser-Gln-Glu-Pro-Pro-
Ile-Ser-Leu-Asp-Leu-Thr-Phe-His-Leu-Leu-Arg-Glu-Val-Leu-
Glu-Met(O)-Thr-Lys-Ala-Asp-Gln-Leu-Ala-Gln-Gln-Ala-His-Ser-
Asn-Arg-Lys-Leu-Leu-Asp-Ile-Ala-NH$_2$. The biologic potencies

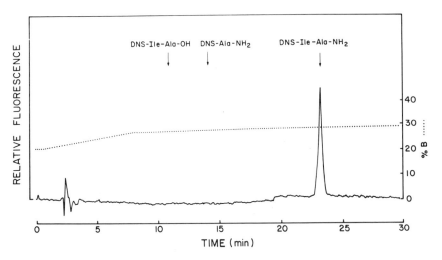

Fig. 2. Reverse-phase HPLC of a mixture of DNS-Ile-Ala-NH$_2$
from synthetic and natural origin (1.5 pmol of each product).
A DuPont Zorbax ODS column (0.46 x 25.0 cm) was eluted
(50°C; 0.8 ml/min) with a mixture composed of 0.012 M phosphoric acid, adjusted to pH 6.5 with triethylamine (A) and
35% (v/v) acetonitrile in n-propanol (B). The excitation
wavelength was 250 nm; emission was measured at 500 nm. A
Hewlett-Packard 1084B chromatograph equipped with a
Schoeffel FS 970 spectrofluoro monitor was used.

of natural CRF and its synthetic replicate determined in the rat anterior pituitary cell culture assay (5) did not differ significantly. Half-maximal responses of ACTH and β-endorphin secretion were at 0.2 to 0.8 nM CRF (5).

Synthetic CRF in the methionine form was found even more potent than CRF in the methionine sulfoxide form. Synthetic CRF (1-27), CRF (1-39) or C-terminally desamidated CRF (in the methionine form) exhibited in vitro less than 0.1% of the potency of CRF in the methionine form.

DISCUSSION

The elucidation of the primary structure of ovine hypothalamic CRF was mainly accomplished with a Beckman 890C spinning cup sequencer modified according to Wittmann-Liebold et al (8,9). This modification was first applied by Hunkapiller and Hood (11) to direct micro-sequence analysis of peptides on a low nmole and subnanomole level.

It was observed that, after modification, the contamination of PTH-amino acid fractions with by-products of the Edman degradation was reduced and that analysis of more than 25 amino acid residues of a peptide was not impaired by carry-over. However, despite extensive purification of Polybrene, sequencer reagents and solvents, micro-sequence analysis was still hampered by background problems in the reverse-phase HPLC identification of PTH-amino acids.

From amino acid analysis, it was concluded that natural CRF was in the methionine sulfoxide form. In view of the observation (13), that methionine residues can be easily oxidized by air oxygen, it was assumed that CRF was oxidized during purification and that the physiologic form of CRF probably contained non-oxidized methionine.

Ovine hypothalamic CRF has significant homologies with frog skin sauvagine (14) and fish urophysis urotensin I (Karl Lederis, personal communication).

ACKNOWLEDGMENTS

We thank Brigitte Wittmann-Liebold for helpful discussions, J. Heil, R. Lee, L. Keadle, J. Vaughan, G. Yamamoto and E. Fung for excellent technical assistance, and L. Wheatley for help in the preparation of this manuscript. This research was supported by NIH Grants AM26741, AM20917, AM18811, AA03504, and HD13527, The Clayton Foundation for Research, California Division, the Charles S. and Mary Kaplan Foundation, Stiftung Volkswagenwerk and Fritz Thyssen

Stiftung. Drs. J. Spiess, J. Rivier, C. Rivier and W. Vale
are Clayton Foundation Investigators.

REFERENCES

1. Guillemin, R. and Rosenberg, B. (1955) Endocrinology
    57:599–607.
2. Saffran, M. and Schally, A.V. (1955) Can. J. Biochem.
    Physiol. 33:408–415.
3. Saffran, M. (1977) in Hypothalamic Peptide Hormones
    and Pituitary Regulation, ed. Porter, J.C. (Plenum,
    New York), pp. 225–235.
4. Burgus, R., Amoss, M., Brazeau, P., Brown, M., Ling,
    N., Rivier, C., Rivier, J., Vale, W. and Villarreal, J.
    (1976) in Hypothalamus and Endocrine Functions, eds.
    Labrie, F., Meites, J. and Pelletier, G. (Plenum, New
    York), pp. 355–372.
5. Vale, W., Spiess, J., Rivier, C., and Rivier, J. (1981)
    Science 213:1394–1397.
6. Rivier, J., Rivier, C., Branton, D., Millar, R., Spiess,
    J., and Vale, W. (1981) in Peptides: Synthesis, Struc-
    ture and Function, eds. Rich, D.H., and Gross, E.
    (Pierce Chemical Company, Rockford) in press.
7. Spiess, J., Rivier, J., Rivier, C., and Vale, W. (1981)
    Proc. Natl. Acad. Sci. (USA) 78:6517–6521.
8. Wittmann-Liebold, B., Graffunder, H. and Kohls, H.
    (1976) Anal. Biochem. 75:621–633.
9. Wittmann-Liebold, B. (1980) in Polypeptide Hormones,
    eds. Beers, Jr., R.F., and Bassett, E.G. (Raven Press,
    New York), pp. 87–120.
10. Spiess, J., Villarreal, J. and Vale, W. (1981) Bio-
    chemistry 20:1982–1988.
11. Hunkapiller, M.W. and Hood, L.E. (1978) Biochemistry
    17:2124–2135.
12. Dwulet, F.E. and Gurd, R.N. (1976) Anal. Biochem. 76:
    530–538.
13. Savige, W.E. and Fontana, A. (1977) Methods Enzymol
    47:453–459.
14. Montecucchi, P.C., Henschen, A. and Erspamer, V.
    (1979) Hoppe-Seyler's Z. Physiol. Chem. 360:1178
    (abstr.).

IDENTIFICATION OF INTERNAL RESIDUES OF LACTOSE PERMEASE OF

ESCHERICHIA COLI BY RADIOLABEL SEQUENCING OF PEPTIDE MIXTURES

Konrad Beyreuther, Barbara Bieseler,
Ruth Ehring and Benno Müller-Hill
Institut für Genetik der Universität zu Köln

D-5000 Köln 41, Germany

ABSTRACT

The present study describes a general method for the rapid
identification of sequence internal radiolabeled residues
of proteins of known sequence such as active site residues
or residues predicted from the DNA sequence of putative
exons. The method employs comparative radiolabel sequence
analysis of peptide mixtures obtained from two different
specific cleavages of radiochemically homogeneous proteins.
The assignment of sequence internal residues is achieved by
correlating the positions of radiolabeled residues determined by Edman degradation to the positions predicted from
the known sequence according to the specificity of the
cleavage reactions. This strategy applied to lactose
permease of E. coli biosynthetically radiolabeled in vitro
and in vivo by chemical modification allowed the identification of the 8 cysteine positions, of the essential
cysteine 148, and the analysis of the accessibility of the
cysteines of in vivo permease towards sulfhydryl reagents.

Abbreviations. TDG, galactosyl-ß-D-thiogalactopyranoside;
NEM, N-ethyl maleimide; PTH, phenylthiohydantoyl; T, tryptic;
CB, cynogen bromide.

INTRODUCTION

At present, the combination of DNA sequencing and N-terminal

protein sequencing is a commonly used, straightforward way
to obtain the sequence of a protein. The amount of protein
needed for such an analysis may be less than 1 fmol (10^{-15}
mol) if radiolabel protein sequencing techniques are
employed. Techniques as rapid and as sensitive as the aforementioned protein sequencing techniques are also needed for
the identification of sequence internal, i.e. more carboxy-
terminal residues than those covered by the former method.
We present a general technique which permits the identification of residues located on peptides beyond the positions
covered by N-terminal protein sequencing provided the
sequence of the protein is known. The procedure is based
on comparative automated sequence analysis of peptide
mixtures radiolabeled under different conditions. On the one
hand, the protein synthesized in vivo was radiolabeled by
posttranslational modification known to affect the active
site (referred to as "in vivo lactose permease"). On the
other hand, the protein was radiolabeled biosynthetically
with a single amino acid during cell-free protein synthesis
(referred to as "in vitro synthesized lactose permease").

The procedure has been worked out with lactose permease of
Escherichia coli an intrinsic membrane protein of known
sequence located in the bacterial cytoplasmic membrane (1-3).
The permease contains an essential cysteine residue which can
be labeled specifically with NEM (4). Recombinant E. coli
plasmids allowing the overproduction of lactose permease(5,6)
have made possible the protein-chemical analysis of the
protein such as the characterization of the primary translation product of the permease gene (2,3), and the experiments reported in this paper on the identification of the
essential cysteine residue of the permease and the accessibility of the cysteine residues of detergent solubilized
lactose permease towards the sulfhydryl reagent NEM.

MATERIALS AND METHODS

The cell-free synthesis of lactose permease was performed in
a coupled transcription-translation system using plasmid
pGM 21 DNA (6) as described previously (2,3). Radiolabeling
with ^{35}S-cysteine was throughout the 30-min synthesis period
under conditions as described in (2,3) and the pelleted
material obtained after 60 min centrifugation at 100 000 x g
(3) was used for further studies. Earlier experiments show
that this fraction contains one predominant radiolabeled

product identified as lactose permease (2,3). The sediment
containing 1 - 2 uCi of biosynthetically labeled permease
was resuspended in 1 ml of 0.1 M ammoniumbicarbonate buffer,
pH 8.2, 5% 2-mercaptoethanol (v/v) and dialyzed against 2 L
of the same buffer for 12 h at 5°C, 4 times against 2 L each
of 0.1 M ammoniumbicarbonate buffer, pH 8.2 for 2 h at 25°C
and finally against 6 M guanidine hydrochloride in 50 mM
potassium phosphate buffer, pH 6.3 for 12 h at 25°C. The
solution was treated with NEM at a final concentration of
0.1 M NEM for 1 h at 25°C and after the addition of 0.1 ml
of 2-mercatoethanol dialyzed against 4 times 2 L of 0.1 M
ammoniumbicarbonate buffer, pH 8.2. Digestion of 35S-(N-ethyl-
succinimido)-cysteine-containing permease with 0.01 mg
TPCK-treated trypsin (total protein in the sample was 0.15 mg)
was performed after heat denaturation at 100°C for 10 min
and terminated after 12 h at 37°C by lyophilization. Cyanogen
bromide cleavage was done in 0.5 ml of 70 % formic acid
using 15 mg of CNBr for 12 h at 25°C under nitrogen.
Lactose permease-containing crude membrane fractions or
cytoplasmic membrane vesicles (7) were prepared from E. coli
strain T 206 harboring plasmid pGM 21 (6).
Solubilization of lactose permease in Lubrol PX (1% w/v) (9)
was done in 0.01 M potassium phosphate buffer, pH 6.3 for 30
min at 4°C and the insoluble material removed by centrifugation at 10 000 x g or 100 000 x g for 30 min at 4°C.
Labeling of lactose permease under the conditions known to
label the essential SH group   of lactose permease (4,8)
was done by treating the membranes with unlabeled NEM in the
presence of the substrate TDG and with radioactive NEM in the
absence or presence of TDG for 30 min. The centrifugation
and washing procedure after the first NEM-treatment was
replaced by dialyses against 0.01 M potassium phosphate
buffer, pH 6.3.
Tryptic digestion of membrane-associated or detergent-
solubilized lactose permease was performed in 50 mM ammoniumbicarbonate buffer, pH 8.2 of heat denatured material
(10 min at 100°C) with TPCK-treated trypsin at a protein:
enzyme ratio of 20:1 in the presence of 10 % (v/v) of
n-butanol which was added after the denaturation step.
The digestion performed for 12 h at 37°C was terminated by
lyophilization. Cyanogen bromide cleavage was done as described
for in vitro permease.
Automated Edman degradation was carried out in a Beckman
Model 890B (updated) sequencer equipped with an undercut cup,
a Sequemat P-6 autoconverter, chemical oil filters (Balzers)

for rough and fine vacuum pumps and a gas purifier (Sulpelco)
to reduce the oxygen and water content to less than 0.1 ppm.
The program used employs 0.2 M Quadrol, single coupling and
single cleavage. Quadrol was supplemented with aminopropyl-
glass and delivered through a sintered glass funnel as
suggested by Frank (10). The peptide mixtures were subjected
to automated Edman degradation in the presence of 3 mg of
the nonprotein carrier polybrene (11) and of 4 mg of hen egg
white lysozyme. Polybrene was subjected to 10 degradation
cycles in the presence of 150 nmol of glycyl-glycine per mg
prior to use and stored in 10 % (v/v) acetic acid in the
cold. Sequence analyses were started after one nondegrading
cycle in the sequencer to reduce the background of radioactivity (cycle 0). The PTH-amino acid derivatives released
from the protein carrier lysozyme were analysed by HPLC on a
Du Pont Zorbax CN column according to (12) and by thin layer
chromatography (13) using a total of 5 - 10 % of the sample.
This allowed to assess the performance of the instrument and
to evaluate repetitive yields and out-of-step recoveries at
least for the protein carrier. The actual reppitive yields
varied between 94 and 96 % and the out-of-step recoveries
between 2 and 3 %. The radioactivity in the fraction obtained
at each cycle was determined in 5 ml Aquasol-2 (NEN) using
90 % of the sample dissolved in 0.1 ml of methanol.

RESULTS AND DISCUSSION

Lactose permease is a strongly hydrophobic protein consisting
of 417 residues of which 71 % are nonpolar. It contains 8
cysteine residues (1).
In order to identify the essential cysteine residue(s) of
lactose permease which is labeled by NEM in the assay of
Fox and Kennedy (4,8) we performed a pilot experiment with
^{35}S-cysteine labeled lactose permease synthesized in vitro.
The material was cleaved with trypsin, staphylococcus aureus
V8 protease, thermolysin and cyanogen bromide. The solubility
problems encountered with the peptides did not allow their
separation by the fingerprint technique. Therefore we tried
direct sequencing of whole peptide mixtures as a possible
way to assign the essential cysteine   of the permease.

In theory, tryptic digestion is expected to yield peptides
which contain the 8 cysteines of lactose permease at different
positions (Table 1). The same should hold true for cyanogen
bromide cleavage (Table 1) and digestion with staphylococcal

Table 1. Sequences covered by expected cysteine-containing
peptides of lactose permease and their cysteine positions

Cysteine residue	T-peptides	Cys pos.	CB-peptides	Cys pos.	T-CB-peptides	Cys pos.
117	75 - 131	43	87 - 145	31	87 - 131	31
148		4	146 - 161	3	146 - 161	3
154	145 - 188	10		9		9
176		32	162 - 267	15	162 - 188	15
234	222 - 259	13		73	222 - 259	13
333	320 - 335	14		10	324 - 335	10
353	345 - 358	9	324 - 362	30	345 - 358	9
355		11		32		11

V8 protease (data not shown). However, all three cleavages
do produce one peptide containing one cysteine beyond position
40. The detection of these cysteine residues might be
difficult by automated sequencing. In practice, washing-out
of peptides, the often encountered repetitive yields of less
than 96%, out-of-step degradation problems and the expected
poor cleavage yields for a membrane-associated protein limit
the number of interpretable sequencing steps to some 40 steps.
Tryptic digestion plus cyanogen bromide cleavage should
produce peptide mixtures which fulfil the requirement that
all cysteines are in positions covered by 40 degradation
cycles (Table 1). However, cysteines 154 and 353 occupy the
same position in the respective peptides. This coincidence
requires the analysis of two peptide mixtures in order to
assign all cysteines of lactose permease by automated
sequence analysis.
Analysis of the peptide mixtures produced by trypsin and
by trypsin plus cyanogen bromide would facilitate the
assignment of cysteines of the permease since five of the
eight cysteines do occupy different positions in the
respective peptides (Table 1). As a result, a shift of
radioactivity associated with these residues should be
observed if the predictions of Table 1 are correct.
We decided therefore to follow this strategy for the identification of the essential cysteine   of lactose permease
isolated from bacterial membranes.
Lactose permease constitutes 5 - 10 % of the membrane protein

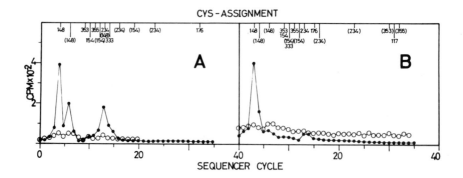

Fig. 1. Radioactivity recovered by automated Edman degradation of specifically radiolabeled in vivo lactose permease
treated with trypsin (A) or trypsin plus cyanogen bromide (B).
Membrane bound permease was pretreated with NEM and TDG and
than reacted with ^{14}C-NEM in the absence of TDG(closed circles)
or in the presence of TDG (open circles). The samples labeled
in the absence of TDG contained 20 000 c.p.m. The other
samples contained 10 500 c.p.m.. Cys assignment as in Fig.3.

in these preparations (crude membranes or Osborn vesicles)
and is the major protein being labeled by the assay
employing radioactive NEM (5). Cysteine residue 148 is shown
to be the only cysteine of membrane bound permease which
reacts with ^{14}C-NEM (Fig. 1). A comparison of Fig. 1A and
1B reveals that the radioactivity recovered at steps 4, 6, 13
and 14 is shifted to position 3 in Fig. 1B suggesting that
incomplete hydrolysis of trypsin is responsible for the
radioactivity released beyond step 4 in Fig. 1A (cf. Tables
1 and 2). The specific radioactivity recovered for cysteine
148 in the experiment described in Fig. 1 is only 5 – 10 %
of the specific radioactivity of the samples analysed
suggesting very poor cleavage yields. Similar yields are
obtained for cysteine 148 of lactose permease synthesized
in vitro in the experiments described in Fig. 3. Such poor
cleavage yields are frequently found for hydrophobic proteins
associated with membranes.

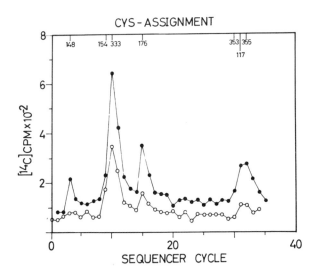

Fig. 2. Radioactivity profile of degradations of detergent solubilized lactose permease treated with cyanogen bromide. ^{14}C-NEM treatment in the absence (closed circles; 107000 c.p.m.) and in the presence of TDG (open circles; 61 000 c.p.m.). Both samples contain equal amounts of lactose permease. Cys assignment according to Table 1.

Four arginine residues at positions 134, 135, 142 and 144 and a lysine residue at position 131 precede cysteine 148 in the permease sequence (1). It can be estimated from Fig. 1A that cleavage at arg 144 accounts for about 40%, at arg 142 for 25%, at arg 135 for 30%, at arg 134 for 5% and at lys 131 for less than 1% of the radioactivity recovered in the experiment.

Membrane vesicles pretreated with cold NEM in the presence of the permease substrate TDG were solubilized in Lubrol PX as described in Materials and Methods and treated with ^{14}C-NEM in the presence (Fig. 2B) and absence of TDG (Fig. 2A). The cysteine residues 117 (step 31), 148 (step 3), 154 (step 9), 176 (step 15), 333 (step 10), 353 (step 30) and 355 (step 32) of detergent solubilized permease are modified to different extents (Fig. 2). Cysteine 234 which does not show up in Fig.2 is expected to appear at step 73 (Table 1). This cysteine is also modified by radioactive NEM under the aforementioned conditions as found after tryptsin plus cyanogen bromide treatment (data not shown).

Are we able to detect all cysteines of lactose permease with
this method? Is the assignment given in Fig. 1 and 2 correct
or do we not cover all cysteine residues of the permease?
The analysis of a set of two different peptide mixtures
listed in Table 1 might not be sufficient for a complete
assignment. This could be for instance the case if the
assumed cysteine positions were incorrect or if neither of
the cleavage procedures does release all peptides covering
the 8 cysteine residues of lactose permease. We therefore
prepared in vitro synthesized lactose permease labeled with
^{35}S-cysteine. The analyses of the tryptic mixture and of the
material treated with trypsin and subsequently with cyanogen
bromide are shown in Fig. 3.
The interpretation of the radioactivity patterns shown in
Fig. 3 is complicated. First, trypsin may not cleave
quantitatively at the first cleavage site preceding the

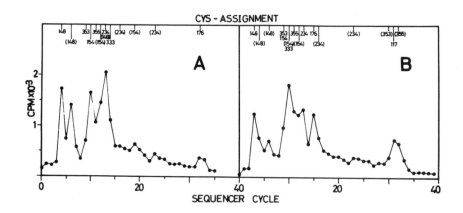

Fig. 3. Radioactivity recovered at each sequencer cycle from
in vitro synthesized permease labeled with ^{35}S-cysteine.
Tryptic peptide mixture (A; 389 000 c.p.m.), peptides after
treatment with trypsin plus cyanogen bromide (B; 474 000
c.p.m.). The cys assignment for positions not included in
Table 1 ( ) refers to partial cleavage products (cf. Table 2).

corresponding cysteine residue as found for in vivo permease
in the experiment described in Fig. 1. Second, the extent of
cleavage is expected to be different at different sites.
Third, problems encountered with the automated sequencing
of peptides and in particular with hydrophobic peptides
contribute to the complex pattern. The latter problem might
be overcome by quantitative evaluation of the data shown in
Fig. 3. Correction for losses due to repetitive yield and
out-of-step degradation reduces the complexity of the
pattern and provides good evidence that all 8 cysteines are
detectable with the method. The assignment shown on top
of Fig. 3 is based on this evaluation (details to be
published). Some positions showing release of radioactivity
above background might either be due to incomplete cleavage
or due to side reaction during cleavage. The latter
complication can not be ruled out but seems to be less
likely since nearly all readioactivity seen in Fig. 3 can
be accounted for on the basis of the cleavage specificities
(Tables 1 and 2). Incomplete cleavage of trypsin influences
the radioactivity pattern if the products are N-terminal
extensions of the peptides listed in Table 1. The degradation
steps which might contain radioactivity released of partial
tryptic peptides are listed in Table 2. The assignment of
cysteines given in parentheses on top of Fig. 3 was done
on the basis of Table 2. For Fig. 3B the additional
assumption was made that incomplete cyanogen bromide cleavage

Table 2. Cysteine positions predicted for partial tryptic
hydrolysis of lactose permease (deduced from (1)) which are
covered by 40 steps of automated Edman degradation

Cysteine residue	Positions
117	—
148	4, 6, 13,14, 17,
154	10, 12, 19,20,23
176	32,34
234	13, 16, 23
333	14, 31
353	9, 18, 34
355	11, 20, 36

of tryptic peptides occured and that the mixture includes
cyanogen bromide fragments with uncleaved internal tryptic
sites.
The method applied to the analysis of internal residues of
lactose permease has provided information on the essential
cysteine residue and on the alterations caused by detergent
solubilization suggesting a direct substrate protection of
several cysteines or more likely different conformations of
permease in the presence and absence of the substrate TDG.
The method is rapid and brings about information of gene
products which is not available by DNA sequencing. The
method can be applied to in vitro and in vivo synthesized
radiolabeled proteins and permits the analysis of proteins
such as membrane proteins which are difficult to analyse by
conventional methods.
It is concluded that neither pure proteins nor pure peptides
are a prerequisite for the localization of sequence internal
residues of proteins of known sequence.

Acknowledgement. We thank K. Neifer, S. Pinto, R. Hanssen and
C. Heibach for technical assistance and the DFG for financial
support through SFB 74.

REFERENCES

1. Büchel, D.E. et al. Nature 283, 541-545 (1980)
2. Ehring, R. et al. Nature 283, 537-540 (1980)
3. Beyreuther, K. et al. in Methods in Peptide and Protein
   Sequence Analysis (Birr, Chr.ed)p.199-212, Elsevier,
   Amsterdam, 1980
4. Fox,C.F &Kennedy,E.P.Proc.Natl.Acad.Sci.USA54,891-899(1965)
5. Teather,R.M. et al. Mol.Gen.Genet. 159, 239-248 (1978)
6. Teather, R.M. et al. Eur.J.Biochem. 108, 223-231 (1980)
7. Osborn, M.J. et al. J.Biol.Chem. 247, 3962-3972 (1972)
8. Fox, F. et al. Proc.Natl.Acad.Sci.USA 57, 698-705 (1967)
9. Beyreuther, K. et al. Biochem.Soc.Trans.8, 675-676 (1980)
10. Frank,G.Hoppe-Seylers Z.Physiol.Chem.360, 997-999 (1979)
11. Tarr, G.E. et al. Anal.Biochem. 84, 622-627 (1978)
12. Johnson, N. et al. Anal.Biochem. 100, 335-338 (1979)
13. Beyreuther, K. in Solid Phase Methods in Protein Sequence
    Analysis (Previero,A.and Coletti-Previero,M.-A. eds.)
    p. 107-119, Elsevier, Amsterdam, 1977

THE APPLICATION OF MONOCLONAL ANTIBODIES

TO THE MICROSEQUENCING OF PROTEINS

Lawrence K. Duffy and Alexander Kurosky

Department of Human Biological Chemistry and Genetics

University of Texas Medical Branch, Galveston, Texas   77550

ABSTRACT

The isolation of peptides through the use of monoclonal antibody immunoadsorbent columns was developed. The α chain of cholera toxin was used as a model for application of this strategy. Monoclonal antibodies to the A subunit of cholera toxin were isolated from mouse hybridoma cell lines. Two monoclonal antibodies with different specificities were used to immunoadsorb a cyanogen bromide digest of either purified native toxin or the α chain of the toxin. The amino-terminal sequences of the isolated fragments were determined by automated sequence analysis. In addition, a S. aureus protease digest was immunoadsorbed with one of the monoclonal antibodies.

INTRODUCTION

Several years ago, at the Second International Conference on Methods in Protein Sequence Analysis, Zabin, Fowler and Brake[1] introduced the use of antibodies as immunological aids in sequencing large proteins. More recently, there has been the development of monoclonal antibodies as "chemical affinity reagents" and the documentation of their use in the purification of proteins by immunoadsorbent methods.   In our studies on the structure of cholera toxin, we have tried to combine these two techniques in order to utilize the high degree of

149

specificity of antigen binding by monoclonal antibodies for the isolation and characterization of peptides.

Cholera toxin is an enterobacterial protein secreted by Vibrio cholerae that is primarily responsible for the severe fluid and electrolyte loss associated with the disease cholera[2]. Intact cholera toxin is composed of two subunits, A and B, which associate noncovalently and are readily dissociable[3]. The A subunit is made up of two chains, α and γ, covalently bonded by a single disulfide[4]. The B subunit is an aggregate of five β chains strongly held together by noncovalent interactions. Primary structural analysis of the β and γ chains have been completed[5,6,7]. The quaternary structure of the toxin can be formulated as $\alpha\gamma\beta_5$. There is good evidence that the NAD^+-glycohydrolase and the ADP-ribosyltransferase activities of cholera toxin reside on the α chain of the A subunit. In order to obtain a high degree of ADP-ribosylation activity by the A subunit, the $\alpha\gamma$ disulfide bond must be reduced[8]. Antisera raised to native (unfrozen) toxin predominantly recognize B subunit determinants; the A subunit determinants appear to be poorly antigenic[9]. However, purified preparations of A subunit were shown to be highly antigenic suggesting that the A determinants are hidden in the intact toxin[9]. This report describes the production of monoclonal antibodies to cholera toxin A subunit and the initial studies of their use to isolate peptides.

METHODS

Purification of cholera toxin and its subunits. The toxin was purified from fermenter cultures of Vibrio cholerae (Inaba strain 569B) by precipitation of the culture filtrate with sodium hexametaphosphate followed by column chromatography on Whatman P11 phosphocellulose[4]. The separation of A and B subunits of cholera toxin was achieved by gel filtration of Sephadex G-75 eluted with 5% formic acid. Precursor A subunit, α chain, and γ chain were also separated on the G-75 column after reduction with 2-mercaptoethanol and alkylation with iodo |1-^{14}C| acetic acid.

Preparation of monoclonal antibodies. Production of monoclonal antibodies to the A subunit of cholera toxin was

as previously reported[11]. Antibodies were collected from
either culture media or from ascites fluid of Balb/c
female mice. They were purified first by ammonium sulfate
precipitation (50% saturation) and then by ion exchange
chromatography on DEAE-Affi-Gel Blue (Bio-Rad) using a
step-gradient of 0.01 M sodium phosphate pH followed with
0.01 M sodium phosphate containing 1.2 M NaCl, pH 8.

Immunoaffinity chromatography. Immunoadsorbent
columns were prepared by coupling the immunoglobulin
fraction (5 absorbance units/ml) to either cyanogen
bromide (CNBr) activated Sepharose (Sigma)[12] or Affi-gel
10 (Bio-Rad)[13]. Protein digests were applied to the
column in either sodium phosphate buffer, pH 8, or
distilled water and continually circulated overnight by
means of a peristaltic pump. One or two ml fractions
were collected and tested for absorbance at 280 nm. When
the baseline stabilized, the column was eluted with 0.5 M
acetic acid. Fractions were monitored by absorbance
(280 nm) or by measurement of relative fluorescence after
reaction with fluorescamine. The column was regenerated
by washing with .01 M sodium phosphate .15 M NaCl,
followed by several column volumes of distilled water.

Protein/peptide cleavages. CNBr hydrolysis was as
described by Lai et al.[14]. The α chain contains four CNBr
peptides (Fig. 3). S. aureus protease hydrolysis of
peptides was carried out for 16 h in 0.15 M NH_4CO_3 at room
temperature with an enzyme/protein ratio of 1:50 (w/w).

Peptide purification and identification. Amino acid
analysis and automated amino acid sequence analysis were
performed as previously described[4,7]. High performance
liquid chromatography (HPLC) peptide separations were
carried out on a Beckman/Altex gradient HPLC using either
an Ultrasphere C-18 column or a SynChropax RP-P column
(SynChrom, Ind.). The gradient system used with the C-18
column was that of Schroeder et al.[15] while the method of
Mahoney and Hermodson[16] was used with the RP-P column.

Competitive inhibition assays. The enzyme-linked
immunoadsorbent assay (ELISA) employed was the indirect
antibody procedure. Inhibition experiments were performed
by incubating varying concentrations of peptide with a
constant amount of antibody in a microtiter plate overnight.

The mixture was then transferred to a microtiter plate precoated with either A subunit or cholera toxin prior to analysis by ELISA.

RESULTS AND DISCUSSION

Two monoclonal antibodies to A subunit ($\alpha\gamma$) of cholera toxin were selected that showed different specificities. Although both antibodies reacted with the α chain neither reacted with the γ chain. One (1-4G7) reacted with intact cholera toxin and α chain CNBr peptide 1 (residues 95-214) while the other (1-4D10) did not react (Table 1). CNBr peptide 3 (residues 1-23) purified by gel filtration and HPLC slightly inhibited monoclonal antibody 1-4D10.

Clone 1-4D10 immunoadsorbent column. CNBr peptides (2 mg) in .01 M sodium phosphate, pH 8, were applied to the affinity column (1X10 cm). The first difficulty encountered was the partial insolubility of the CNBr digest after lyophilization. This was overcome by adjusting the pH to approximately 10 with NaOH and titrating back to pH 8 with HCl. As seen in Fig. 1, a small peak was eluted from the column with 0.5 M acetic acid. The amino acid composition of this pool was found to be: /Asx_5, Ser_2, Glx_2, Pro_2, Gly_2, Ala, Ile, Leu_2, Tyr, Lys_2, Arg_2, Hse/.

Automated sequence analysis (Beckman 890B) showed the amino terminal sequence of the eluted peptide to be: /Asn-Asn-Asp-Lys/. This sequence was identical to the amino-terminal sequence of the α chain. The sequence results confirmed that the peptide eluted from the 1-4D10 column was CNBr 3. The yield of this peptide based on amino acid analysis was approximately 10%. The low yield

Table 1   Reactivity of Monoclonal Antibodies with Cholera Toxin, and its Component Chains, and CNBr Peptides.

Antibody Clone	Cholera Toxin	α Chain	γ Chain	β Chain	CNBr Peptides			
					1	2	3	4
1-4G7	+	+	—	—	+	ND	—	—
1-4D10	—	+	—	—	—	ND	(+)	—

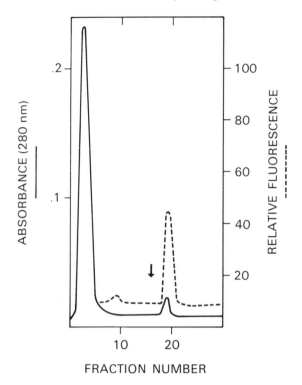

Fig. 1.   Immunoadsorbent chromatography of the CNBr digest of cholera toxin with antibody clone 1-4D10.   Addition of 0.5 M acetic acid indicated by arrow.

is probably related to the low amount of antibody coupled to the column due to a small antibody titer. For large scale purification of peptides, this initial experiment indicated that larger column volumes may be needed than that previously reported for proteins. Another disadvantage was the fact that the unbound pass through pool contained sodium phosphate and chloride salts. A possible improvement would be the use of more volatile buffers such as N-methylmorpholine and trifluoroacetic acid.

The soluble portion of a S. aureus digest (4 mg/ml H_2O) was also applied to the 1-4D10 column and subsequently eluted with 0.5 M acetic acid. A single peak was obtained (data not shown) which had the following

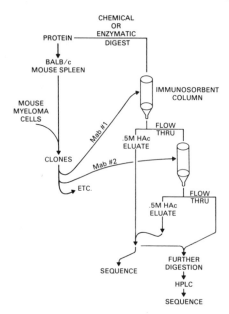

Fig. 2.   Schematic representation of peptide isolation by
a cascade arrangement of monoclonal antibody immunoadsorbent
columns.

composition: /Asx$_2$, Thr, Ser, Glx$_2$, Pro, Gly$_2$, Val$_2$, Met,
Ile, Leu, Tyr, Phe, Lys, Arg/. Sequence analysis showed
the presence of 3 peptides in this pool with amino-terminals
Asn, Tyr, and Met. The predominant peptide had an amino
terminal sequence that began /Tyr-Phe-Asp-Arg-Gly/. This
sequence corresponded to residues 30-34 in the α chain
sequence. The yield of this peptide was also less than 10%
of the total nanomoles applied.

     Clone 1-4G7 immunoadsorbent column. The peptides which
did not bind to the antibody 1-4D10 column were subsequently
applied to a 1-4G7 column. Elution with 0.5 N acetic acid
resulted in a single peak (data not shown) which had the
following amino acid compositions: /Asx$_{13}$, Thr$_3$, Ser$_6$,
Glx$_{11}$, Pro$_{10}$, Gly$_{14}$, Ala$_{11}$, Val$_7$, Ile$_6$, Leu$_9$, Tyr$_8$, Phe$_4$,
His$_6$, Lys, Arg$_9$/. The amino terminus was identified to be
phenylalanine, the same as that of peptide CNBr 1. The
yield was 1%. This is probably related to the insolubility
of this large CNBr peptide (residues 95-214). The results

of these initial experiments showed that the use of
monoclonal antibodies to purify peptides will be a useful
aid in sequencing proteins. For example, a cascade
approach as described by Springer[17] for the isolation of
cell surface antigens would be possible for protein with
more than one antigenic determinant. Depending on the
size of the peptides obtained further digestion of the
fragment followed by repeated passage down the immunoadsorbent column or by the use of HPLC would generate new
peptides for sequencing (Fig. 2).

In addition to primary structural information, the use
of monoclonal antibody immunoadsorbent columns will allow
rapid determination of the antigenic regions of proteins as
they are being sequenced. In the case of the A subunit of
cholera toxin, the fact that the monoclonal antibody which
reacted with native cholera toxin bound peptide CNBr 1 indicated that this region (residues 95-214) is at least in
part exposed to the environment. It was striking that
monoclonal antibody 1-4D10, which recognized the amino-
terminal region (residues 1-37) in the α chain, does not
react with native toxin. This suggested that this region is
buried in native tosin possibly due to interaction with
B subunit.

ACKNOWLEDGMENT

We wish to thank Linda Merryman, Ann Waguespack, Horace
D. Kelso, and Billy Touchstone for expert technical
assistance and Dr. R. Denney for help and advice in

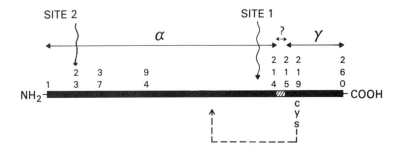

Fig. 3.   Schematic representation of the primary structure
of cholera toxin A subunit, illustrating possible antigenic
sites. Met residues occur at positions 23, 37, 94, and 215.

producing the monoclonal antibodies. Supported by Grant
29039 from the National Institute of General Medical
Sciences, by Grant CA 17701 from the National Center
Institute and by the J.W. McLaughlin Foundation.

REFERENCES

1. Zabin, I., Fowler, A.V., and Brake, A.J. (1977). In
    Solid Phase Methods in Protein Sequence Analysis,
    Previero, A. and Colleti-Previero, M.A. eds., North-
    Holland Publishing Co., New York, pp. 257-263.
2. Finkelstein, R.A. (1973). CRC Crit. Rev. Microbiol. 2,
    553-623.
3. LoSpalluto, J.J., and Finkelstein, R.A. (1972). Biochem.
    Biophys. Acta 257, 158-166.
4. Duffy, L.K., Peterson, J.W., and Kurosky, A. (1981)
    FEBS Lett., 126, 187-190.
5. Lai, C.-Y. (1977). J. Biol. Chem. 252, 7249-7256.
6. Kurosky, A., Markel, D.E., and Peterson, J.W. (1977).
    J. Biol. Chem. 252, 5257-5264.
7. Duffy, L.K., Peterson, J.W., and Kurosky, A. (1981).
    J. Biol. Chem., in press.
8. Lai, C.-Y. (1980). CRC Crit. Rev. Biochem. 9, 171-206.
9. Finkelstein, R.A. (1975). In Current Topics in Micro-
    biology and Immunology, Vol. 69, Springer-Verlag, New
    York, pp. 137-195.
10. Markel, D.E., Hejtmancik, K.E., Peterson, J.W., and
    Kurosky, A. (1979). J. Supramolec. Struct. 10, 137-149.
11. Kurosky, A., Duffy, L.K., and Denney, R.M. (1981). In
    Protides of Biological Fluids, Peeters ed., Pergamon
    Press, Oxford, Vol. 29, in press.
12. Bureau, D., and Daussant, J. (1981). J. Immunol. Meth.
    41, 387-392.
13. Caldwell, H.D., and Kuo, C.C. (1977). J. Immunol. 118,
    437-441.
14. Lai, C.-Y., Cancedda, F., and Chang, D. (1979). FEBS
    Lett. 100, 85-89.
15. Schroeder, W.A., Shelton, J.B., and Shelton, J.R. (1980).
    Hemoglobin 4, 551-559.
16. Mahoney, N.C., and Hermodson, M.A. (1980). J. Biol.
    Chem. 255, 11199-11203.
17. Springer, T.A. (1981). J. Biol. Chem. 256, 3833-3839.

Solid Phase Sequencing

SENSITIVITY ENHANCEMENT IN SOLID-PHASE PROTEIN SEQUENCE ANALYSIS: AN EXAMINATION OF METHODOLOGY

MARCUS J. HORN and ALEX G. BONNER

SEQUEMAT INC.

109 School St., Watertown, Mass., U.S.A.

One of the early impacts of the genetic engineering revolution has been a marked increase in interest in amino acid sequence analysis of proteins and peptides, especially on the so-called "micro" level. The scope of the term "micro-sequencing", however, has yet to be defined. As such, the question of sensitivity in protein sequence analysis is continually being raised.

For this report, we have investigated the level of sensitivity accessible by solid-phase protein sequence analysis. We also have attempted to pin-point factors within the sequencing methodology which contribute to or detract from sensitivity enhancement. These factors fall into three distinct areas:

1. Detection of the phenylthiohydantoin (PTH) derivatives.
2. Background contribution of the sequencing methodology.
3. Repetitive yield of the methodology.

By minor modifications of the sequencing methodology, based on the above, routine solid-phase protein sequence analysis has been extended to the 100-300 picomole level.

SENSITIVITY

While the word sensitivity is ofttimes used in reference to protein sequence analysis, it is more relevant to examine the minimum or minimal amounts of material that may be successfully sequenced in the solid-phase sequencer. As a generalization, it appears at present that the maximum degree

of detectability for protein sequence analysis can be obtained by sequencing intrinsically radio-labelled proteins, or
by using radio-labelled phenylisothiocyanate (PITC).

Figure 1 shows typical sequence analysis results from
an intrinsically radio-labelled protein. The protein (2 picomoles by specific activity), labelled with ^{35}S-methionine
and ^{3}H-lysine, was coupled to NH_2-aryl glass, in the presence
of a carrier cytochrome c. Attachment yield was 75%, based
on analysis of the radioactivity in the coupling washings.
As can be seen in Figure 1, sequence analysis proceeds de-
finitively, and with high repetitive yield, on a sample of
approximately 1.5 picomoles. Based on these and other data,
the amount of material to be sequenced does not seem to be
the limiting factor in solid-phase protein sequence analysis.
A sample can be sequenced --- whether the resulting PTH derivatives can be detected and identified is the more important question.

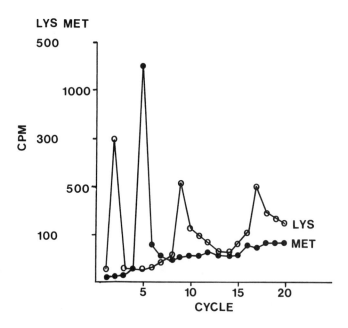

Figure 3. Determination of methionine and lysine residues
in an intrinsically labelled protein following 20 steps of
solid-phase Edman degradation.

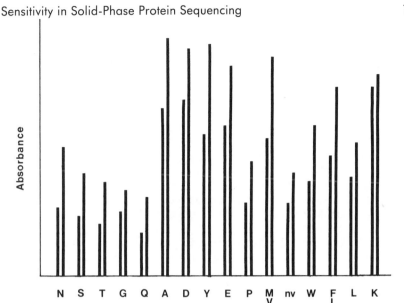

Figure 2. Comparative relative UV-absorbance of PTH-amino acids at 254 nm and 269 nm. Absorbance at 254 nm is the left-hand member of each pair. Standard single-letter codes are used for each amino-acid. "nv" = norvaline.

DETECTION

Perhaps since most laboratories have fixed-wavelength detectors, absorbance at 254 nm is used routinely for detection of the PTH derivatives arising from Edman degradation. The absorption maximum for the phenylthiohydantoin amino acids, however, is nearer 269 nm[2]. The relative absorbances for the various PTH derivatives at 254 and 269 nm are shown in Figure 2. The absorbance yields at 254 nm are increased for all derivatives when detection is at 269 nm --- from a 7-8% increase for lysine to a nearly two-fold increase for asparagine and glutamine. In addition, any background from solid-phase sequence analysis detected at 254 nm is diminished when detected at 269 nm.

There has been much discussion[3,4,5] concerning the need for "high-purity" reagents and solvents to improve the lower limits in protein sequence analysis. Laursen, in the original description of the solid-phase sequencer[6], remarked on

the ability of the solid-phase instrument to perform well
using reagent-grade chemicals alone. While the criteria for
reagent-grade chemicals have obviously declined since then,
the solid-phase methodology still does not require "ultrapure" reagents and solvents for sequence analysis at a low
level of material.

As a matter of fact, "high-purity" chemicals have actually proved detrimental in certain cases. Dichloroethane
is used routinely in the solid-phase sequencer and in the
automatic-conversion device[7]. When a variety of sources of
dichloroethane were evaluated as to their effect on the recovery of PTH-phenylalanine compared to a standard[5], the
reagent-grade material actually gave the best results.
Figure 3 presents a summary of the results of these tests.
We postulate that the highly purified material contains
free radicals generated by the repurification, or that some
stabilizing agent has been removed. Based on these data,
one should avoid the blanket statement that high purity reagents and solvents improve the yield in sequence analysis.
Sequencer reagents and solvents are best evaluated on an
application basis.

BACKGROUND

One of the main sources of background in solid-phase
protein sequence analysis arises from the breakdown of the
backbone of the solid support. By proper preparation and
washing of the solid support, the majority of "bleed" from
glass and polystyrene resins can be eliminated.

In standard practice, however, excess resin amino-groups
are blocked with isothiocyanates[6], and excess resin isothio-
cyanate-groups are blocked with amines[8] --- both blocking
reactions give rise to thiourea groups on the solid support.
The "cracking" of thioureas in the presence of acid and
heat (as happens in the cleavage step of the Edman degradation)
is a general method of producing isothiocyanates. This reaction (Figure 4) can lead to generation of background in
solid-phase Edman degradation. Proper choice of blocking
agents easily eliminates interference in the high-performance
liquid chromatograph (HPLC) analysis.

Ethanolamine has been used as a blocking agent for isothiocyanate resins[8]. In addition to the breakdown reaction
previously mentioned, we have observed another reaction of
interest when ethanolamine is used for resin blocking. During experiments involving pulse-labelling[9] with fluorescent

PTH-PHENYLALANINE RECOVERY FOLLOWING
DISSOLUTION IN DICHLOROETHANE

BRAND A	GLASS-DISTILLED	9%
BRAND B	HPLC-GRADE	16%
BRAND C	HIGH PURITY	10%
BRAND D	ACS REAGENT	90%

Figure 3. Comparative recoveries of PTH-phenylalanine
following dissolution in dichloroethane from various sources.

or radio-labelled isothiocyanates for low-level sequence
analysis, it was noted that the foremost incorporation of
the labelled reagent was into the support, and not into the
polypeptide (Figure 5). The support adduct, besides having
prevented labelling of the polypeptide, generated additional

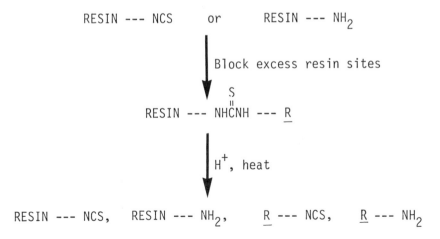

Figure 4. Breakdown of resin-bound thioureas to yield
possible background contributions during solid-phase Edman
degradation.

background products. N,N-dimethylethylenediamine (DMEDA)
has proved a satisfactory alternative blocking agent for
isothiocyanate supports.
    Since reagents and solvents used in solid-phase protein
sequence analysis are not deposited or coated by evaporation
on the supporting medium (as is done in liquid- or gas-liquid
phase sequence analysis), these components are not major
contributors to background problems. A background artifact
is occasionally present, however, migrating near PTH-alanine
in the HPLC chromatogram. This peak, a pyridinium salt,
arises from an ion-exchxnge interaction between the acidic
hydrogen ions on the glass or polystyrene backbone and the
pyridine-containing buffer. When the buffer[6] is replaced
by a 10-25% solution of triethylamine in methanol, similar
to that used by Machleidt and Machleidt[10], the interfering
peak is removed (Figure 6).

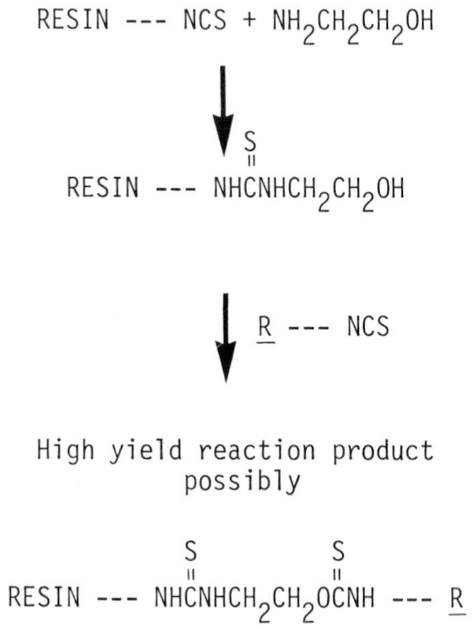

$$RESIN --- NCS + NH_2CH_2CH_2OH$$

$$\downarrow$$

$$RESIN --- NH\overset{S}{\overset{\|}{C}}NHCH_2CH_2OH$$

$$\downarrow \quad \underline{R} --- NCS$$

High yield reaction product
possibly

$$RESIN --- NH\overset{S}{\overset{\|}{C}}NHCH_2CH_2O\overset{S}{\overset{\|}{C}}NH --- \underline{R}$$

Figure 5. Reaction of ethanolamine-blocked DITC-glass with
isothiocyanates.

.OO1 —

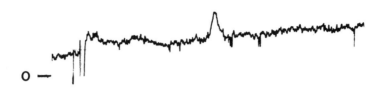

O —

BACKGROUND

Figure 6. Typical uncorrected background chromatogram (HPLC) of one step of solid-phase Edman degradation on DITC-glass blocked with N,N-dimethylethylenediamine. Diphenylthiourea (DPTU) is only evident artifact. DPTU does not interfere with identification of any PTH-derivative.

REPETITIVE YIELD

Matsueda and Margolies[11] have reported repetitive yields 97% and 98% in the solid-phase Edman degradation of synthetic peptides. These degradations were performed on an unmodified solid-phase sequencer, with standard reagent grade chemicals. Similar results have been noted by others[12]. With identical reagents and instruments, however, repetitive yields for peptides and proteins coupled intact to solid supports are generally lower, in the range of 90-94%[13]. Obviously, the solvents and reagents are not the major contributors to the decrease in repetitive yield.

Our current study indicates that an interaction between the solid support and the polypeptide is responsible, to an extent, for the noted decrease in repetitive yield. It appears that the higher the loading of polypeptide on the support, the higher the repetitive yield. As has been shown[14], however, overloading of the support generates overlap in sequence analysis.

The intrinsically labelled protein shown in Figure 1 was used as a model system. The sequence determination, as detailed previously, was performed after immobilization of

the protein in the presence of a large excess of carrier
C. krusei cytochrome c, a cytochrome with a free amino terminus. Calculations show a repetitive yield of 95-97%. A
similar immobilization, using comparable amounts of carrier
horse heart cytochrome c, a cytochrome with an acetylated
amino terminus, showed a 92-94% repetitive yield. Apparently,
something available in the C. krusei cytochrome c, but not
available in the horse heart cytochrome c, probably an amino
group, aids in scavenging the interfering reactant.

To obtain a further indication of what the interfering
reaction was, underivatized porous glass beads (100 mg) were
incubated with anhydrous trifluoroacetic acid at 50°C for
one hour. After evaporation of the acid, 150 nanomoles of
C. krusei cytochrome c in buffer[8] were added to the glass
beads, and stirred at room temperature for 2.5 hours. After
exhaustive washing of the glass beads, 18 nanomoles of the
cytochrome c could be sequenced for 10 steps. Again, there
apparently was a reaction between the glass backbone and the
polypeptide.

Kent, et al.,[15] have proposed a mechanism involving
trifluoroacetic acid and the hydroxyl groups of the polystyrene backbone to explain a blocking of the amino termini
during solid-phase peptide synthesis. A similar mechanism
may indeed be involved in the inactivation of amino-groups
during solid-phase protein sequence analysis, or there may
be a direct coupling of the amino-terminal residue to the
solid support. In either case, the solution proposed by
Kent, et al., for peptide synthesis can be adapted to solid-
phase Edman degradation. The introduction of a large excess
of primary or secondary amine to the bound polypeptide
following the trifluoroacetic acid cleavage step should, and
does, competitively scavenge reactants that might otherwise
react with the newly-generated amino-terminal residue. A
similar procedure is likewise effective in the liquid-phase
sequencer, minimizing polypeptide interactions with the
glass cup[13].

In practice, a buffered solution of benzylamine or
β-alanine is introduced into the sequencing program following
the wash after the cleavage step. Excess of the amine is
washed out prior to the addition of buffered PITC solution.
Improvements in repetitive yield have averaged 1-2% per
cycle, with increases as high as 3% noted. Optimal conditions
have yet to be established.

The proposed interaction, and its subsequent remedy,
appears to be peptide dependent, with hydrophobic peptides
and proteins showing less interaction, less improvement after

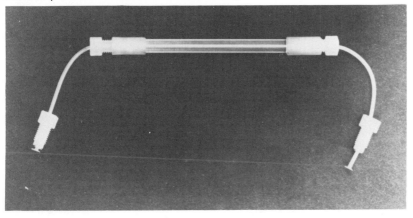

Figure 7. Miniaturized reaction column for use in low-level solid-phase Edman degradation.

amine treatment, and likewise, higher repetitive yields. In this regard, Sandberg, et al.,[16] have reported the solid-phase sequence analysis of 81 consecutive amino acid residues in a highly hydrophobic polypeptide chain. The average repetitive yield over the 81 residues was 98.5%. Other hydrophobic peptides were also sequenced with repetitive yields in excess of 95%.

MINIATURIZATION

Standard procedures for solid-phase protein sequence analysis have routinely used 35-50 mg of polystyrene resins, or 150-200 mg of glass supports per sample. Reaction columns for these supports have ranged in size from 2mm(I.D.)x100mm to 3mm(I.D.)x200mm. Incumbent upon any improvement in methodology for sequencing at picomolar levels or lower is a miniaturization of the reaction column as well as a reduction in the amount of solid support employed. For low level sequence analysis, we now use columns of 0.3-0.5mm inner diameter and 70mm length; interconnections are made via external threads on the glass column (Figure 7). These columns have dead volumes of less than 1 microliter, and may be used for 5-15 mg of glass supports or 2-4 mg of polystyrene resin.

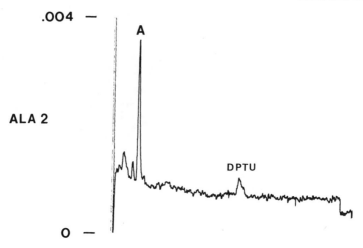

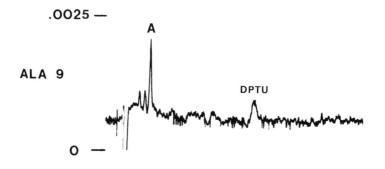

Figure 8. HPLC tracings of steps 2 and 9 in the solid-phase Edman degradation of C. krusei cytochrome c. Approximately 120 picomoles of protein were sequenced. The above analyses represent 40% of the sequenced material.

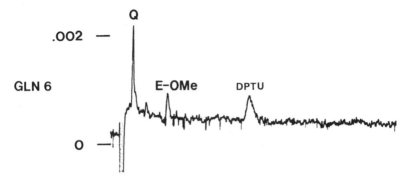

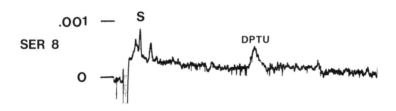

Figure 9. HPLC tracings of steps 6 and 8 in the solid-phase
Edman degradation of C. krusei cytochrome. See Figure 8.

        Reduction in the size of the columns allows for reduction
in the volume of reagents and solvents needed for reaction
and washing. With chemical consumption reduced, as well as
the amount of solid support used, any contributions to background from these factors are further minimized.
        Use of the smaller columns has also dictated a decrease
in the standard cycle time --- the solid-phase sequencer
program may easily be shortened to about 45 minutes by using
one-half the times specified in a standard sequencer program[7].
Additional fine tuning of parameters allows a program time
of 20-25 minutes for the sequence analysis of 500 picomoles
or less of polypeptide[13].

With application of the previously-mentioned factors
(except for the buffered amine pretreatment), 375 picomoles
of C. krusei cytochrome c was reacted with DITC glass[3]. By
amino acid analysis, 300 picomoles (80%) was coupled to
13 mg of DMEDA-blocked DITC-glass. Approximately 40% (5 mg)
was sequenced, and 40% of each cycle was analyzed by HPLC.
A comparison of recoveries of alanine in steps 2 and 9
(Figure 8) indicates a repetitive yield of 92% per cycle.
Glutamine shows little deamidation, while serine is recovered
in 20-22% yield (Figure 9). In a subsequent experiment[13],
using the buffered amine pretreatment, the repetitive yield
was raised to 94-95% per cycle.

REFERENCES

1. Horn, M.J. (1981) Submitted for publication.
2. Edman, P. and Henschen, A. (1975) in Protein Sequence
   Determination, Needleman, S.B., ed., Springer-Verlag,
   New York, p.238.
3. Hunkapiller, M.W. and Hood, L.E. (1980) Science 207, 523.
4. Hewick, R.M., Hunkapiller, M.W., Hood, L.E. and Dreyer,
   W.J. (1981) J. Biol. Chem. 256, 7990.
5. Niall, H.D. (1980) in Methods in Peptide and Protein Sequence Analysis, Proceedings of the 3rd International
   Conference on Solid Phase Methods in Protein Sequence
   Analysis, Birr, Chr., ed., Elsevier/North Holland Biomedical Press, Amsterdam, p.123.
6. Laursen, R.A., (1971) Eur. J. Biochem. 20, 89.
7. Horn, M.J. and Bonner, A.G. (1977) in Solid Phase Methods
   in Protein Sequence Analysis, Proceedings of the 2nd
   International Conference, Previero, A. and Coletti-Previero,
   M.A., eds., Elsevier/North Holland Biomedical Press,
   Amsterdam, p.163.
8. Machleidt, W., Wachter, E., Scheulen, M. and Otto, J.
   (1973) FEBS Lett. 37, 217.
9. Bonner, A.G. and Horn, M.J. (May 8, 1979) U.S. Patent No.
   4,153,416.
10. Machleidt, W. and Machleidt, I. (1977) in Solid Phase
    Methods in Protein Sequence Analysis, Proceedings of the
    2nd International Conference, Previero, A. and Coletti-
    Previero, M.A., eds., Elsevier/North Holland Biomedical
    Press, Amsterdam, p.233.
11. Matsueda, G.R. and Margolies, M.N. (1979) FEBS Lett. 106,
    89.

12. Kent, S.B.H., Riemen, M., LeDoux, M. and Merrifield, R.B. (1982) This volume.
13. Horn, M.J. (1981) Manuscript in preparation.
14. Horn, M.J. (1973) Dissertation Boston University, Boston.
15. Kent, S.B.H., Mitchell, A.R., Engelhard, M. and Merrifield, R.B. (1979) Proc. Natl. Acad. Sci, USA 76, 2180.
16. Sandberg, L.B., Alvarez, V.L., Wolt, T., Hansen, B. and Leach, C.T. (1981) Submitted for publication.

PROGRESS IN SOLID-PHASE SEQUENCING:

EXTENDED DEGRADATIONS OF PEPTIDES AND PROTEINS

WERNER MACHLEIDT and HELMUT HOFNER

Institut für Physiologische Chemie, Physikalische Biochemie und Zellbiologie der Universität, Goethestrasse 33, D-8000 München 2, FRG.

INTRODUCTION

Large peptides and proteins which contain internal lysine residues can be coupled to isothiocyanato-glass with yields usually better than 50%. This simple coupling reaction proceeds equally well in the presence of high concentrations of detergent and in organic solvents (1).It has been applied to proteins that have been eluted from stained SDS gels (cf. J.E. Walker, this volume). Using the liquid-phase sequencer or, most recently, the gas-phase sequencer, degradations over 80-100 steps have been reported (cf. the papers of G. Frank and M. Hunkapiller, this volume). Can the solid-phase sequencer be improved to a similar degree of performance? In this paper we want to discuss some of the major problems that will have to be solved before.

METHODS AND INSTRUMENTATION

Immobilization of Proteins

Bovine cytochrome b (2) was coupled to isothiocyanato-glass of 177 Å mean pore size (1) in 2-5% SDS adjusted to pH 8 with solid $NaHCO_3$. 50 nmole of protein were reacted with 90 mg of support for 5 h at 25°C. Glass-bound cyto-chrome b was deformylated by incubation in 0.5 M HCl in methanol for 2 h at 25°C. Acidolytic cleavage of Asp-Pro-bonds in the immobilized protein was performed in 80%

173

formic acid for 15 h at 25°C; cyanogen bromide cleavage of glass-bound cytochrome b in the same solvent using a 100-fold excess of the reagent over methionine.

Solid-Phase Sequence Analysis

A non-commercial solid-phase sequencer was used which includes a device for automated conversion and subsequent injection of the samples into a high pressure liquid chromatograph operating on-line to the sequencer (3). By the use of two additional valves the reaction column is closed at its outlet and pressurized with nitrogen to about 2 bar. The liquid content of the column can be purged into the sample vessel and the support dried with nitrogen. A typical sequencer cycle is given in table 1.

Identification of PTHs by HPLC and handling of quantitative data by the programmable computer portion of a SP 4100 (Spectra Physics) was done as described earlier (4).

The OPA-reagent contained 0.8 g o-phthalaldehyde and 2 ml of mercaptoethanol in 1 l of borate buffer pH 10.4. It was delivered to the reaction column for 10 min, followed by a 15 min wash with PA(MeOH (see below), both 0.6 ml/min at 25°C.

TABLE 1. 60 MIN SEQUENCER CYCLE
used with a 0.16 x 7 cm column at 50°C

Function	Time (min)
PA/MeOH (0.6 ml/min)	5
PA/MeOH (0.6 ml/min)+PITC (0.03 ml/min)	10
Rest	1
Column pressurized	10
Relax	1
MeOH (0.6 ml/min)	15
Column purged with nitrogen	3
Relax	1
TFA (0.3 ml/min) + collect	1
Column pressurized	5-10
Column purged with nitrogen + collect	2
Relax	1

PA/MeOH: 50mM n-propylamine (Sigma) in methanol (Merck,p.a.); MeOH: methanol; PITC: phenylisothiocyanate (Fluka puriss.p. a.); TFA: trifluoroacetic acid (Fluka purum, redistilled).

RESULTS AND DISCUSSION

n-Propylamine as a Scavenger in the Coupling Step

Originally we used 1-2% triethylamine in methanol instead of the usual coupling buffer. Repetitive yields up to
95% have been obtained with this solvent (3). Using different batches of methanol (all Merck, p.a.), the repetitive
yields were as low as 85%. They were raised to 96-97% using
the same methanol but replacing the triethylamine by n-
propylamine. Obviously, the primary amine scavenges contaminants of methanol that lead to irreversible blocking
reactions. Another possible effect is the neutralization of
reactive species formed on the support during the cleavage
step (cf. M.J. Horn, this volume). In contrast to the scav-
engers used in liquid-phase sequencing (5,6),the propylamine
is present during the reaction with PITC which is delivered
in an 8-fold excess over the amine.The resulting reaction
products are washed out, leaving only two moderate peaks
that do not interfere with PTH analysis.

Maximum Repetitive Yield

We have found that the repetitive yield (r.y.) in extended degradations of peptides and proteins is strongly
dependent on the position of anchor residues to the solid
support and the nature of residues between them. Maximum
r.y. have been obtained within the sequence 1-15 of horse
myoglobin (attached to isothiocyanato-glass) preceding the
first anchor point at Lys-16. The r.y. between Gly-5 and
Gly-15 was 97.1% using a 5 min TFA cleavage step (fig. 1).
Adding back the overlap portions, the total r.y. is 98.5%.
Only 1.5% per cycle are lost by irreversible blocking
reactions or washed out. This shows clearly the high
efficiency of solid-phase sequencing in the presence of a
scavenger.

Factors Limiting Extended Degradations

Unfortunately, this maximum r.y. is not maintained in
extended degradations of large peptides and proteins. Part
of it is dispersed as overlap over the following steps, and
most of it is lost by the so-called non-specific cleavage
of peptide bonds.

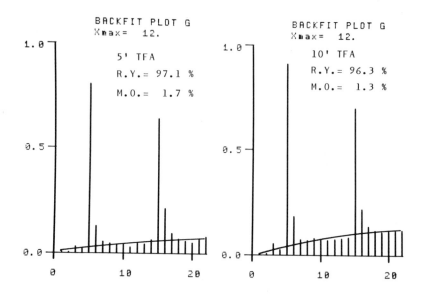

Fig. 1. Relative yields of PTH-Gly from the solid-phase degradation of horse myoglobin as determined by HPLC.
R.Y.= repetitive yield; M.O.= mean overlap per step.

Overlap. Experiments with different reaction times
have shown that the coupling step is virtually complete.
Overlap is due to incomplete cleavage. Incomplete cleavage
may result from an insufficient reaction time, but in solid-
phase sequencing also from incomplete and inhomogenous
penetration of TFA into the pores of the support. Penetration was significantly improved by the use of a pressurized
reaction column (see Methods). We tried to find the shortest
cleavage time that leads to an acceptable degree of overlap. For unknown reasons, the per-step overlap was rather
high in the begin of a degradation and approached a lower,
constant value after 10-15 cycles. This constant overlap
was about 2% for 5 min TFA cleavage and 1% for 10 min
cleavage time (fig. 1). A mean overlap of 1% per cycle
seems to be acceptable in extended degradations (50% overlap after 50 steps). Under these conditions - 10 min
cleavage - the limiting factor is the non-specific cleavage
of peptide bonds.

Non-specific Cleavage of Peptide Bonds

Brandt et al. (7) have shown that the so-called non-specific cleavage of peptide bonds during liquid-phase Edman degradation is not random but rather occurring at certain sensitive peptide bonds, preferably at aspartyl bonds, and, due to N-O-acyl shift, before serine and threonine residues. Our experience with solid-phase Edman degradation using TFA as cleaving agent confirms these findings.

Non-specific cleavage of peptide bonds leads to background and lowers the repetitive yield according to the losses of in-frame sequence. In solid-phase degradations, non-specific cleavage may cause discontinuous drops of yield at the anchor points of the protein to the solid support. All cleavage products that do not contain an anchor point will be washed out. Consequently, gaps will grow within the sequence between two anchor points (e.g. between Asp-20 and Ile-28 in sperm whale myoglobin attached via its lysine residues). When the last anchor residue preceding a gap is reached (e.g. Lys-16 in myoglobin), a yield-drop will occur due to the missing portion of in-frame sequence. Experimental data for myoglobin (without use of a scavenger) agree well with this concept: The r.y. between Leu-2 and Leu-9 was 95.0%, then dropped to 94.2% between Leu-9 and Leu-29, indicating the gap between Asp-20 and Ile-28, and rised to 94.8% again between Leu-29 and Leu-40.

A serious yield-drop is to be expected when an anchor point is reached which is followed by a long non-anchored sequence containing several cleavage-sensitive bonds. This situation was found during the degradation of deformylated cytochrome b (fig. 2). The r.y. was 96% between Met-1 and Met-11. Then a marked yield-drop occurred at the Lys-12 anchor point, and the degradation proceeded with a r.y. of 93.4%. According to the DNA sequence (B.G. Barrell, Cambridge, personal communication), Lys-12 is followed by a long sequence containing no anchor residues but many cleavage-sensitive bonds. It was not possible to determine a continuous sequence over more than 30 residues. The really limiting factor was not the low yield as such, but the difficulty of discriminating the low sequence-specific signals from the high level of background.

Both high background and yield-drops result from the non-specific cleavage of peptide bonds. Extended degradations

of large peptides and proteins on solid supports would
become more frequent, if the degree of non-specific cleavage could be reduced considerably. It remains to be established whether non-specific cleavage can be further reduced
by physical means like the use of minicolumns (cf. the
papers of M.J. Horn and J.E. Walker, this volume) or porous
layer beads as a sequencing support. Scrupulous exclusion
of water from the cleavage step will not solve the problem,
as the (most important) cleavage of aspartyl bonds is
rather suppressed by the presence of water (7). Ideally,
cleavage agents other than TFA would have to be found which
catalyze the Edman-type cleavage reaction much more efficiently than the non-specific cleavage of peptide bonds.
Tarr has suggested the use of boron trifluoride (8), but
safety restrictions preclude the use of the gas in a normal
solid-phase sequencer. As Brandt et al. have shown (7), non-
specific cleavage can be diminished by derivatization of
sensitive residues prior to the degradation. Derivatization
reactions can be easily performed on solid supports. Our
preliminary results indicate that glycination of carboxyls
lowers the background as well as the yield-drops in solid-
phase degradations. The resulting derivatives of Asp and
Glu can be identified by HPLC using a normal PTH program.
A suitable blocking group for the hydroxyl amino acids has
still to be found.

OPA-Blocking of Amino Groups before Proline

Temporary reduction of background may be achieved by
blocking all amino groups when a proline residue is reached.
Bhown et al. (6) have used fluorescamine for this purpose
in the spinning-cup sequencer. This reagent did not work in
the solid-phase instrument, presumably because it is hydrolyzed before it reaches the bottom of the reaction column.
Therefore we replaced it by o-phthalaldehyde (OPA). OPA-
blocking suppressed the background completely (fig. 3) until
it rised to its maximum value again within the following
10-15 cycles. After OPA-blocking the overlap is almost
completely abolished. Repeated OPA-blocking at several Pro
residues will allow to shorten the cleavage time.

The degradation of immobilized bovine cytochrome b
after acidolytic cleavage of a single Asp-Pro-bond (figs. 2,
3) is a typical moderately extended degradation with the
solid-phase sequencer. 60 residues were determined without
blanks. A mean r.y. of 96% was maintained over the whole

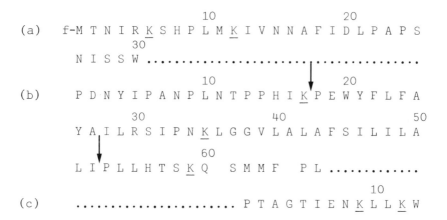

```
                        10                   20
(a)    f-M T N I R K S H P L M K I V N N A F I D L P A P S
                _         _
                30
       N I S S W ....................................
                        10          ↓        20
(b)    P D N Y I P A N P L N T P P H I K P E W Y F L F A
                                         _
                30                  40                   50
       Y A I L R S I P N K L G G V L A L A F S I L I L A
         ↓           _
                        60
       L I P L L H T S K Q   S M M F   P L .............
                      _
                                                   10
(c)    ..................... P T A G T I E N K L L K W
                                              _     _
```

Fig. 2. Partial sequences obtained by solid-phase Edman
degradation of bovine cytochrome b.
(a) N-terminal sequence of the deformylated protein
(b) Sequence after acidolytic cleavage of an Asp-Pro-bond
    (Asp-252 - Pro-253 in the DNA-derived sequence)
(c) Sequence after cyanogen bromide cleavage and OPA-block
    (Pro-367 - Trp-379 in the DNA-derived sequence)
Anchor points are underlined; arrows indicate OPA-blocks.

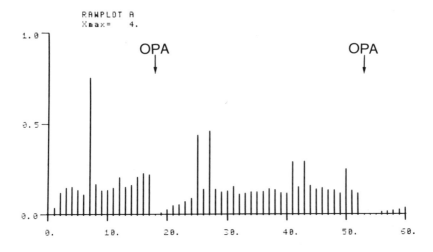

Fig. 3. Yields of PTH-Ala from the degradation of acidolyzed
cytochrome b (fig. 2,b) as determined by HPLC (1.0 on the
ordinate is equivalent to 5 nmole).

degradation including two OPA blocks before Pro-18 and
Pro-53. The limiting event was a yield-drop at the Lys-59
anchor point which is followed by a long non-anchored sequence containing several cleavage-sensitive bonds (predicted from the DNA sequence).

Another application of the OPA block is a selective
sequencing of peptides beginning with proline within mixtures. Using this technique, the C-terminal cyanogen bromide
fragment of bovine cytochrome b was sequenced from the
mixture of fragments that remained attached to the support
after cyanogen bromide treatment of the immobilized protein
(fig. 2). In step 12 the C-terminal Trp was released and
identified in the amino acid analyzer.

ACKNOWLEDGEMENTS

We wish to thank Dr. B.G. Barrell, Cambridge, who kindly
provided the DNA sequence of bovine cytochrome b prior to
publication.
The skillful technical assistance of Ms. G. Behrens and
U. Borchart is greatly appreciated.
The investigations were supported by the Deutsche Forschungsgemeinschaft, Sonderforschungsbereich 51 'Medizinische Molekularbiologie und Biochemie', München (B/31).

REFERENCES

1. Machleidt,W.& Wachter,E. (1977) Methods Enzymol. 47,
    263-277.
2. v. Jagow,G.,Schägger,H.,Engel,W.D.,Machleidt,W.,Mach-
    leidt,I. & Kolb,H.J. (1978)·FEBS Lett. 91,121-125.
3. Machleidt,W. & Hofner,H. (1980) in Methods in Peptide
    and Protein Sequence Analysis (Birr,Chr.,ed.) pp. 35-47,
    Elsevier/North Holland, Amsterdam
4. Machleidt,W. & Hofner,H. (1981) in High Pressure Liquid
    Chromatography in Biochemistry (Henschen,A.,Hupe,K.-P. &
    Lottspeich,F.,eds.),de Gruyter, Berlin, in press.
5. Frank,G. (1979) Hoppe Seyler's Z.Physiol.Chem. 360,
    997-999.
6. Bhown,A.S.,Bennett,J.C.,Morgan,P.H. & Mole,J.E. (1981)
    Anal. Biochem. 112,158-162.
7. Brandt,W.F.,Henschen,A. & v.Holt,C. (1980) Hoppe Sey-
    ler's Z.Physiol.Chem. 361,943-952.
8. Tarr,G.E. (1977) Methods Enzymol. 47,335-357.

THE USE OF DABITC IN AUTOMATED SOLID-PHASE SEQUENCING

Johann Salnikow*, Arnold Lehmann[+] and
Brigitte Wittmann-Liebold[+]

*Technische Universität Berlin, Institut für
Biochemie und Molekulare Biologie, Franklin-
straße 29, D-1000 Berlin 10
and
[+]Max-Planck-Institut für Molekulare Genetik,
Abt. Wittmann, Ihnestr. 63-73, D-1000 Berlin 33
West Germany

INTRODUCTION

The critical factor limiting detailed chemical characterization of physiological important proteins and peptides in the
past decade has often been the availability of pure material.
Consequently, there is a tendency in protein chemical technology including sequencing towards more sensitive methods
with the aim to extend the range of structural studies to
these scarce molecules. Although radioactive labelling of
the Edman reagent increases the sensitivity range by an order of magnitude, this method is often hampered by the insufficient purity and specific activity of the commercially
available compound and the rather cumbersome identification
procedure. A different approach has been used by Chang et al.
(1) by incorporation of a high absorptive chromophore - an
azo-group - into the Edman reagent. The resulting DABITC

Abbreviations:
DABITC, 4-N,N-dimethylaminoazobenzene 4'-isothiocyanate, DMF,
dimethyl formamide, EDC, 1-ethyl-3-dimethylaminopropyl carbodiimide, PITC, phenylisothiocyanate, TFA, trifluoroacetic
acid.

reagent leads to colored thiohydantoins whose detection limit approaches 5 picomoles permitting microsequencing. The
reagent, however, does not yield clean-cut degradation steps
unless used in conjunction with PITC in a double coupling
procedure (2). Adaptation of this microsequencing method to
the solid-phase technique has been successfully attempted
(3-5). Here we describe improvements of this technique with
respect to peptide attachment, programming of the sequencer
and automated conversion.

MATERIALS AND METHODS

Peptide Attachment to Aminopropyl Glass. Peptides are coupled underivatized to aminopropyl glass with EDC according to
the following procedure (6):
  ↓ dissolve up to 50 nanomoles of dried peptide in 100µl
    anhydrous TFA in a micro test tube
  ↓ incubate for 15 min at room temperature, then remove the
    TFA on the rotary evaporator and dry in vacuo over KOH
    pellets (total 20-30 min)
  ↓ add 2mg EDC freshly dissolved in 200 µl DMF and sonicate
    for 1 min
  ↓ add ca. 50 mg APG, deaerate and incubate at 40°C for
    60 min with gentle stirring
  ↓ wash with 2x2ml methanol and 1 ml ethyl ether and dry.
Solid-Phase Sequencing. A LKB 4020 solid-phase instrument
with a two-column system in the microsequencing mode was
equipped with an additional valve prior to the column select
valve permitting nitrogen purge of the reaction column. The
sequencing column (2 mm inner diameter) was shortened to
about 15 mm. A conversion unit was incorporated by instanta-
neous dilution of the TFA eluate with water to about 30%
with the help of a peristaltic pump (Fig.1). 2 min collecting

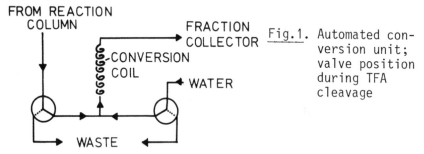

FROM REACTION COLUMN

FRACTION COLLECTOR

CONVERSION COIL

WATER

WASTE

Fig.1. Automated con-
version unit;
valve position
during TFA
cleavage

time will suffice to fill 2/3rd of the conversion coil. The
thiazolinones are converted at 50°C for the approximate
period of one Edman cycle and then transferred by the next
charge to the fraction collector.
The degradation program has been designed in accordance with
the double coupling procedure of Chang et al. (2) with DABITC
followed by PITC (Fig. 2). Both coupling processes are repeated separately with intermittent washing and nitrogen
purge steps. The DABITC cycle is performed in short reaction
intervals for economical reasons, i.e. to save reagent. The
TFA cleavage step is limited to 4 min with 2 min collecting
as outlined above: the bulk amount of colored thiazolinone
appears with the front of the cleaving acid. A short collect
signal just prior to the TFA cleavage serves to segment the
already converted thiazolinone in the coil from the next
sample.
The thiohydantoins are dried, dissolved in 200 µl n-butyl
acetate and after mixing with 500 µl water extracted with
the organic phase. Identification was performed by micro
thin layer chromatography (2) and high performance liquid
chromatography (7); Ile and Leu were differentiated in an
additional system (8).
Sequencing Reagents. DABITC, recrystallized: 0.5% solution
in redistilled DMF (v/v), prepared freshly every day. PITC,
redistilled: 5% solution in redistilled DMF (v/v). Buffer:
0.4 M N-ethyl morpholine-trifluoroacetate pH 9.0: pyridine:
DMF = 20:40:40. All reagents were of analytical grade and
freshly purified if necessary.
A detailed description of all methods is documented in (6).

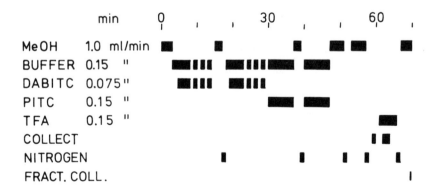

Fig. 2. Degradation program.

RESULTS AND DISCUSSION

Peptide attachment to amino group containing supports via
bifunctional isothiocyanate or homoserine lactone activation
has been proven a reliable method, which, on the other hand,
is limited to trypsin and cyanogen bromide derived protein
fragments. The search for a general procedure led to the use
of carbodiimides, which have been established as the classical condensing agents in peptide synthesis since long time.
As has been shown by Wittmann-Liebold et al. (9), peptide
attachment with EDC under specified conditions constitutes
a satisfactory method even in the absence of NH_2-terminal
protecting groups, the sole requirement being a free COOH-
terminal carboxyl group. Carboxyl side chains originating
from aspartic and glutamic acid, respectively, react to a
lesser extent thus permitting the identification of the corresponding thiohydantoins.
Optimization studies with a variety of peptides have shown,
that the assay described in "Methods and Materials" for amino propyl glass represents the best conditions with respect
to incubation time and EDC concentration (6). Similar conditions have been determined for aminopolystyrene with the
most significant difference in coupling yield being the attachment effeciency of peptides with COOH-terminal Lys, which
react rather sluggish with the glass support (6). Aminopolystyrene resin, however, possesses two distinct disadvantages:
first, the build-up of high back pressures during pumping
requires manyfold dilution with inert glass beads for satisfactory flow rates, second, the hydrophobic matrix exerts
strong interactions with the DABITC reagent resulting in
high background contamination, which, in turn, interferes
with the identification procedures. Therefore, the glass derived support appears as the most suitable matrix for the
solid-phase microsequencing technology using DABITC.
The sequencing program has been tested with some short and
long peptides (Fig.3). Smaller peptides like Leu-Leu-Val-Tyr
and bradykinin could be sequenced right up to the last amino
acid (arginine, which did not elute from the HPLC column,
was verified by micro thin layer chromatography); 26 degradation steps of 4,5 nanomoles oxidized insulin-B-chain could
be identified by high performance liquid chromatography,
with the thin layer chromatographical technique the sequence
up to residue 30 could be deduced.

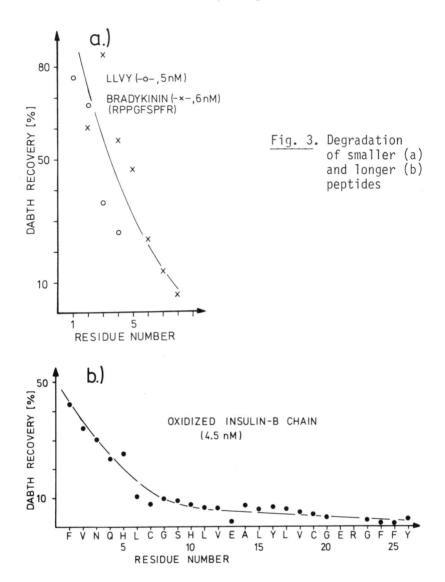

Fig. 3. Degradation
        of smaller (a)
        and longer (b)
        peptides

The sensitivity of detection can be increased by suppression
of background contamination, which, in turn, is a function
of available dead volumes in the system. Consequently, di-
minishing these volumes by using microbore tubing for all
connections and miniaturizing the reaction column will improve microsequencing. A microcolumn (Fig.4), which will

Fig.4.

Microcolumn

accommodate 5-25 mg of amino propyl glass ob-
viating dilution with additional glass beads
can easily be fabricated from a piece of large
bore teflon tubing (1,5 mm inner diameter) by
drawing out one end in the heat yielding a
conical constriction. The other large bore end
serves for column filling. Flanging of both
ends after threading through the appropriate
fittings will permit a facile and tight joining to the teflon lines with the help of suitable polypropylene connectors. The glass beads
are secured in the column by placing a porous teflon disc (Zitex H 662-123, product
of Chemplast, USA) between the bottom connectors. Dead space above the glass bead column
is minimized by insertion of a flanged piece
of small bore teflon tubing. The microcolumn
displays good flow characteristics and, besides, has the advantage of an inexpensive
one-way item.

Limiting factors impairing the sensitivity of the technique
are at present incomplete attachment yields as well as the
rapidly deteriorating degradation yields (Fig.3). The reduced yields are apparently not due to acid labile peptide
linkages since hydrolysis and amino acid analysis of TFA
eluates for detached peptides proved negative. Precycling of
the immobilized peptides with all reagents except isothiocyanate - a procedure, which has a beneficial effect on yields
in the liquid phase sequencer - proved as well ineffective.
Loss of peptide during degradation by labilization of the
alkali sensitive amino propyl glass bond on exposure to the
sequencing buffer has been tested by preincubation of the
support for 1 hour at 50°C in the buffer prior to peptide
attachment; a yield improvement, however, was not observed.
Significant higher yields in liquid phase sequencing can be
obtained by flushing the sequencing cup after the cleavage
step shortly with dilute ammonia thus neutralizing any traces
of acid still left in the lines and connections. A similar
effect can be achieved in solid-phase sequencing by addition
of very little tertiary amine to the washing solvent without
necessitating an additional program function, i.e. inclusion
of 0.1-0.2% triethylamine in the methanol used. Triethylamine trapped in the glass beads will be blown off by the
following nitrogen flush prior to the TFA cleavage thus preventing salt formation with the acid. As can be seen from

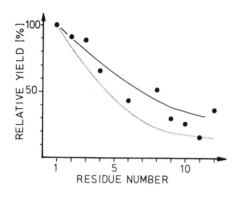

Fig.5. Degradation of
2 nm oxidizided
insulin B-chain
in the microcolumn;
washing solvent:
MeOH containing
0.1% triethylamine.
The lower curve
(··············) is from
Fig.3b. For compa-
rison the first Ed-
man step is set to
100%.

Fig.5 some yield improvement can be obtained by this measure.

In summary, due to the limiting factors described, in the
present version of the solid-phase microsequencing technology using DABITC a general sequencing range of 1-20 nanomole
of peptide appears appropriate although in some cases a sequence could be deduced with considerable less material. The
successful sequence determination of peptides up to 34 residues from such diverse sources like ribosomal proteins, viral
proteins and plant enzymes (6) shows that the solid-phase
technique using DABITC is a powerful tool in microsequencing.

ACKNOWLEDGEMENTS

We greatly appreciate the technical assistance of Mrs. Gisela
Haeselbarth. This work was supported by the Deutsche Forschungsgemeinschaft.

REFERENCES

1. Chang, J.Y., Creaser, E.H. and Bentley, K.W. (1976) Biochem. J. 153, 607-611.
2. Chang, J.Y., Brauer, D. and Wittmann-Liebold, B. (1978)
   FEBS lett. 93, 205-214
3. Hughes, G.J., Winterhalter, K.H., Lutz, H. and Wilson,
   K.J. (1979) FEBS lett. 108, 92-97
4. Chang, J.Y. (1979) Biochim.Biophys.Acta 578, 188-195
5. Wittmann-Liebold, B. and Lehmann, A. (1980) in "Methods
   in Peptide and Protein Sequence Analysis", Birr, Chr.ed.,
   Elsevier/North-Holland, pp. 49-72

6. Salnikow, J., Lehmann A. and Wittmann-Liebold, B. (1981)
   Anal.Biochem., in press.
7. Chang, J.Y., Lehmann, A. and Wittmann-Liebold, B. (1980)
   Anal. Biochem. 102, 380-383
8. Yang, C. (1979) Hoppe-Seyler's Z. Physiol. Chem. 360,
   1673-1675
9. Wittmann-Liebold, B. and Lehmann, A. (1975) in "Solid-
   Phase Methods in Protein Sequence Analysis", Laursen,
   R.A. ed., Pierce Chemical Company, pp. 81-90

IMPROVED AUTOMATIC CONVERSION FOR USE WITH A

LIQUID-PHASE SEQUENATOR

Michael N. Margolies,[*] Andrew Brauer,[*] Christine Oman,[*] David G. Klapper,[†] and Marcus J. Horn[**]

[*]Depts. of Surgery and Medicine, Massachusetts General Hospital and Harvard Medical School, Boston, MA; [†]Dept. of Bacteriology and Immunology, Univ. of North Carolina School of Medicine, Chapel Hill, NC,; and [**]Sequemat, Inc., Watertown, MA USA

The spinning cup sequenator designed by Edman and Begg in 1967 (1) incorporated automation of the coupling and cleavage reaction; conversion of the anilinothiazolinone (ATZ)-amino acid obtained at each cycle of Edman degradation to the corresponding phenylthiohydantoin (PTH)-amino acid was performed manually as a separate step using aqueous acid. The conversion step was automated for the spinning cup instrument by Wittman-Liebold (2,3) and subsequently adopted in other laboratories (4) using trifluoroacetic acid (TFA) as the conversion reagent. An automated conversion device was devised for a solid-phase sequencer by Birr (5). The advantages of automating the conversion step have been enumerated (2-8) and are summarized here: 1. The background levels of PTH-amino acids obtained during automated conversion are reproducible from cycle to cycle, although they gradually accumulate during the degradation at varying rates. 2. The yields for a specific PTH-amino acid are generally reproducible under a given set of conditions, in contrast to certain PTH-amino acids obtained following manual conversion where the yields vary depending on time elapsed prior to conversion. 3. As the automated system may be maintained secure from oxygen, it was predicted that improved yields would occur relative to manual conversion because of prevention of oxidative desulfuration of phenylthiocarbamyl (PTC)-amino acids. 4. Automated conversion results in prompt conversion of the unstable ATZ-amino acids to the more stable PTH derivatives, thus improving the yields of serine and

threonine. 5. The degree of deamidation of asparagine and glutamine is reduced during automatic conversion; 6. The recovery of
PTH-amino acids ordinarily found in the aqueous phase following
ethyl acetate extraction after manual conversion is more reliable
using automated conversion.

The use of methanol-HCl for the conversion reaction was
originally suggested by Tarr (9) who noted that the conversion in
methanol-HCl appeared to be more rapid than in aqueous acid, as
the intermediate in the methanol-HCl conversion is an ester. This
suggestion was subsequently adapted by Horn and Bonner (7), who
constructed an automated conversion device for the solid-phase
sequencer which depended upon the use of HCl-methanol. The
greater volatility of this conversion reagent permits more rapid
drying of the sample following conversion. Automated conversion
was subsequently employed for the solid-phase sequencer in conjunction with an on-line HPLC identification system (8). In this
report we describe our experience with an automated conversion
device employing methanol-HCl, in conjunction with a liquid-phase
sequenator. The results are compared quantitatively to the results
for manual conversion using aqueous acid.

In order to optimize the yields of PTH-amino acids obtained
during methanol-HCl conversion, we characterized the intermediates in this conversion pathway in order to monitor the extent of
conversion. As shown in Figure 1, the ATZ is converted to an
intermediate PTC-amino acid during aqueous acid conversion. The
formation of the PTC-amino acid is rapid, the rate-limiting step

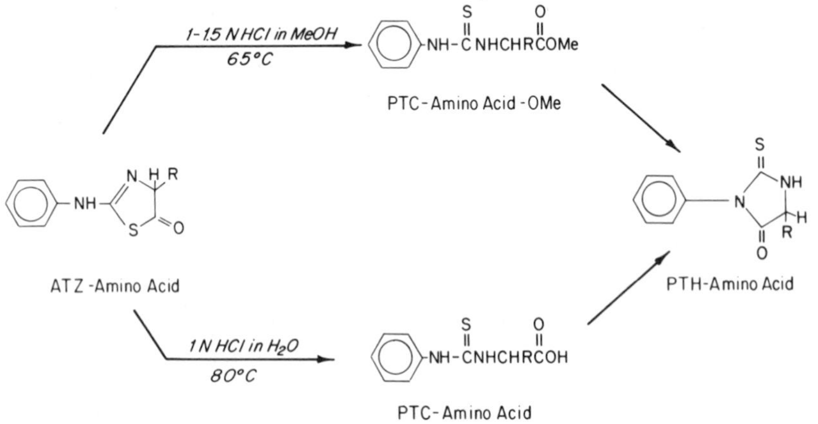

Figure 1. Conversion of the ATZ-amino acid to the PTH-amino
acid occurs via the PTC-amino acid during aqueous acid conversion
(lower portion of figure). The intermediate form during conversion
in methanol-HCl is the PTC-amino acid methyl ester (upper portion
of figure).

being the subsequent cyclization to the PTH (10,11). When conversion occurs in methanol, however, the intermediate is the PTC-amino acid methyl ester which is formed rapidly with the subsequent cyclization to the PTH being rate-limiting. The PTC-amino acid methyl ester intermediates were characterized by HPLC in time studies of conversion of methanol-HCl (vide infra).

MATERIALS AND METHODS

Automated Edman degradation was carried out in a Beckman 890 C sequenator equipped with a cascade refrigeration cold trap (Beckman) and in a Beckman 890 B sequenator as well. The spinning cup sequenators were otherwise unmodified except that automated conversion was carried out with the model P-6 autoconverter (Sequemat, Watertown, MA, USA). A dilute Quadrol program was used in all instances (12). The S1 wash (benzene:ethylacetate) was increased from 600 sec to 650 sec to remove excess Quadrol. Polybrene was added in the case of peptide degradations. PTH-amino acids and the PTC-amino acid methyl ester intermediates were identified by high pressure liquid chromatography (HPLC) on 4.6 mm x 25 cm Zorbax-ODS columns (Dupont) at 46° on a Waters liquid chromatograph (Figure 2). The initial buffer was 0.044 M NaAc, pH 4.15, 10% acetonitrile; the final buffer was 90% acetonitrile in the same buffer. The elution program employed curve 6 of the Waters model M-720 systems controller with variation in the slope and flow-rate (1.8-2.6 ml/min). PTH-amino acids were quantitated using a Hewlett-Packard model 3385 integrator or by measurement of peak heights.

In order to identify the PTC-methyl ester intermediates, butyl chloride extracts delivered into the sequenator fraction collector were dried and then multiple aliquots were converted manually using 1.0-1.5 N HCl-methanol at 65°. The aliquots were converted for various lengths of time and rapidly dried under nitrogen prior to HPLC identification. In order to obtain representatives of all the commonly encountered amino acids, and in subsequent experiments to compare the efficiency of manual and automated conversion, a variety of proteins and peptides were degraded (see Table I). These were selected so that all of the PTH-amino acids were represented relatively near to the amino-terminus in one or more of the peptides so as to avoid the statistical problems associated with increasing background and overlap. In the case of degradations involving proline, additional cleavage time was introduced into the sequenator program (13). The conversion reagent was prepared by the addition of acetyl chloride (Sequemat) to methanol (Baker, "Instra-analyzed") at -20° or by bubbling HCl gas into methanol. The conversion reagent was prepared at least

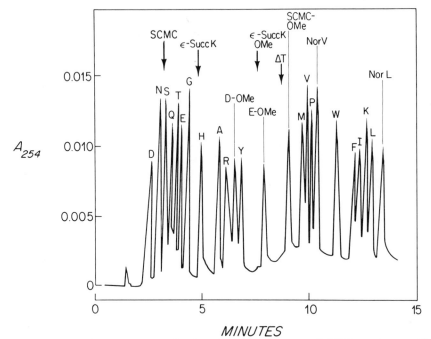

Figure 2. High pressure liquid chromatograms of PTH–amino acid
standards (0.5 nmol) encountered following methanol-HCl conversion. The one letter code for amino acids is used (16): SCMC,
carboxymethyl cysteine; ε-Succ-K, ε-succinyl lysine; D-OMe, aspartic acid methyl ester; ε-Succ-K-OMe, ε-succinyl lysine methyl
ester; E-OMe, glutamic acid methyl ester; ΔT, dehydrothreonine;
SCMC-OMe, carboxymethylcysteine methyl ester; NorV, norvaline;
NorL, norleucine.

weekly; at the end of seven days the molarity of HCl in methanol
in the "autoconverter" had decreased from 1.5 to 1.3.

In order to confirm the identity of PTC-amino acid methyl
esters identified from Edman degradation of proteins of known
sequence, the corresponding PTC-amino acid methyl esters of
three of the amino acids (leucine, alanine, and glycine) were
synthesized. The free amino acids were converted to the corresponding methyl esters followed by coupling with PITC to form
PTC-methyl esters.

RESULTS AND DISCUSSION

During conversion in HCl-methanol, the PTH-amino acids of

TABLE I

Proteins and Peptides Used for Evaluation
of Conversion Methods

Sperm whale apomyoglobin.

Light chains from murine anti-arsonate hybridomas (immuno-globulins) 44-10, 3D10, and 1210.7

Light chains from murine anti-digoxin hybridoma 26-10.

CNBr peptide (46 residues) from anti-digoxin hybridoma heavy chain 26-10.

Tryptic peptides from succinylated light chains of murine hybridomas

Arginine vasopressin.

aspartic acid, glutamic acid, ε-succinyl lysine, and carboxymethylcysteine are converted to their methyl esters. The extent of conversion using either manual conversion (1.0 N HCl-methanol, 65°, 10 min) or automated conversion (1.0-1.5 N HCl-methanol, 55-65°, 7-13 min) was greater than 95%. Each of these methyl ester derivatives were separated by HPLC without modification of the program used previously. As the PTC-methyl ester of valine converts relatively slowly to the PTH and is only partially separated from PTH-Norleucine by HPLC (see Table II), calculations of repetitive yield and relative yields are less accurate using PTH-Norleucine as an internal standard. Therefore, PTH-Norvaline was used as an internal standard in place of PTH-Norleucine in these experiments.

We attempted to characterize the chromatographic behavior of the PTC-amino acid methyl ester intermediates for all of the common amino acids as a prerequisite to an estimation of the extent of conversion. This proved successful for 15 PTC-methyl esters (see Table II). An example of the data (see Materials and Methods) generated in time course studies of conversion using methanol-HCl is shown in Figure 2 for leucine conversion. As shown in the upper left, the PTC-methyl ester intermediate of leucine appears rapidly; the rate-limiting step is the subsequent conversion to PTH-leucine, analogous to the kinetics for conversion in aqueous acid (10,11). The HPLC chromatograms corresponding to aliquots taken at different conversion times (10 sec, one min, and five min) are also illustrated in Figure 2. Experiments identical to that for PTH-leucine were done for each of the PTH-amino acids. The results are summarized in Table II. We were successful in identifying and characterizing 15 PTC-methyl ester intermediates. We failed to identify these compounds for aspartic

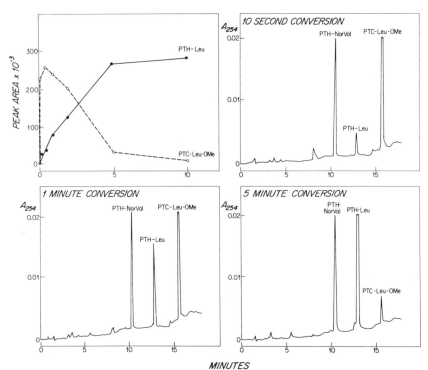

Figure 3.   Manual conversion of ATZ-leucine to PTH-leucine obtained at cycle 2 from the degradation of myoglobin using 1.0 N HCl in methanol, 65°.   Upper left: relative amounts of PTC-leucine methyl ester and PTC-leucine formed following various times of conversion. ATZ-leucine was not detected. Sample HPLC chromatographs are shown in the remaining panels for certain individual time intervals.   PTH-Norvaline was used as an internal standard.

acid, asparagine, arginine, and carboxymethylcysteine. In the case of glycine, leucine, and serine, the time course of conversion was also studied using aqueous acid (1.0 N HCl, 80°, 10 min), in order to identify the PTC-amino acid intermediates (see Table II).   In general, the rates and extent of conversion in 10 min using aqueous acid at 80° was equivalent to those using HCl-methanol at 65°.   As observed previously for the pathway in aqueous acid (11), the conversion in HCl-methanol is also slowest for glycine (14), which makes the identification of the PTC-methyl ester of glycine a useful diagnostic measure to determine the extent of conversion. To lend credence to the identification of the PTC-methyl esters

TABLE II
HPLC Separation of PTH-Amino Acids and PTC Intermediates Found During Conversion[a]

Retention Time (Minutes)

	PTH-amino acid	PTH-amino acid methyl ester	PTC-methyl ester amino acid	PTC-amino acid		PTH-amino acid	PTH-amino acid methyl ester	PTC-methyl ester amino acid	PTC-amino acid
Asp	2.25	6.30			Tyr	6.55		9.95	5.00
Asn	2.60				ΔThr	8.35			
SCMC	2.80	9.00			Met	9.65		12.00	
Ser	2.90(2.35)[b]		4.30	1.45	Val	9.85		13.30	
Gln	3.15		3.60		Pro	10.10		(8.20)[c]	
Thr	3.30		(5.65)[c]		Nor Val	10.35			
Glu	3.50	7.75	9.65		Trp	11.35		15.00	
Gly	3.85		5.35	2.00(1.65)[d]	Phe	12.10		15.90	
ε-succ Lys	4.15	7.25			Ile	12.30		16.00	
His	4.60		5.65		Lys	12.50		14.70	
Ala	5.55		8.35		Leu	12.75		15.70	
Arg	5.90				Nor Leu	13.10			

[a]Conversion in 1.0 N HCl-methanol, 65°, was used to generate PTH-amino acid methyl esters and PTC-amino acid methyl esters. Conversion in 1 N HCl, 80°, was used to generate PTC-amino acids.

[b]PTH-serine is accompanied by an additional earlier eluting peak found consistently during automated conversion in methanol-HCl.

[c]Identification of the PTC-methyl esters of threonine and proline are tentative.

[d]An additional peak was identified (?ATZ) appearing transiently during a time course study of manual conversion (aqueous HCl).

195

based on HPLC chromatograms from Edman degradations, in the
case of three amino acids (glycine, leucine, and alanine), the
corresponding PTC-amino acid methyl esters were synthesized
from the free amino acids. In each case they co-eluted on HPLC
with the putative methyl ester intermediates identified by conversion of sequenator samples. Examination of Table II indicates that
the PTC-amino acid methyl esters are less polar than the corresponding PTH-amino acid, except in the case of proline, where the
putative PTC-methyl ester elutes earlier on reverse-phase chromatography. In contrast, the PTC-amino acids were more polar than
the corresponding PTH. In order to detect the less polar PTC-
methyl ester intermediates the HPLC program was extended
several minutes beyond the retention time of PTH-Norleucine in
these experiments.

The instability of the ATZ of serine and threonine make these
residues ideal candidates to investigate improvement of yields
through prompt automated conversion. Inman and Appella (1975)
reported that the half-lives of the ATZ of serine and threonine
were only 2.3 and 13 hours, respectively, at 23°. A time-course
study of serine conversion employing the ATZ serine obtained from
cycle three of the degradation of sperm whale apomyoglobin is
shown in Figure 4. In two successive sequencer experiments, the
ATZ was treated differently prior to manual conversion in methanol-HCl. In the experiment depicted in the upper panel the butyl
chloride extract containing ATZ-serine was delivered directly into
methanol-HCl in the sequenator fraction collector. This was
subsequently dried down at 4° and thereafter subjected to manual
conversion (1.0 N HCl-methanol, 65°) for varying time periods.
This maneuver was based upon the results reported previously by
Horn and Bonner using solid-phase sequence analysis (6), where the
yield of PTH-serine was improved by employing a methanol wash
following TFA cleavage to deliver the ATZ to the fraction collector. They hypothesized that the hydroxyl group of serine had
been trifluoroacetylated, followed by the elimination of TFA in hot
aqueous acid, producing the dehydroalanine derivative and possibly
polymeric forms. They proposed that in the presence of methanol-
HCl, transesterification at the hydroxyl occurred, yielding free
serine ATZ and thus giving improved yields of PTH-serine. In the
second experiment (lower panel) the butyl chloride extract was
delivered into the fraction collector without methanol-HCl pretreatment and dried down prior to subsequent conversion. The
serine yield improved from 14% to 28% of theoretical yield when
the ATZ was delivered into methanol-HCl.

Inspection of Figure 4 (arrow) indicates that at zero time
significant amount of PTH-serine is already present without exposure to the conversion acid. This was observed not only for serine

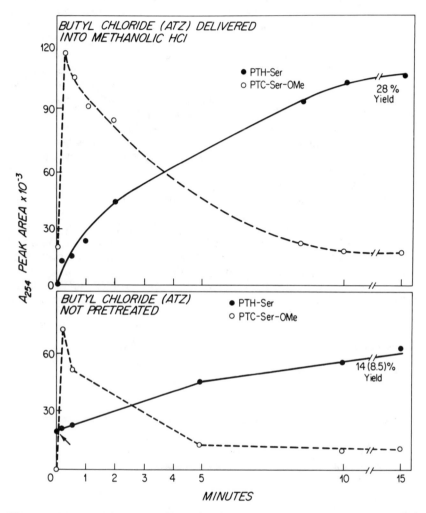

Figure 4. Manual conversion of ATZ–serine to PTH–serine obtained from cycle 3 of the degradation of myoglobin (1.0 N HCl-methanol, 65°). Aliquots were withdrawn at various time intervals and identified by HPLC. Upper panel: butyl chloride containing the ATZ-amino acid was delivered into methanol-HCl in the sequenator fraction collector (4°) where it was dried prior to manual conversion. Lower panel: butyl chloride extract was dried down in fraction collector in the absence of reagent prior to manual conversion. The arrow indicates the amount of PTH–serine already present at zero time ("pre-conversion").

but for many other PTH-amino acids obtained from the spinning
cup sequenator.   This so-called "pre-conversion" was variable in
amount.   Presumably it occurred because of the presence of water
in the cleavage acid with conversion occurring during the cleavage
step.   Another possible explanation is the presence of HFBA
carried over in the butyl chloride if drying following cleavage is
incomplete.   It must be emphasized that zero time samples should
be examined when one is evaluating the results of automated
conversion.   As the ATZ of serine is so unstable with time, it is
likely that the low yield of PTH-serine experienced using manual
conversion in aqueous acid represents material converted by the
above mechanisms and already present prior to the actual manual
conversion step.   Indeed, as shown in Figure 4, manual conversion
(in this instance in methanol-HCl) does not add greatly to the yield
of PTH-serine unless the sample is obtained immediately from the
fraction collector.

     Based upon the improvement in PTH-serine yield, the pro-
gram for automated conversion was modified as shown in Figure 5,
so that the butyl chloride extract was delivered into the conversion
acid in the autoconverter chamber without a preliminary drying
step.   Under these circumstances, the yield of PTH-serine during
autoconversion was improved.   It was reported previously that

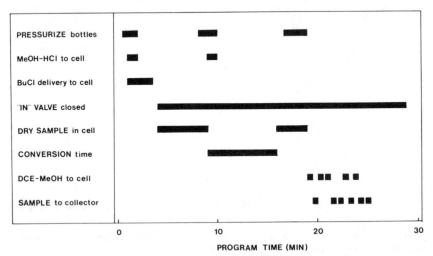

Figure 5.   Revised program for automated conversion in P-6
autoconverter (Sequemat) used in conjunction with a spinning cup
sequenator.   In comparison to the earlier program (6), the butyl
chloride extract (4 ml) is delivered into 1.5 N HCl-methanol (0.5
ml) at 55-65° and dried.   Further conversion occurs for seven
minutes after the second delivery of methanol-HCl.

serine losses may also be minimized by lowering the temperature at which the ATZ was dried down prior to conversion (7,8). The yield of PTH-serine obtained using methanol-HCl automated conversion may be improved slightly by lowering the temperature to 55°, but the conversion of the other ATZ-amino acids is affected to such an extent that conversion time must be prolonged to approximately 13 minutes in order to reduce the level of PTC-methyl ester intermediates which may cause confusion in the

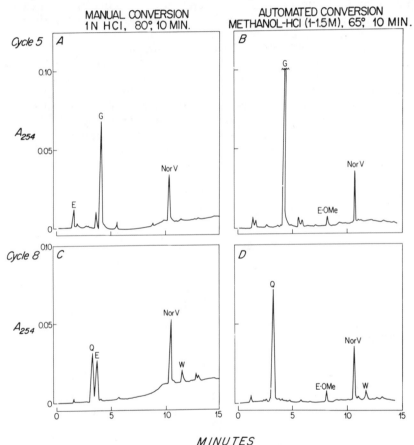

MANUAL CONVERSION
1N HCl, 80° 10 MIN.

AUTOMATED CONVERSION
METHANOL-HCl (1-1.5M), 65° 10 MIN.

MINUTES

Figure 6. Comparison of the yields of PTH-amino acids quantitated by HPLC obtained during Edman degradation of myoglobin using manual conversion in aqueous acid vs. automated conversion in methanol-HCl. Upper panels: cycle 5 (glycine). Lower panels: cycle 9 (glutamine). PTH-Norvaline is included as an internal standard.

interpretation of HPLC chromatograms. For the results shown in Figures 6 and 7 and Table III, the conditions of autoconversion were 1.0-1.5 N HCl, 5-8 min, 65°. The first ten cycles of degradation of myoglobin are a useful test system to measure the extent of conversion. The yield of PTH-serine in cycle three, the extent of glycine conversion in cycle five as measured by the ratio of PTH-glycine to PTC-glycine methyl ester, the extent of valine conversion (cycles one and ten), and the degree of deamidation of glutamine at cycle eight permit adjustment of conditions until an acceptable compromise in the yields of these residues is obtained.

For each of the proteins and peptides listed in Table I, we compared the yields of PTH-amino acids using the automated conversion system (methanol-HCl) to manual conversion in aqueous HCl. As shown in Figure 6, automated conversion results in an improved yield of PTH-glycine and in significant reduction in the degree of deamidation, here exemplified by PTH-glutamine. In Figure 7 are shown similar comparisons of for the yield of PTH-serine which is significantly improved using the automated conversion system. An additional early eluting peak detected at 254 nm was found associated with PTH-serine following automated conversion. The yield of PTH-threonine, however, was not consistently improved although in general the ratio of PTH-threonine/PTH-dehydrothreonine was enhanced.

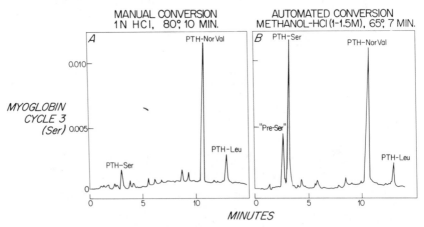

Figure 7. Comparison of the yields of PTH-serine quantitated by HPLC obtained during Edman degradation of myoglobin using conversion in aqueous acid vs. automated conversion in methanol-HCl. In the automated conversion system, a serine-associated peak ("Pre-ser") is obtained in addition to PTH-serine. The sensitivity of the detector was doubled for the first 4 1/2 min of the chromatogram.

TABLE III

Relative Yields (%) of PTH-Amino Acids
Obtained from Spinning Cup Sequenator[a]

	Manual Conversion 1 N HCl, 80° 10 min	Automated Conversion 1.0-1.5 N HCl-Methanol, 65° 5-9 min
Ala	90±10	90±10
Arg	35±25	75±10
Asn	75 (+25 as Asp)	93±5 (+7[b] as Asp-OMe)
Asp	55±10	80[b]±10
Glu	70±10	100[b]
Gln	45±10 (+45 as Glu)	88±8 (+12[b] as Glu-OMe)
Gly	45±20	70±15
His	35±25	75±15
Ile	90±10	90±10
Leu	100	100
Lys	50	60
Met	90±10	90±10
Phe	90±10	90±10
Pro	21±10	50±15
Ser	5±5	27±8
Thr	19±10 (+ ΔT)[c]	30±15 (+ ΔT)[c]
Trp	90	100
Tyr	65±15	65±15
Val	100	95±5

[a] Yields are quantitated from HPLC chromatograms from degradation of peptides and proteins listed in Table I.

[b] Recovered as the methyl ester.

[c] A variable amount of PTH-dehydrothreonine is detected at 313 nm.

In Table III are summarized the yields of each of the PTH-amino acids obtained in both manual conversion (1 N HCl) and automated conversion using methanol-HCl. The yields of serine, glycine, proline, arginine and histidine were consistently improved using automated conversion. The results compare satisfactorily to those reported using TFA (4). In addition to improved yields of certain residues, there was a consistent decrease in deamidation of asparagine and glutamine.

The characteristics and advantages of automated conversion using methanol-HCl include the following: 1. The conversion is more rapid than in aqueous acid, permitting milder conditions with resultant decrease in destruction of labile anilinothiazolinones; 2.

The PTH-methyl esters of aspartic acid, glutamic acid, ε-succinyl lysine, and carboxymethylcysteine are easily identified and separated by HPLC; 3. Delivery of the anilinothiazolinone directly into methanol-HCl results in prompt conversion to the more stable PTC-methyl ester intermediates and improved yields, particularly for serine; 4. The yield of glycine, proline, arginine and histidine are also improved compared to manual aqueous acid conversion, as well as the degree of deamidation of asparagine and glutamine; 5. Quantitation of the PTC-methyl ester intermediates by HPLC permits monitoring of the extent of conversion and thereby modification of the conversion conditions.

ACKNOWLEDGEMENTS

Supported by NIH grants CA 24432 and HL 19259.

REFERENCES

1. Edman, P. and Begg, G., European J. Biochem. 1:80-91, 1967.
2. Wittman-Liebold, B., Hoppe-Seyler's Z. Physiol. Chem. 354: 1415-1431, 1973.
3. Wittman-Liebold, B., Graffunder, H., and Kohls, H., Anal. Biochem. 75:621-633, 1976.
4. Hunkapiller, M.W. and Hood, L.E., Biochemistry 17:2124-2133, 1978.
5. Birr, C. and Frank, R., FEBS Lett 55:61, 1975.
6. Horn, M.J. and Bonner, A.G., In: Previero, A. and Coletti-Previero, M.A. (eds). Solid-Phase Methods in Protein Sequence Analysis. INSERM Symposium V, Elsevier/North Holland Biomedical Press, New York, pp. 163-176, 1977.
7. Henschen-Edman, A. and Lottspeich, F., In: Birr, C. (ed). Methods in Peptide and Protein Sequence Analysis. Elsevier/North Holland Biomedical Press, New York, pp. 105-114, 1980.
8. Machleidt, W. and Hofner, H., In: Birr, C. (ed). Methods in Peptide and Protein Sequence Analysis. Elsevier/North Holland Biomedical Press, New York, pp. 35-47, 1980.
9. Tarr, G.E., Anal. Biochem. 63:361, 1975.
10. Edman, P., Acta Chem. Scand. 10:761-768, 1956.
11. Ilse, D. and Edman, P., Aust. J. Chem. 16:411-416, 1963.
12. Brauer, A., Margolies, M.N., and Haber, E., Biochemistry 14:3029-3033, 1975.
13. Brandt, W., Edman, P., Henschen, A., and von Holt, C., Hoppe-Seyler's Z. Physiol. Chem. 357:1505-1508, 1976.
14. Tarr, G.E., In: Hirs, C and Timasheff, S. (eds). Methods in Enzymology, Vol. XLVII. Academic Press, Inc., New York,

pp. 335–357, 1977.
15. Inman, J.K. and Appella, E., In: Laursen, R.A. (ed). Solid
Phase Methods in Protein Sequence Analysis. Pierce Chemical, Rockville, IL, pp. 241–253, 1975.
16. IUPAC-IUB Commission on Biochemical Nomenclature, J.
Biol. Chem. 243:3557–3559, 1968.

A STUDY OF THE EDMAN DEGRADATION IN THE ASSESSMENT OF THE PURITY OF SYNTHETIC PEPTIDES.

Stephen B.H. Kent*, Mark Riemen, Marie LeDoux, and R.B. Merrifield

The Rockfeller University

New York, NY 10021

INTRODUCTION

The level of amino acid omission in the chain assembly of peptides synthesized by the solid phase method can be determined by quantitating "preview" in the Edman degradation of the peptide, while it is still attached to the resin (Ref. 1). Previous reports indicated a minimum level of 0.3 to 0.5% preview per step(Refs. 2,3). The improved synthetic methodology for solid phase peptide synthesis recently developed in this laboratory (Refs. 4-6) had been shown, by direct analysis of the crude products after cleavage from the resin, to give rise to less than 0.05% amino acid omission per step in the assembly of simple model peptides (Ref. 5).

In order to use the quantitation of preview in the Edman degradation to evaluate levels of omission as low as this in longer and more complex peptides, we wanted to know: was the reported minimum level of preview a real measure of synthetic efficiency, or was it an artefact of the Edman degradation?

The level of preview is cumulative, and
this acts as a means of amplifying the
sensitivity of the method. For this reason, we
set out to apply the Edman degradation to the
total sequence analysis of long peptides as
synthesized on these improved resins.

RESULTS

Conditions of the Edman Degradation and
Conversion Reaction.

All sequencing was performed using a
Sequemat Mini 15 solid phase sequencer. The
standard program supplied with the instrument was
used without modification. Conversion of the
ATZ-derivatives to PTHs was performed manually
with 2N HCl in methanol at 65 degrees Centigrade
for 10 minutes. After evaporation of the reagent
under a stream of nitrogen the residue was taken
up in methanol containing PTH Tyr as an internal
standard.

Under the conversion conditions used here
most derivatives used in synthesis were
essentially totally stable. The PTH-derivatives
of Asn and Gln were converted to the side-chain
methyl esters in about 50% yield, as were
Asp(OBzl) and Asp(cHxl) to lesser extents.
Ser(Bzl) and Thr(Bzl) were significantly destroyed in the conversion reaction, but still
gave acceptable yields of the PTH derviatives for
the determination of preview.

HPLC Separation of the PTH-Derivatives.

PTH derivatives of free and side chain-
-protected amino acids were separated and
quantitated by reverse phase HPLC (Table I),
under conditions based on Bhown, et al. (Ref. 7).
Although not all the PTH-derivatives were
completely separated, the system proved adequate
for all but a few rare sequences for the

evaluation of preview.

TABLE I. Separation of the PTH derivatives.

PTH	Time(min)	PTH	Time(min)
His	7.5	Ala	12.9
His(DNP)	21.4	Pro	19.9
Trp(Form)	23.0	Val	20.6
Arg(Tosyl)	25.7	Cys(4MeBzl)	34.2
Asp(OBzl)	28.1	Met	20.4
Asp(cHxl)	31.1	Ile	25.1
Asp(OMe)	14.1	Leu	26.2
Asn	7.1	Tyr	16.1
Thr(Bzl)	28.7	Tyr(BrZ)	38.1
Ser(Bzl)	27.7	Phe	24.4
Glu(OBzl)	19.0	Gly	8.9

Notes: C18 MicroBondapak, 5-85%B over 40min, 2.0mL/min. Buffer composition in Ref. 7.

Minimum Levels of Preview.

A simple model sequence was used to determine if any artefactual preview was inherent to the Edman degradation under the conditions used. The peptide, Leu-Ala-Gly-Val--Leu-Ala-Gly-Val-Phe (LAGVLAGVF), was synthesized on Phe covalently attached to 1% cross-linked polystyrene resins, both by stepwise addition of the amino acids and by the addition of LAGV segments (we thank Virender K. Sarin for the latter). The Boc-LAGV used in the latter case was analyzed by ion exchange chromatography and contained no detectable impurities in which an amino acid was omitted (detection limit <0.02mole% each).

Sequencing of these peptides gave extremely low levels of preview, less than 0.01% per cycle in the case of the segment synthesis (Table II). A very low level of background was seen in these experiments, typically about 0.2% total. This was

TABLE II. Preview in model peptides.

Peptide	Preview (%)	@ (cycle#)	Average (%/step)
LAGVLAGVF	0.73	@ #8	0.09
LAGVLAGVF (segments)	<0.1	@ #8	0.01
LAGV-His(DNP)- LAGVF	0.12	@ #8	0.02
LAGV-His-LAGVF	<0.1	@ #8	0.01
LAGV-Pro-LAGVF	<0.1	@ #8	0.01
AT-His(DNP)-TFV	1.0	@ #2	0.5
	1.7	@ #5	0.3
AT-His-TFV	2.9	@ #2	1.4
	2.1	@ #5	0.4

Notes: peptides were synthesized by stepwise solid phase methodology on Pam-Resins (Ref. 5). Corresponding peptides containing His were generated from this His(DNP) peptide-resin. Single letter code, with T=Thr(Bzl).

quite constant from step to step. Preview was corrected for this background by subtracting the amount of PTH found two cycles before from the amount found in the cycle prior to the residue under investigation.

Investigation of natural bovine insulin B chain for preview gave ambiguous results. Varying levels of apparent preview, from 0.3 to 1% per cycle, were found superimposed on rather high backgrounds. These experiments were unsatisfactory and the question of the level of preview in natural peptides and pure peptides of complex structure is still being investigated.

Artefactual Preview in Sequences Containing Histidine or Proline.

It has been reported that artefactual preview is generated by premature cyclization and cleavage under the basic conditions of the PITC coupling reaction, at His (Ref. 8) and at N-methyl amino acids(Ref. 9). Because we had seen increments in the observed level of preview at His residues in synthetic peptides from nerve growth factor (see Table III), we decided to investigate this by synthesis and sequencing of suitable model peptides. The results are shown in Table II. No increase in preview was apparent on sequencing through His(Im-DNP), His, or Pro. Investigation of models of the sequence in which the problem was first observed (Thr(Bzl)-His--Thr(Bzl)-Phe-Val; see last two entries in Table II) showed no increase in preview at or after the His residues. This is in complete agreement with the report of Niall (Ref. 10) who failed to detect premature cyclization of His in a number of peptides in the spinning cup sequencer. It was concluded that a synthetic problem had led to the original observation in our work.

Total Sequence Analysis of Synthetic Peptides.

Synthetic fragments of nerve growth factor, acyl carrier protein, hepatitis B surface antigen, and CCK, and total synthetic beta-endorphin, thymosin alpha-1, and cecropin A, were completely sequenced. The results are given in Table III.

Apart from the incomplete coupling of the histidines discussed above and single incomplete couplings in the thymosin alpha-1 and hepatitis B surface antigen syntheses, only a low constant level of preview was observed (approximately 0.12% per cycle). Despite the fact that this is higher than observed for the unambiguously synthesized model peptide, it is not clear whether this level is an inherent artefact of the Edman degradation performed on peptides containing many trifunctional amino acids, or whether it is a true measure of the level of

Table III. Sequencing of synthetic peptides.

Peptide	Preview(Cycle#) (%)	Average (%/step)	Rep. Yield
NGF fragment	1.1 (#09)	0.12	
	6.6 (#12)	0.55	--
ACP(65-74)	1.1 (#09)	0.12	97%
HbSAg(135-155)	9.1 (#08)	---	
	9.0 (#16)	<0.1	97.5%
Cys-OPCCK	3.1 (#08)	0.38	89.5%*
thymosin-α_1	4.5 (#15)	0.3	
	14.8 (#21)	---	
	15.4 (#25)	---	97.5%
β_h-endorphin	3.6 (#30)	0.12	97%
cecropin A (1-33)	3.3 (#31)	0.11	95%

Notes: NGF=nerve growth factor; ACP=acyl carrier protein; OPCCK=C-terminal octapeptide of CCK. All peptides were prepared by stepwise solid phase peptide synthesis on Pam-Resin, except *=(4MeBHA)-Resin. Data for thymosin (Ref.12) and ACP(65-74) (Ref.13) have been reported previously.

chronic omission of amino acids in the stepwise solid phase syntheses.

Repetitive Stepwise Yields

In all these studies, very high repetitive stepwise yields (96-98% per step) were observed (Table III). Side reactions we had identified and overcome to improve the yield of synthesis on resin supports included: acidolytic cleavage of the benzyl ester link between the C-terminal amino acid and the resin(Ref. 4); and, blocking of the growing peptide chains by trifluoro-acetylation of the α-amino groups(Ref. 6). These seemed to be analagous to problems observed in the solid phase Edman degradation. We expected that the Pam-resins would show very high repetitive stepwise yields due to their 100-fold greater acid stability and/or the absence of extraneous functionalities that lead to term-inating side reactions.

In order to distinguish these two possible mechanisms for the very high sequencing efficiency observed, the peptide LAGVLAGVF was assembled on a benzyl ester resin and on a Pam resin support. Sequencing in both cases gave very low cumulative preview (<0.9%). Very high repetitive stepwise yields were observed on the Pam-Resin (98% per step), while substantially lower yields were found on the benzyl ester resin (90% per step). However, hydrolysis and amino acid analysis of the sequenced resins gave the same low residual amount of blocked peptide (8.5, 8.8% respectively, corresponding to 1% blocking per step).

This indicates that the improved sequencing yield of the peptidyl-Pam-Resin compared with the benyzyl ester resin is due solely to decreased acidolytic losses. This was expected for the Pam-resin and has also been reported by Matsueda, et al. (Ref. 3).

However, this does not address the question of the very high sequencing yields of the protected peptides bound to these resins compared with that found for free peptides covalently attached to sequencing resins by acid stable linkages (88-92% per step). In these cases, the residual blocked peptide entirely accounts for the lowering of the stepwise sequencing yield (unpublished observations).

Here, the difference in level of terminating side reactions is probably related to the level of extraneous functionalities on the resins used, and to the side chain functionalities being in the unblocked/protected state. It is clear from the studies reported in this work that the Edman degradation has an efficiency of at least 98% per step under the conditions used, in the absence of interfering side reactions. This suggests that the use of clean sequencing resins and the blocking of side chain functionalities (based on an understanding of their involvement in side reactions) will lead to similarly high repetitive stepwise yields in the sequencing of free peptides and proteins.

Miscellaneous Studies

Methionine Sulfoxide-Containing Sequences. The behavior of methionine sulfoxide under the conditions of the Edman degradation was investigated using the peptides LAGV-Met-LAGV--Resin and LAGV-Met(O)-LAGV-Resin. These peptides were prepared concurrently on the same batch of LAGV-Resin and had the expected amino acid analyses. The Met-containing peptide showed a repetitive stepwise yield of 98% with 0.87% cumulative preview. The Met(O)-containing peptide, on the other hand, showed a 50% drop in yields of the PTH derivatives of Leu, Ala, and Gly after the Met(O) cycle. Nothing absorbing at 254nm was detected in the Met(O) cycle. Cumulative preview was again 0.94%.

These data indicate 50% termination of the chains at the Met(O) residue under the conditions of the Edman degradation used in this investigation. Mechanisms analagous to the cyclization involved in the cynaogen bromide cleavage can be written to rationalize the termination in the sequencing, but as yet these are purely hypothetical.

Deprotection of the Side Chain in His(DNP)-Containing Peptides. Thiophenol has been used to remove the DNP group from the imidazole side chain while the protected peptide is still resin-bound (Ref. 11). Sequencing of the peptide

LAGV-His(DNP)-LAGVF-Resin showed no trace of free His (<0.1mole%), while after treatment of the same resin with thiophenol (2% v/v in DMF, 30min at RT) only free His was detected. This indicated quantitative (>99.9%) removal of the Im-DNP group under the recommended conditions.

REFERENCES

1. H.D. Niall, G.W. Tregear, J. Jacobs, in "Chemistry and Biology of Peptides", J. Meien-hofer, Ed., Ann Arbor Press, 1972, pp.695-699.
2. G.W. Tregear, et al., Biochemistry, 16 (1977), 2817-2823.
3. G.R. Matsueda, E. Haber, and M.N. Margolies, Biochemistry, 20 (1981), 2571-2580.
4. A.R. Mitchell, B.W. Erickson, M.N. Ryabtsev, R.S. Hodges, and R.B. Merrifield, J. Am. Chem. Soc., 98 (1976), 7357- 7362.
5. A.R. Mitchell, S.B.H. Kent, M. Engelhard, and R.B. Merrifield, J. Org. Chem., 43 (1978), 2845-2852.
6. S.B.H. Kent, A.R. Mitchell, M. Engelhard, and R.B. Merrifield, Proc. Natl. Acad. Sci., 76 (1979), 2180-2184.
7. A. S. Bhown, et al., J. Chromatog., 148 (1978), 532-535.
8. W.A. Schroeder, Meth. Enzymol., XI (1967), 445-461.
9. J.Y. Chang, FEBS Lett., 91 (9178), 63-68.
10. H.D. Niall, Meth. Enzymol., XXVII (1978), p.942, ff.
11. M.C. Lin, B. Gutte, D.G. Caldi, S. Moore, and R.B. Merrifield, J. Biol. Chem., 247 (1972) 4768-4774.
12. T.W. Wong and R.B. Merrifield, Biochemistry, 19 (1980), 3233-3238.
13. S.B.H. Kent and R.B. Merrifield, in Proc. European Peptide Symp, Gdansk, 1980.

SOLID PHASE SYNTHESIS OF POLYPEPTIDES AND OLIGONUCLEOTIDES USING A SOLID PHASE SEQUENCER.

ALEX G. BONNER, MARCUS J. HORN, RICHARD S.NEVES
Sequemat/Genetic Design Inc., 111 School St.,
Watertown, MA 02172
and STEPHEN B.H. KENT, Molecular Genetics Inc.,
10320 Bren Road East, Minnetonka, MN 55343

ABSTRACT

Recent experiments with a Sequemat solid-phase sequencer indicate the feasibility of adapting this instrument to perform the solid phase synthesis of polypeptides and oligonucleotides. Only programming and minor plumbing changes are necessary to reversibly convert the Sequemat sequencer into a solid phase synthesizer. Microbore columns were used with polystyrene, acrylamide and silica resins diluted with unfunctionalized glass beads to provide back pressures of 150-190 psig. The usual scale of the reactions provided 0.001 to 0.1 mmole of polypeptide or oligonucleotide. The cycle time was 1-1.5 hours per synthetic step with each step providing only 25-50 ml total waste volume.

DISCUSSION

Automatic, solid phase chemical methods are widely used for protein sequence analysis (1,2), peptide synthesis (3,4,5), and oligonucleotide synthesis (6,7,8). Comparisons of these methods revealed so many similarities that it seemed practical to design or adapt one instrument to be used interchangeably for all three chemistries shown in Figure 1.

Other than the chemistry, the only major differences between automation of these methods are the type of solid support, the type of reaction vessel and the scale of the reaction.

215

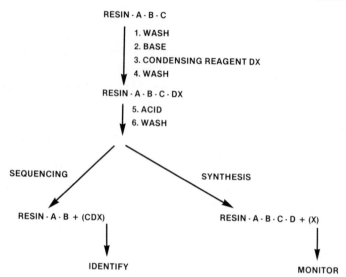

FIGURE 1. General Chemistry of Solid Phase Reactions.

All automatic systems for solid phase chemistry require an
inert reagent/solvent storage and delivery system, a flexible, cyclic programmable controller and a reaction vessel
(shaker or column) for the resin/solid support. A fraction
collector is necessary for sequence analysis and useful for
synthesis chemistry where costly intermediates can be collected, regenerated and used again. The system must be
adaptable to variable amounts and types of solid supports
and it is desirable to control the temperature of the reaction vessel.

The existing design of the Sequemat solid phase sequencers
is ideally suited to meet the above requirements. As shown
in Figure 2, minor alterations involved changing the sequencer waste/collect valve into a recycling valve and adding a
valve for a third solvent which could fill the entire recycle loop. The sequencer collect/partition valve becomes
the synthesizer waste/collect valve and the sequencer col-
umn-2 function is used as a bypass function to flush lines
to waste without passing through the column. Injections of
the synthetic intermediates is accomplished by using one of
the syringe pumps, or by installing a syringe loaded, loop
injection valve, or by use of an automatic sample injector.

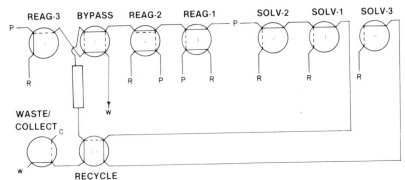

FIGURE 2.   Valve Diagram of the Modified Sequencer.

The Sequemat sequencer uses microbore columns as reaction
vessels and various amounts of resin are loaded in columns
which vary in length and/or diameter. Derivatized silica,
porous glass and Kel-F graft polystyrene can be packed directly. Derivatized polystyrene and acrylamide resins which
swell in various solvents are diluted with glass beads or
Kel-F beads as described by Laursen(1) to prevent high back
pressures. Small analytical columns (50 x 0.5mm) or various
preparative columns (250 x 3mm) can be attached in series.
The standard Sequemat system can be used with up to 200 mg
of "swelling" resins and minor modifications should make it
possible to use up to gram amounts of resin for peptide
synthesis. Small scale oligonucleotide synthesis with 1-5 mg
of resin are handled easily with the analytical sized columns.

The recycling capability is made possible only by the inclusion of a small bubble trap before the inlet of the recip-
rocating pump. The pump is relocated close to the column to
minimize dead-volume to about 1.0 ml. We have performed high
yield peptide synthesis (99.7% per step) by slow, continuous
delivery of the BOC amino acid preformed symmetrical anhydride.
However, the recycle function provides rapid delivery of the
entire volume of the active species and continuous, pulsing
flow to penetrate the resin. The recirculating solution
could be monitored continuously by flow photometry for the
advantages mentioned by Birr (5).

As shown in Figure 2, the daisy-chain of valves provides a
single column inlet and eliminates any problem with reagent
cross contamination. Also, use of a column provides for
efficient, rapid washing of the resins so that cyclic program times are 1-1.5 hours and waste volume is only 25-50
ml per step.

RESULTS

For peptide synthesis we have usually used symmetrical an-
hydrides of BOC amino acids preformed with dicyclohexylcarbodiimide or diisopropylcarbodiimide. The starting
resins were prepared as described by Mitchell(9). Figure 3
shows the "preview" sequence results of the synthetic product Leu-Ala-Gly-Val-Resin. Total synthesis cycle time was
60 minutes with 20 minutes for coupling, 15 minutes for deprotection and 2 minutes for neutralization.

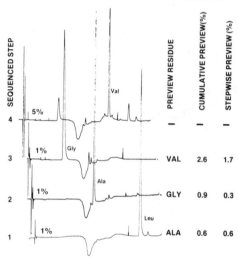

FIGURE 3.   Preview Sequence Analysis of Leu-Ala-Gly-Val-
Resin synthesized on the modified sequencer.

For DNA synthesis we have primarily followed procedures described by Matteucci(7) for the phosphoramidite chemistry.
The phosphotriester chemistry methods were essentially those
described by Miyoshi(10).   Figure 4 shows the analytical results for a dGCGCGC synthesis. The starting resin was amino
acrylamide type with a succinyl linkage to the 3' hydroxyl
of 5' DMT (N^4-benzoyl) deoxycytidine. The first condensation used protected dG monomer and two successive condensa-
tions used protected dGC dimers in 5-fold excess over resin
equivalents. Mesitylene sulfonyl 3-nitrotriazole was used
as condensing reagent in 3-fold excess over monomer or dimer.
Deprotection reagent was 10% TCA in chloroform for 90 sec.
After cleavage from the resin and purification by DEAE cellulose homochromotography the isolated, purified yield was
estimated at 20% (42 OD units). Cycle time was 90 minutes
with 60 minutes for condensation recycling.

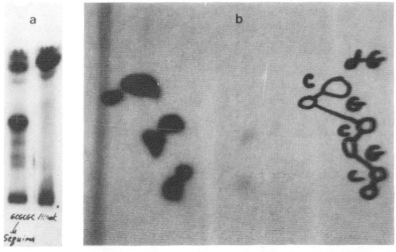

FIGURE 4.   DEAE Cellulose homochromotography (a) of P^{32}
kinased sample. Wandering spot sequence anal.
(b) of product dGCGCGC.

REFERENCES

1. Laursen, R.A., Eur.J.Biochem. 20(1971)89.
2. Laursen, R.A., Machleidt, W., (1980) Methods Biochem.
   Anal., 201.
3. Merrifield, R.B., Stewart, J.M., Jernberg, N., Anal.
   Chem. 38(1966)1905.
4. Barany, B., Merrifield, R.B., (1979) In: "The Peptides",
   Vol. 2,(Gross,E., Meinhofer,J., eds.)Academic Press, NY.
5. Birr, C., (1978) "Aspects of the Merrifield Peptide
   Synthesis", In: "Reactivity and Structure",(K.Hafner,ed.)
   Vol. 8, p.77, Springer-Verlag, Berlin and NY.
6. Miyoshi,K., Huang, T., Itakura, K., Nucl.Acids Res.
   8 (1980)5491.
7. Matteucci, M.D., Caruthers, M.H., JACS 103(1981)3185.
8. Alvarado-Urbina, G., et.al., Science 214(1981)270.
9. Mitchell, A.R., Kent, S.B.H., Erickson, B.W., Merrifield,
   R.B., (1976) Tetrahedron Letters 3796.
10. Miyoshi, K., Miyake, T., Hozumi, T., Itakura, K., Nucl.
   Acids Res. 8(1980)5473.

ABBREVIATIONS: BOC, t-butyloxycarbonyl; DMT, dimethoxytrityl;
dG, 5'-DMT-N^2-isobutyryl-deoxyguanosine 3'-(p-chlorophenyl)
phosphate; dGC, the dideoxynucleotide of dG and N^4-Benzoyl-
deoxycytidine 3'-(p-chlorophenyl) phosphate; TCA, trichloroacetic acid.

Manual Sequencing

MANUAL BATCHWISE SEQUENCING METHODS

George E. Tarr

Department of Biological Chemistry
University of Michigan
Ann Arbor, MI   48109

INTRODUCTION

Until recently the rate-limiting step in protein structural determination was the purification of peptides, so sequencing efforts focused on long degradations of large fragments. With the advent of HPLC methods the sequencer has been confronted with a host of suitable material mostly of small to moderate size. Improvements in procedure for automatic instruments like the spinning cup have extended their applicability, range and sensitivity, but have only slightly increased their speed: one day's worth of HPLC runs can easily produce enough peptides to occupy an automatic sequenator for half a year.

The manual sequencing methods I have been advancing over the past several years (1,2) have been improved again, especially with regard to batchwise processing, in order to cope with the HPLC output. Two strategies have evolved: a partitioning method capable of degrading any small peptide (2-12 residues) to the C-terminus, and a film method for larger peptides and extended runs. Both methods operate conveniently on 10 samples at a time in the 0.1-10 nmole range and both have cycle times under one hour. A modification of the former method provides for a quick and quantitative N-terminal analysis.

223

HARDWARE

The manual sequencing station previously described (2) was changed slightly to include an all-glass vacuum manifold bearing size 18 Rotaflo valves (Corning) to which were attached by vacuum tubing SC-24 Mininert valves (Pierce) lacking sliders and with enlarged bores. These valves, modified as in (2), also served as closures for the 40 ml screw cap vials used as reaction and drying chambers. The standard aluminum block of the block heater was drilled out to 28 mm to accomodate one of these vials within 4 of the existing 15 mm holes.

Sequencing and extraction vessels were 6x50 mm culture tubes. To aid in liquid transfer by decantation, each sequencing tube was given a tooth (Fig. 1): portions of the rims of two tubes were softened in a flame, lightly joined, pulled out several mm, snapped apart and fire polished. Up to 10 such tubes were readily accomodated in one 40 ml vial.

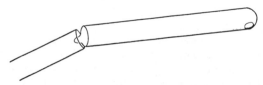

Fig. 1 Sequencing tube with tooth pressed against a second 6x50 mm tube for transfer of upper phase. The bottom should not be raised much above horizontal to avoid decanting the lower phase if, as shown here, that phase happens not to spread out well on the glass. The film method is also aided by the tooth, but no care need be taken as to angle of tilt.

REAGENTS

Quality of PITC was assessed by vacuum drying 1 μl and running half of the residue at 0.1 OD in the PTH system (3). Pierce PITC was usually free of significant contamination except for a modest amount of DPTU (diphenylthiourea); vacuum distillation of Aldrich PITC gave material of equal quality. Reagent grade EtOH was resistilled. Aldrich TEA (triethylamine) and DMF (dimethylformamide) were good sans purification, as were Burdick and Jackson MeCN, hep (heptane) and MeOH, but B and J EA (ethyl acetate) was contaminated with peroxides. Removal of peroxides by distillation, with or without prior filtration through alumina, was generally adequate, but to insure continued

quality a small amount (about .01%) EtSH (ethanethiol) was
always added to any EA used in sequencing procedures. HCl
and TFA also sometimes became deleterious in a manner
preventable with a trace of EtSH.

Working stocks of coupling and cleavage reagents were
stored in 1 ml Reactivials (Pierce) or in 13x100 mm screw
cap culture tubes under N_2 and at 0° except when in use.
The valve cap closures were modified as previously described
(1). All except conc HCl were removed under N_2 barrier with
syringe.

N-TERMINAL ANALYSIS

The previously recommended procedure (2) called for
partitioning an aqueous coupling buffer with a series of
hep:EA mixtures progressively more polar to a final 2:1 or
1:1. Cleaner analyses and better discrimination of
byproducts and coupled peptide could be achieved with non-
aqueous (or nearly) coupling systems (20% TEA in 90% EtOH or
DMF) which were vacuum dried to remove solvent, buffer, and
PITC before partitioning. Three washes of a small aqueous
layer with hep: EA 7:1 now readily removed all DPTU and
other byproducts exept for any PTU that may have been
generated from NH_3, hence samples were routinely redried
from TEA before coupling. PTH analyses not suffering from
interference by PTU would obviate this step. To leave the
most apolar PTC-peptides in the lower phase, even with such
low EA concentrations, it was necessary to exchange residual
TEA with a more polar amine, so a small amount of TMA
(trimethylamine) was incorporated into the first wash
solvent (see Table 1). For the purposes of N-terminal
analysis the cleavage and conversion reactions were
combined, with the incubation time in 1N HCl/MeOH at 65°
extended to 20 min to accomodate slowly cleaving residue
pairs. Recoveries for most residues were 90-95%; the lowest
was for Ser at 70-75%. The method is summarized in Fig. 2.

As the major reactive species in the coupling is not
PITC but its activated product with TEA, coupling was faster
and more consistent if the full coupling mixture was 'aged'
at RT for 5-10 min before use. 'Aged' medium could then be
stored for several hours at 0° without losing effectiveness
or unacceptably raising the analytical background.

PARTITIONING METHOD FOR SHORT PEPTIDES

The coupling and wash procedure described for N-

Table 1. Wash tests for the partitioning method

First wash	hep:EA 15:1		hep:EA 15:1 .25% TMA		hep:EA 7:1 .25% TMA	
Second wash	7:1		7:1		7:1	
	W1	W2	W1	W2	W1	W2
PTC-HL	0	tr	0	0	0	0
PTC-GLA	0	0	0	0	0	0
PTC-FF	42	35	tr	0	1	0
PTC-II	67	27	3	0	4	0

Results expressed in % extraction of PTC-peptide into 250 µl
wash solvent from 2.5 µl water. The separate phases were
dried and converted for PTH analysis.

terminal analysis precedes the cleavage and extraction
steps. For reasons given before (2) conc HCl is better than
TFA. The published extraction step involved a roughly
neutral pH created by pyridine. Loss of extraction
efficiency as salt accumulated and wash-out of some non-
polar peptides prompted a re-examination of these
conditions. The rationale behind these tests was that a

Fig. 2 N-terminal analysis

1. Set-up
   Large peptides: make TFA film (precycle) in 6x50 tube.
   Redry from 2-4 µl TEA.
2. Couple
   Add 10 µl 'aged' 90% EtOH:TEA:PITC 7:2:1. 10 min 50°, N_2.
   Vacuum dry 3 min 50°.
3. Wash
   Add 2.5 µl H_2O and 250 µl hep:EA 15:1/.25-.5% TMA.
   Centrifuge briefly.
   Vortex well, centrifuge 30 sec, decant upper phase.
   Wash twice with hep:EA 7:1.
   Add internal standard and vacuum dry.
4. Cleave/convert
   Add 20 µl 1N HCl/MeOH, 20 min 65°. Vacuum dry.

Notes: Step 1, no carrier added to large peptides. Step 3,
decant carefully as shown in Fig. 1; with large peptides,
more polar washes (e.g., 2:1) are safe and may give cleaner
analyses, but such washes are rarely necessary.

peptide ought to behave differently from any ATZ (anilinothiazolinone) or PTH because it is a zwitterion. Extraction from an acidic aqueous phase should remove all derivatives except for those of His and Arg, with the peptide held in the lower phase by its positive charge. At basic pH, His and Arg derivatives should become extractable, while the peptide will remain hydrophilic because of its negative charge; pHs much above neutrality were not tested, however, because of the instability of Arg derivatives.

Some of the data obtained is shown in Table 2. As predicted, PTH-Asp was readily extractable from an acidic aqueous phase (water added to the residual acid following the cleavage step) but not from a weakly basic phase; PTHs-His and -Arg behaved oppositely. On the other hand, apolar peptides like Trp·Met·Asp·Phe·amide and ε-PTC-Phe·Gly·Lys were not easily extracted under either condition. The most effective buffer for extracting PTH-Arg-- HFA/TMA-- clearly has properties beyond those attributable to its nominal pH. This is partly due to its extraction into the organic phase, which raises the pH of the aqueous phase. Note the behavior of PTH-Asp; also total Arg recovery values were somewhat lowered in that system relative to other neutral buffers (but more pronounced in EA than in MeCN, so that effect is not simple). However, a comparison of HAc and HPr at neutral pH showed no benefit of HPr in terms of the differentiation of PTH-Arg and the most extractable peptide --Leu·Trp·Met·Arg·Phe·Ala. Evidently,

Table 2. Extraction tests for partitioning method

Upper phase	EA	hep:EA 1:2	φ:MeCN 2:1	EA	EA	hep:EA 1:3	φ:MeCN 2:1	EA
Lower phase	H_2O	H_2O	H_2O	.1Ac TMA pH 5	.1 HFA/TMA pH 7			.1PO$_4$ NH$_3$ pH 7
PTH-Asp	96	89	89	60	6	2	4	33
PTH-Gln	96	79	84	89	92	86	81	91
PTH-His	6	1	1	83	88	85	85	90
PTH-Arg	2	0	1	5	59	19	30	16
WMDF·amide	2	0	0	6	4	1	2	9
FGK·εPTC	10	3	6	15	11		6	
LWMRFA	0	0	0		29	2	6	8

All samples were dried from conc HCl to simulate post-cleavage conditions. Values are % extraction into 60 μl upper phase from 5 μl lower phase. φ = benzene; HFA = hexafluoroacetone.

the increased ion-pairing power of HPr over HAc affects both
PTH and peptide equally. Many solvent mixtures were tested:
all gave the same relative extraction of these two compounds
from HFA/TMA buffer.

The effective use of double pH extraction in actual
sequencing is complicated by the hydrolytic instability of
ATZs, which are by far the major product from cleavage by
conc HCl despite its aqueous component. The dilute aqueous
acid of the first extraction hydrolyzes ATZs more or less
completely in a few minutes at RT, and the resulting PTC-
amino acids resist extraction at neutrality or above. One

Fig. 3 Partitioning method for small peptides

1. Set-up
   Redry samples in 6x50 toothed tubes from 2-4 μl TEA.
2. Couple
   Either 90% EtOH or DMF:TEA:PITC 7:2:1 or 12:3:2. 'aged'
   5-10 min RT; 10 μl, 5 min 50°, under N_2.
   Vacuum dry 2-3 min 50° or 5-10 min RT.
3. Wash
   As for N-terminal analysis.
4. Cleave
   Add 5 μl conc HCl, 5 min RT or 1.5-2 min 50°, under N_2.
   Vacuum carefully, full vacuum (<70 mtorr) for 3-5 min RT.
5. Extract
   a. General procedure: Add 10 μl MeOH, 2 min 50° or 10 min
      RT. Redry. Add 5 μl H_2O, 40 μl hep:EA 1:5, vortex well
      and centrifuge 30 sec. Decant into 6x50 tube. Extract
      with 40 μl .02M HFA/.016TMA in hep:EA 1:5 or 1:10.
   b. When low Arg recovery can be tolerated: MeOH treatment
      best but not essential if no Arg; change buffer to
      HAc/TMA pH 6-7.
   c. Peptides lacking His and Arg: No MeOH treatment. Do
      two extractions with hep:EA 1:5.
6. Convert
   Use 2 μl 1N HCl/MeOH, 10 min 65°.

Notes: Step 2, the first ratio is in current use-- no advantage is implied. Steps 2 & 5, although the reaction vessel is
routinely flushed with N_2, tests under air for extended
periods revealled no loss of PTC-peptide or PTH. Step 5, the
extraction solvents for the acidic and neutral media can be
ϕ:MeCN 2:1 and 3:2, respectively. Step 6, if PTC-G·Me does
not interfer with analysis of PTH-Ala or any other, then
temperature may be lowered to 50° for improved recovery of
PTHs Asn, Gln, Ser and Thr.

solution is to change the ATZs into something more stable
first: MeOH was found to react rapidly and quantitatively
and to give no adverse side effects (half-life of ATZ-Met at
23° about 30 sec). The resulting PTC-amino acid methyl
esters extracted much like the PTHs.

Fig. 3 shows a scheme for sequencing by this method.
Extraction alternatives that avoid the MeOH treatment or the
toxicity of HFA are also given. The dangers of peroxides in
EA can be avoided and the recovery of PTH-Arg increased by
substituting benzene:MeCN, but this represents an increase
in toxicity. n-Butyl chloride:MeOH mixtures showed poor
physical properties during extraction, but some such
suitable alternative could undoubtedly be found.

The results of a typical sequencing run using this
method is shown in Table 3. This tryptic peptide ends in
two hydrophobic residues and an Arg, just the sort of thing
that is difficult to retain in a sequencing vessel while
extracting the Arg derivative in reasonable yield in its
turn. The recoveries shown are the usual as experienced in
the three separate laboratories now employing this method.
It is not clear why repetitive efficiencies generally do not
rise much above 90%, but amino acid analysis of the material
remaining after completing a run suggests that at least half
of the inefficiency is due to chemical blockage of N-terminus.

Table 3. Example of sequencing by the partition method

				Residue					
Cycle	Ala	Leu	Asp	Tyr	Asn	Glu	Tyr	Phe	Arg
1	4.11								
2	0.20	3.91							
3		0.27	3.05						
4			0.89	2.63					
5			0.74	0.58	1.21				
6			0.21	0.11	0.46	1.60			
7			0.21	*	0.11	0.60	1.47		
8			0.08	*	0.03	0.13	0.58	1.29	
9				*			0.10	0.46	0.44
10				*			0.05	0.12	0.44
11									0.19

Recoveries in nmoles PTH. Sequencing by the general procedure
given in Fig. 3 using hep:EA 1:5 for both extraction steps.
Total repetitive efficiency 91% includes in-phase (87%) and
out-of-phase. Yields of Asp for cycles 5-8 belong mostly
with the Asn; Tyr values for cycles 7-10 have been included
entirely under the second Tyr.

FILM METHOD FOR LARGER PEPTIDES

Films analogous to, but less perfect than, those found
in the spinning cup could be produced in round-bottomed
tubes simply by vacuuming away the coupling or cleavage
agents. As in the automatic instrument, addition of
Polybrene considerably enhanced the retention to peptides.
Furthermore, this readily solvated carrier improved the
speed, completeness, and consistency of the coupling
reaction. A second polymer-- poly(methylthiocarbamyl-amino-
ethylcarbamido-ethylene) or MTC-ED-PAA-- also possessed
these qualities to a fair degree and, furthermore,
protected the PTC-peptide from oxidation. As EtSH in the
wash solvents and cleavage acids also served the latter
function, this polymer does not seem uniquely helpful yet.

The progenitor of the film method described here was an
extension of the aqueous pyridine system (2) in which the
large peptide or protein was precipitated or deposited as a
rough film by withdrawing not quite all of the aqueous phase
into hep:EA while slowly rotating the nearly horizontal
tube. Subsequent washes removed the remaining water. Films
were made after cleaving by drying the TFA with an N_2
stream. This procedure required much judgment and practice
and it was readily applied to only one sample at a time, but
it was capable of handling proteins as large as 50,000 MW
and of successfully degrading 20 nmoles of smaller ones well
into the 30s. Although the present method is not yet as
successful there appears to be no intrinsic reason why it
will not soon become at least as good.

After a coupling procedure like that given for the
partitioning method, or after cleaving with TFA, the
extractive effectivenss of a given solvent decreased
drastically with each successive application, so a series of
increasingly polar solvents was required for both steps. In
both cases, solvents more polar than acetone or MeCN
resulted in catastrophic loss of peptide. The wash
procedure given in Fig. 4, which stops at the polarity of
acetone, leaves a variable amount of DPTU, yet does not
effectively retain the non-polar peptides given in Table 1,
although more polar peptides, long or short, behave well.
Attempts to utilize the exchange of TMA for TFA to improve
retention were unsuccessful.

The results of some extraction tests are given in Table
4. For somewhat better and more consistent extraction, .1%
HAc was added to each solvent; the stronger acids HFo and
TFA were not as effective. The procedure shown in the last

Table 4.   Extraction tests for film method

	First EA	Second EA:acetone 2:1 →————————————→					acetone	
		EA:acetone 1:2 →————→			acetone			
First / Second / Third		acetone			MeCN		acetone	
		1X	2X	3X	2X	3X	2X	3X
PTH-Asp	53	70	85	93	82	96	90	97
PTH-Gln	68	78	90	95	85	96	94	99
PTH-His	69	76	84	86	83	90	90	94
PTH-Tyr	70	72	88	94	79	97	91	98
PTH-Arg	36	64	77	82	74	86	87	92
ATZ-Arg		86						
WMDF·amide	2	7	11	12	11	18	13	14

Set up as films with 20 μg Polybrene/4 μg MTC-ED-PAA,
vacuum dried from TFA; 40 μl each solvent, each with
.1% HAc. Extraction values are cumulative.

Fig. 4   Film method for large peptides

1. Set-up
   Add 50 μg Polybrene to each sample, vacuum dry, precycle
   with TFA. Redry film from TEA.
2. Couple
   Same as for partition method, but couple 8 min 50°.
3. Wash
   To dry film add 300 μl hep:EA 15:1, rock tube and decant.
   Repeat with EA, then 2-3x with acetone.
   Add internal standard and dry.
4. Cleave
   Add 8 μl TFA, 5 min 50°, under N_2.
   Vacuum dry about 2 min at RT.
5. Extract
   Add 50 μl EA:acetone 2:1/.1% HAc, rock and decant into
   6x50 tube containing 10 μl MeOH.
   Repeat with acetone/.1% HAc.
   (or, use 35 μl, extract 1x with 2:1 and 2x with acetone.)
6. Convert

Notes: Step 1, proteins generally precipitated in the tube
from aqueous solution with minimal acetone or from 88% formic
acid with EA in order to free from associated amino acids or
small peptides. Step 4, if the bond is known to be slow (G,
W,P followed by H,T,K,D,E) then cleave 10-20 min. Step 5,
the second extraction procedure is more efficient, but more
tedious; MeOH in extraction tube improves the yields of PTH-
Ser and Thr by reacting with the dehydration-prone ATZs.

column was chosen for simplicity. Downward adjustments in polarity can be made for non-polar peptides, but the usefulness of this ploy is limited by the level of discrimination currently experienced in the wash step.

A scheme for this method is shown in Fig. 4. As with the partitioning method, EA can undoubtedly be avoided if desired. Runs of 20 have been attained, with the limit caused more by out-of-phaseness than by low repetitive efficiency or wash-out. With the current deficiencies rectified, the simplicity of the necessary procedures will make the method attractive for multiple-sample automation. Indeed, construction of the first prototype is underway.

CONCLUSION

The partitioning method and associated N-terminal method operate well and reliably. Important changes in the near future seem unlikely except perhaps for improvements in the coupling and/or cleavage conditions to decrease the blocking reaction. As the HPLC has a prodigious ability to produce peptides in the appropriate size range, even a protein chemist adverse in principle to manual sequencing may find virtue in these methods. While film technique is still flawed, it would seem to have the greater future because of its simplicity of operation and its ability to manage peptides of any size.

ACKNOWLEDGEMENTS

I wish to thank Shaun Black, Valerie Fujita, and Ryuji Kobayashi for testing these methods. This work was supported by U.S. Public Health Service grants MH21539, GB22556, AM10339 and by National Science Foundation Grants SER 77-06923 and PCM 76-14947.

REFERENCES

1. Tarr, G.E. 1975 Anal. Biochem. 63, 361-370.
2. Tarr, G.E. 1975 Methods in Enzymol. 47, 335-357.
3. Tarr, G.E. 1981 Anal. Biochem. 111, 27-32.
4. Tarr, G.E., J.F. Beecher, M. Bell, D.J. McKean 1978 Anal. Biochem., 84, 622-627.

MICRO QUANTITATIVE EDMAN MANUAL SEQUENCING

Clyde Zalut
Department of Medicine
Children's Hospital Medical Center
Boston, MA 02115

INTRODUCTION

The ability to quantitatively as well as qualitatively determine amino acid residues from protein/peptide sequencing would provide an important tool for the study of proteins, which up to recently has not been utilized.[1] Edman clearly noted this; and was surprised so little emphasis was placed on the quantitative aspects of sequencing when so much additional information could be obtained by its realization.[2] The method to be described attempts to fulfill this intention by having a technical ease which facilitates reproducibility and by incorporating a simple, sensitive, accurate system for quantitation; thus not only protein chemists but anyone dealing with proteins could avail themselves of the potential of quantitative sequencing. In addition the method has the flexibility to effectively treat a broad scope of sequencing problems such as rapidly deblocking and sequencing pyroglutamyl polypeptides and preventing material losses during extractive procedures (without special carriers).

Finally the capacity of this method to sequence mixtures quantitatively would not only allow for characterization of samples containing a variety of proteins as envisioned by Edman, but would furnish a powerful means for primary structure determination and the study of protein chemistry in general.[1,2,3]

233

METHODS
Back Hydrolysis/Amino Acid Analysis and HPLC Analysis

Anilinothiazolinone or phenylthiohydantoin (PTH)-amino acids were back hydrolyzed using 0.2 ml of 0.1% $SnCl_2$ under vacuum for 18-24 h at 150°C.[4] HPLC analysis of PTH-amino acids was achieved by using a Dupont Zorbax ODS column and a quasi-isocratic buffer program.[5]

Sequenced PTH-amino acids could be quantitated down to approximately 300 pmol using back hydrolysis and the amino acid analyzer, and to 30 pmol using HPLC analysis.

Sequencing Procedures

All sequencing experiments are performed with a simultaneously run control sample of insulin and the addition of an internal standard (Nle) to both the experiment and the control samples. Since the method outlined in Scheme 1 does not have an extraction step after the coupling reaction (as in most other procedures), the addition of Nle at each sequencing cycle acts as an important and complete internal quantitative monitoring device; thus results are not considered quantitatively reliable unless the Nle yields are at least 80% of theory.

The information to follow relates to Scheme 1.

Coupling Reaction (Steps 1-5)

Usually 6-8 samples can be run simultaneously. It is advised to use PITC directly and not in solution, where we have found it less effective. Coupling reaction time and temperature should not be changed without thorough study. We have found that certain amino acids such as phosphoserine and phosphothreonine need higher temperature and longer time for complete reaction (30 min, 55°C, DMAA/PYR), but for most other amino acids the conditions given are satisfactory.

Consecutive sequencing using our method is not effective with Quadrol buffer (0.1 or 0.25M) since its buffering capacity is impaired by the lack of an extraction step directly following the coupling reaction.[1] However, for solely N-terminal determinations it is the best buffer with our method (fewest interferring by-products for HPLC analysis).

Drying time can range from 1/2 to 2 h depending on quantity and type of coupling buffer used. For denaturing buffers (guanidine-HCl and SDS) the drying time is long and the sample tends to bump (one remedy is to lay the samples on their sides).

If in doubt about the completeness of the coupling reaction repeat the coupling step after step 5.

SCHEME 1 – MANUAL SEQUENCING METHOD *

Coupling Reaction

1. Place 1-25 nmol polypeptide and an equivalent amount of Nle in a 10x75 mm pyrex test tube. The total sample should be dry or in no more than 25 μl solution
2. Add 0.1 ml of coupling buffer (e. g., DMAA/PYR)
3. Add 10 μl PITC (80,000 nmol)-Parafilm tubes
4. Keep 15 min at 45°C (Temp-Block)-Gently stir occasionally
5. Dry sample thoroughly under vacuum in a heated desiccator (45°C) with P_2O_5 and NaOH

Cleavage Reaction

6. Add 0.1 ml TFA-Parafilm tubes
7. Keep 10 min at 45°C (Temp-Block)-Gently stir occasionally
8. Dry in an unheated desiccator with NaOH present, first at low vacuum (water aspirator) and then under high vacuum

Extraction

9. Add 1 ml anhydrous ether, mix, then add 0.1 ml H_2O Remix lightly-Parafilm tubes-Centrifuge. Two layers should form; otherwise, add another 0.1 ml H_2O
10. Remove ether layer into a 10x75 mm pyrex tube (pasteur pipette) and repeat ether extraction (1 ml). Vortex well Centrifuge and combine ether extracts
11. Blow off residue ether from the H_2O phase with N_2; add Nle, then dry sample (N_2 or vacuum)-No heat. After drying, the next sequencing cycle can be started
12. Dry ether extracts under N_2 then proceed to step 13 or 14

Work-up and Analysis

13. Back hydrolysis for Amino Acid Analysis-Add 0.2 ml constant boiling HCl with 1% $SnCl_2$ to the dried ether extract and hydrolyze at 150°C for 22 h under vacuum
14. Conversion for HPLC-Either of two methods were used:
(a) 20% TFA conversion
Add 0.1 ml 20% TFA to step 12 residue-N_2 purge-Parafilm sample-Stir gently then incubate at 55°C for 25 min Dry sample under high vacuum with P_2O_5 and NaOH
(b) Dry conversion
Place dried residue (step 12) at 80°C for 30 min

* All reagents are sequenator grade
Fresh anhydrous ether is used (tested for peroxides)
For HPLC analysis use freshly distilled PITC, which could be used at least a month if kept at -20°C under N_2

Cleavage reaction (steps 6-8)
     More TFA can be used if needed for solubility. Care
should be taken to dry the sample gradually (usually 30 min).
Extraction (steps 9-12)
     The extraction with ether is the most critical step!
If a peptide (usually small and/or hydrophobic) tends to
extract in ether (determined by amino acids present in the
back hydrolysis analysis) a "reverse extraction" should be
used--that is, water added first with slight stirring
followed by ether. This should reduce peptide extraction
unless the cause is due to inadvertant removal of some
aqueous phase, which can be prevented by more care in the
extraction technique.
Back Hydrolysis (step 13)
     Poor quantitative results were found when using
shorter hydrolysis times (e. g., 4 h).
Conversion (step 14a or b)
     Both dry and 20% TFA conversion techniques each have
their good points.[6,7] However, if improved yields of
PTH-Ser and PTH-Thr could be obtained using dry conversion
it would easily be our choice not only for our method but
also for automatic sequencing where it could readily be
incorporated under program control.
Extended sequencing
     It is suggested that when Ser, Gln and Trp become
exposed as the N-terminal amino acid (from trial runs)
further sequencing should not be delayed.
     Between sequencing steps the peptide is stored at -20°C
after the addition of the internal standard followed by
drying with N_2 or vacuum.
Pyroglutamyl sequencing[8]
     10 ul of the blocked peptide solution (20-40 nmol) in
0.1N HCl is reacted with 100 ul of enzyme solution (6 mg
pyroglutamyl amino peptidase to 1 ml phosphate buffer) for
3 h at 50°C, then directly sequenced without drying using
the method in Scheme 1.
     We strongly advise running a simultaneous control
sample (e. g. pyroGlu-Ala) to check enzyme effectiveness.
Guanidine-HCl or SDS as a coupling buffer
     Use 100 ul 8M guanidine-HCl followed by an equal
amount of DMAA/PYR (or 0.1M Quadrol) then sequence as usual;
or use 100 ul of a 1% SDS in DMAA/PYR buffer solution
followed by the usual procedure. Care should be taken
during drying after the coupling reaction.[1]

RESULTS AND DISCUSSION

A thorough demonstration of the usefulness of the above method has already been described.1 A summary of these results along with some additional information follow.

The method has proven effective on free amino acids, small peptides, large proteins, and blocked N-terminal peptides. Even histidine and arginine can be directly determined without additional procedures.

It has been possible to sequence 10 to 15 steps on amounts from 10 to 20 nmol and to quantitate the results without the need of sophisticated equipment or complicated, elaborate techniques (the major item in the procedure is a 10x75 mm pyrex test tube and the HPLC system uses only a single pump without a programmer).

At the 1 nmol level only 3 to 5 steps could be achieved with back hydrolysis due to limited sensitivity of our amino acid analyzer. However, HPLC and a sensitive detecting system (Waters 440, 254nm UV-detector) should allow for more extended and/or pmol range quantitative sequencing, Fig. 1.

For primary structure and related studies it is a great advantage to use both back hydrolysis and HPLC analysis as complimentary systems.

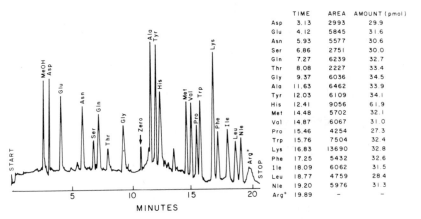

	TIME	AREA	AMOUNT (pmol)
Asp	3.13	2993	29.9
Glu	4.12	5845	31.6
Asn	5.93	5577	30.6
Ser	6.86	2751	30.0
Gln	7.27	6239	32.7
Thr	8.08	2227	33.4
Gly	9.37	6036	34.5
Ala	11.63	6462	33.9
Tyr	12.03	6109	34.1
His	12.41	9056	61.9
Met	14.48	5702	32.1
Val	14.87	6067	31.0
Pro	15.46	4254	27.3
Trp	15.76	7504	32.4
Lys	16.83	13690	32.8
Phe	17.25	5432	32.6
Ile	18.09	6062	31.5
Leu	18.77	4759	28.4
Nle	19.20	5976	31.3
Arg*	19.89	—	—

Fig. 1. Microquantitative HPLC PTH-amino acid analysis. A. Program I — PTH-amino acid standard, 30 pmol each amino acid except His and Arg, 60 pmol each. Starting and elution buffers, 22.5 and 46.5% acetonitrile, respectively, in 0.01 M sodium acetate, pH 5.4. Flow rate 0.9 ml/min. Temperature 60°C. UV absorbance monitored at 254 nm. Full-scale 0.005 A. This analysis was made as if the sample were an unknown with quantitation estimated relative to a standard run made earlier. The baseline was zeroed before Ala to reduce the refractive index effects caused by the change in buffers (22.5—46.5%) appearing at that time. * Because of background noise Arg was not machine integrated, but could be determined manually by peak height — 62 pmol.

TABLE 1

MIXTURE ANALYSIS OF A TRYPTIC DIGEST OF S-AMINOETHYLATED
HUMAN HEMOGLOBIN B-CHAIN*

	Quantitative amino acid assignments(nmol)[a,b,c]				
Cycle	1	2	3	4	5
Fragment #					
1(N-term)	Val(26)	His(10)	Leu(27)	Thr(16)	Pro(13)
2	Ser(26)	Ala(24)	Val(23)	Thr(16)	Ala(20)
3	Val(26)	Asn(13)	Val(23)	Asp(17)	Glu(18)
4	Leu(26)	Leu(22)	Val(23)	Val(21)	Tyr(20)
5	Phe(30)	Phe(20)	Glu(18)	Ser(10)	Phe(19)
6	Val(26)	Lys(13)			
7	Ala(26)	His(10)	Gly(20)	Lys(17)	
8	Lys(18)				
9	Val(24)	Leu(22)	Gly(20)	Ala(19)	Phe(19)
10	Gly(37)	Thr(11)	Phe(22)	Ala(19)	Thr(20)
11	Asn(3)	Lys(4)			
12	Leu(26)	His(10)	Val(23)	Asp(17)	Pro(13)
13	Leu(26)	Leu(22)	Gly(20)	Asn(12)	Val(20)
14	Val(26)	Leu(22)	Ala(24)	His(10)	His(8)
15	Glu(30)	Phe(20)	Thr(16)	Pro(13)	Pro(13)
16	Val(26)	Val(25)	Ala(24)	Gly(23)	Val(20)
17	Tyr(14)	His(3)			

a. Approximately 40 nmol taken for digestion
b. Quantitative assignments are made from information
    obtained at, before or after any one cycle
c. All data is as obtained uncorrected. Amino acids
    present but not shown did not exceed 3 nmol
 * Partial structure-human hemoglobin B-chain:
                    1   2
 1 H-Val-His-Leu-Thr-Pro...Lys/Ser-Ala-Val-Thr-Ala...Lys/
                    3   4
18 Val-Asn-Val-Asp-Glu...Arg/Leu-Leu-Val-Val-Tyr...Arg/
                    5   6   7   8
41 Phe-Phe-Glu-Ser-Phe...Lys/Val-Lys/Ala-His-Gly-Lys/Lys/
                    9   10
67 Val-Leu-Gly-Ala-Phe...Lys/Gly-Thr-Phe-Ala-Thr...Cys/
         11   12   13
94 Asn-Lys/Leu-His-Val-Asp-Pro...Arg/Leu-Leu-Gly-Asn-Val...
                        14   15
112 Cys/Val-Leu-Ala-His-His...Lys/Glu-Phe-Thr-Pro-Pro...Lys/
                    16   17
133 Val-Val-Ala-Gly-Val...Lys/Tyr-His-OH 146

It is felt that one of the most important aspects of quantitative sequencing is mixture analysis, to which the method described here is especially adapted. Quantitative information on protein mixtures could be used for: their general characterization, primary structure determination without time consuming component isolation and purification, to follow chemical and enzymatic modifications, and to detect amino acid substitutions in genetic variants.

In a recent study on the glycosylation of human hemoglobin A, quantitative mixture sequencing (using Scheme 1) was successfully applied to locate lysine 17 as a reactive site for glycosylation.[9] In another experiment we easily, quantitatively followed 5 sequencing steps from a mixture of 17 fragments obtained from a tryptic digest of the S-aminoethylated human hemoglobin B-chain (see Table 1). Thus with only five cycles we could monitor over half of the 146 amino acid residues in the molecule. Shortly we hope to complete the entire primary structure determination of E_4 (a bovine embryonic enamel protein) using mainly mixture sequencing.

ACKNOWLEDGEMENTS
I wish to thank Wm. Henzel, J. Mazer, R. Shapiro and R. Wong for technical assistance, and E. Merler and Wm. Gray for invaluable suggestions and patience. Thanks also to F. Bunn for helping the work on hemoglobin.

REFERENCES
1. Zalut, C., Henzel, W.J. and Harris, H.W. Jr. (1980) J. Biochem. Biophys. Methods 3, 11-30
2. Edman, P. and Henschen, A. (1975) in Protein Sequence Determination: A Sourcebook of Methods and Techniques (Needleman, S.B., ed.) pp 232-279 Springer-Verlag, N. Y.
3. Gray, W.R. (1968) Nature (London) 220, 1300-1304
4. Mendez, E. and Lai, C.Y. (1975) Anal. Biochem. 68, 47-53
5. Zalut, C. and Harris, H.W. Jr. (1980) J. Biochem. Biophys. Methods 2, 155-161
6. Guyer, L.R. and Todd, C.W. (1975) Anal. Biochem. 66, 400-404
7. Wittman-Liebold, B., Graffunder, H. and Kohls, H. (1976) Anal. Biochem 75, 621-633
8. Podell, D.N. and Abraham, G.N. (1978) Biochem. Biophys. Res. Commun. 81, 176-185
9. Shapiro, R., McManus, M.J., Zalut, C. and Bunn, H.F. (1980) J. Biol. Chem. 255, 3120-3127

Sequencing by Mass Spectrometry

PROTEIN SEQUENCING BY MASS SPECTROMETRY

H.R. Morris, G.W. Taylor, M. Panico, A. Dell,
A.T. Etienne, R.A. McDowell and M.B. Judkins.

Department of Biochemistry,
Imperial College of Science and Technology,
London SW7. U.K.

INTRODUCTION

Mass spectrometry is playing an increasingly important part
in the expanding field of biomolecular structure determination. Indeed, over the past ten years the technique has
graduated from the organic chemistry service laboratories
to reach a position where today total protein sequence
information may be derived by MS techniques alone (1). Mass
spectrometry is probably best known for its contribution to
the structure elucidation of biologically active substances
where some unusual structural feature is either known or
suspected to be present. Some examples of this type are
shown in Fig. 1, which shows the structures, determined in
this laboratory, of the enkephalins, Leukotriene D (Slow-
Reacting Substance of Anaphylaxis), γ-carboxy glutamic acid,
antifreeze glycopeptides 'AF8' and locust Adipokinetic
Hormone, (1,2). This work was carried out on materials of
unknown structure usually at the low nanomole level, and
clearly shows the power of a technique which, unlike
"classical" methodologies, is completely independent of
compound class.

Protein/Peptide Analysis

Proteins and peptides probably represent the most difficult
examples of structural analysis for mass spectrometry. They
are in general polar, highly complex molecules, (often
multiply charged at pH7) whose complexity may be further
compounded by post-translational modification (e.g. glycosylation, phosphorylation etc.). Further their molecular

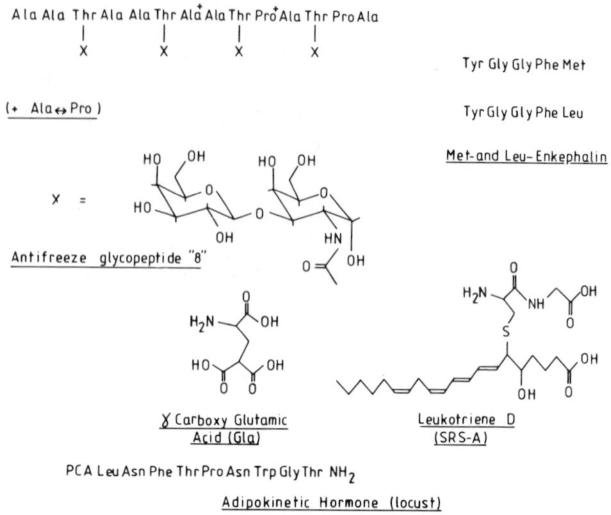

Ala Ala Thr Ala Ala Thr Ala Ala Thr Pro Ala Thr Pro Ala
        | | | |
        X   X   X   X

(+ Ala ↔ Pro)

x =

Antifreeze glycopeptide "8"

Tyr Gly Gly Phe Met

Tyr Gly Gly Phe Leu

Met- and Leu-Enkephalin

γ Carboxy Glutamic
Acid (Gla)

Leukotriene D
(SRS-A)

PCA Leu Asn Phe Thr Pro Asn Trp Gly Thr NH₂

Adipokinetic Hormone (locust)

Fig.1. Important biological substances whose structures
were determined by MS.

weights are well in excess of the mass range of conventional mass spectrometers (1000 mu). The reasons for studying
protein structures in general are many and varied – proteins
playing both structural and control (enzymatic/hormonal)
roles throughout living systems – but why use mass spectrometry when both classical and recombinant DNA sequencing
methods are available? The answer lies in the unique
characteristics of mass spectrometry in terms of 1) sensitivity, 2) specificity and 3) mixture handling capability.
Because MS is independent of sample class, it is particularly
well suited to the analysis of peptides or proteins with
unusual structural features.

The method most widely used to generate amino acid
sequence is electron impact (EI) MS analysis of volatile
N-acetyl N,O permethyl derivatives (1). These derivatives
are volatile in the ion source – an essential prerequisite
for EI analysis- and most importantly, fragment in a defined
and specific way (Fig. 2). Fragmentation occurs at each amide
bond resulting in N-terminal sequence ions; the
mass difference between the sequence ions corresponds to the

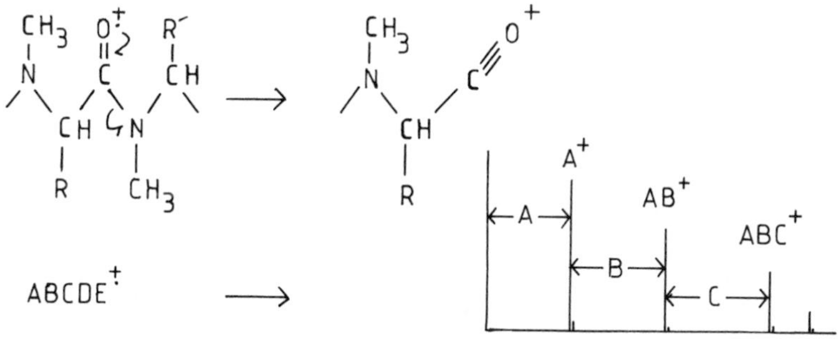

Fig. 2. Formation of N-terminal sequence ions from N-acetyl-N,O-permethyl peptides.

mass of each residue within the peptide. To obtain spectra, peptides of suitable size for MS analysis must first be generated. We introduced the strategy of non-specific protease digestion to accomplish this, using elastase, thermolysin or subtilisin. Following a single ion exchange step or HPLC on the digest, mixtures of up to five peptides are derivatised and inserted directly into the ion source. The mixture is fractionally volatilised over a temp. range of 100-350°C and, observing the concomitant increase and decrease in ion intensity as each peptide volatilises, the "pure" spectrum (and thus the structure) of each component can be determined. Using this method, we have determined the total sequence of dihydrofolate reductase from a methotrexate-resistant organism. Sequences of chloramphenicol acetyltransferase, azurin and ribitol dehydrogenase were determined in a similar manner in conjoint MS-classical studies (1). Using a modification of the above strategy, the first few N-terminal residues of a protein may readily be determined without the need for full structural analysis. A review of protein sequencing by MS may be found in Ref.1.

Post-Translational Modifications
The strength of MS in protein sequence analysis lies in its ability to handle unusual or modified residues. Perhaps the simplest example of this is found in N-terminally blocked peptides/proteins. Because the peptide is deliberately "blocked" during derivatisation (i.e. acetylation/permethy lation), the problems encountered by classical sequencing

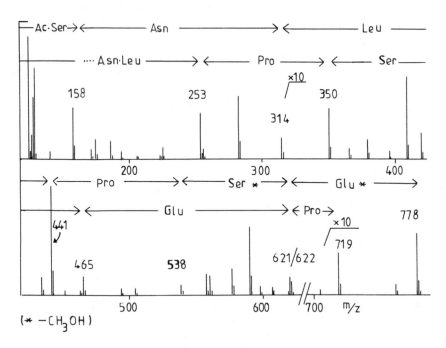

Fig.3. The blocked N-terminal sequence of glutamate dehydrogenase from Neurospora was determined as Ac-Ser-Asn-Leu-
Pro-Ser-Glu-Pro, from the spectrum of the N-^{2}H-acetyl-N,O-
permethyl derivative.

strategies do not arise. Recent examples of blocked N-
terminals determined in this laboratory include: a calcium
binding protein from pig intestinal mucosa (ions at m/z 158,
211,243 clearly define the N-terminal sequence as Ac-Ser-
Ala-Gln-)(3); Delta haemolysin - an extracellular cytotoxic
protein with a formyl group (m/z 174,259,:f-Met-Ala) (4);
Profilin (m/z 128:Ac-Ala-)(5) and Glutamate dehydrogenase
from Neurospora. In the latter case, a complex spectrum
was obtained (Fig.3) from which the N-terminal heptapeptide
was determined as Ac-Ser-Asn-Leu-Pro-Ser-Glu-Pro (6). In
each case, isotopically labelled reagents (d$_6$-acetic anhydride) were used to define the presence of any natural N-
acetyl blocking group.

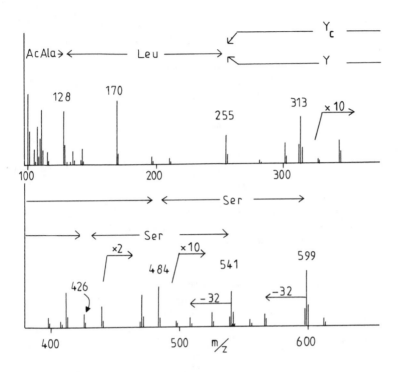

Fig. 4 . EI spectrum of the N-acetyl N,O permethyl derivative of a peptide derived from Prothrombin, from which the
sequence Ala-Leu-Y$_c$-Ser was deduced. The residue Y$_c$ was
determined as arising from γ-carboxyglutamic acid; Y is the
decarboxylated form.

        Because MS provides an exact physicochemical measure-
ment (i.e. mass) it is ideally suited to the analysis of new
structures, where no reference compounds exist with which to
make classical interpretation. This was clearly shown in the
structure determination of the new amino acid -γ-carboxyglu-
tamic acid (Gla)- in the blood clotting zymogen Prothrombin.
The first clue as to the presence of an unusual structure
was the isolation of peptides with anomolous electrophoretic
mobility. Following derivatisation (N-acetyl-N,O permethyl)
the peptides were subjected to EIMS. It was clear from the
spectra obtained that an unusual species was present. For
example, in the spectrum shown in Fig. 4 sequence ions were
present for the peptide Ac-Ala-Leu-Y$_c$-Ser; the mass difference

of "Y_c" (and its related signal 58 mass units lower) did
not correspond to any known amino acid. It was deduced that
"Y_c" represented a derivative of a new amino acid γ-carboxyglutamic acid -Gla- and that Y was the decarboxylated product.
Classically this residue was assigned as glutamic acid. In
retrospect this can be rationalised as the result of facile
decarboxylation in the Dansyl-Edman procedure. Gla has now
been identified and positioned in both Prothrombin and
Factors X_1 and X_2 by this method. (1).

In a similar manner, residues such as N-methyl lysine
give rise to novel mass differences in the spectrum, and
present few problems for MS. Studies on glycoproteins/
glycopeptides which come under this heading are described
later, since they are better considered following a description of Fast Atom Bombardment (FAB), which is having a considerable impact in this area.

Soft Ionisation/Fast Atom Bombardment
Although EIMS has played an important role in our peptide
studies, there are some problems where other methods of
ionisation must be used in order to obtain full structural
information. In EI, a considerable amount of internal
energy is transferred to the molecular ion, often resulting
in appreciable fragmentation, with little or no current carried by the high molecular weight species. To obtain molecular
weight information it is necessary in many cases to use
"softer" methods of ionisation. Chemical ionisation is one
such method - the ionising species being a charged reagent
ion (e.g. CH_5^+, NO^+, H_3O^+ from CH_4, NO, H_2O) instead of an
electron beam. Fragmentation occurs via a protonated pseudo-
molecular ion ($M+H^+$) and follows a different pattern from EI
leading to additional C-terminal ions (cf. N-terminal data by
EI). The value of 'CI' as a complementary technique was well
illustrated in studies on the peptide portion of antifreeze
glycopeptide from antarctic fish blood. N-terminal sequence
data was determined in the usual manner by EI; the intensity
of the spectrum above m/z 1000 was too low to allow the full
sequence to be deduced. CI however gave C-terminal sequence
ions (but again no molecular ion) from which the full peptide
sequence was deduced as Ala-Ala-Thr-Ala-Ala-Thr-Pro-Ala-Thr-
Ala-Ala-Thr-Pro-Ala; some heterogeneity was observed. The
carboxydrate was O-glycosidically linked at Thr.

In the above example, both EI and CI methods failed to
produce a molecular ion, although the data produced gave the

full sequence. This was not the case in studies of the peptide antibiotics Echinomycin and Vancomycin. In both cases
only partial data could be obtained by EI; a molecular
weight was essential to define the structure. A molecular
ion (m/z 1100) was obtained on underivatised Echinomycin by
the discovery of an 'in beam   EI technique. The sample,
loaded on a quartz tip, was placed directly in the electron
beam under EI conditions. It rapidly volatilised and ionised
producing a molecular ion with little fragmentation. This
spectrum, together with NMR/protein chemical techniques/FD,
was used to correct the previously determined structure of
the antibiotic. (7).

   This technique was unfortunately found to to be very
sensitive to structure - in model studies we obtained excellent results for some peptides but no spectrum at all for
others with the same amino acid composition but different sequence. More reliable ways were sought for future
molecular weight determinations.

   In the vancomycin study we turned to field desorption
(FD) MS. Here the sample is loaded onto a tungsten wire
activated with 10μm needles of polymerised benzonitrile. The
sample is desorbed from the wire under the influence of a
high field gradient; very little energy is imparted into the
molecule, which desorbs as the molecular species with little
fragmentation. The FD spectrum of a Vancomycin derivative
showed an intense molecular ion at m/z 1277 defining for the
first time the molecular weight of the substance. The lack
of meaningful fragmentation in FD greatly reduces its value
in day to day peptide sequencing. FD is however routinely
used in this laboratory to determine molecular weights of
polar, thermally labile species; some recent data on Vitamin
B_{12}, oligosaccharides and glycopeptide antibiotics will be
outlined later.

   Recently, a new and powerful mode of MS ionisation has
been introduced into mass spectrometry - Fast Atom Bombardment (FAB)MS. Here the sample is loaded in a glycerol matrix
onto a metallic target and bombarded with accelerated argon
atoms (8). Ionisation occurs, and a long lived ion beam is
generated which is then mass analysed. The energetics of the
process are such that intense quasimolecular ions ($M+H^+, M-H^-$)
are observed; fragmentation, to give structurally useful
information, also occurs in many cases. We have examined a
variety of peptides by this technique at the low μg level.

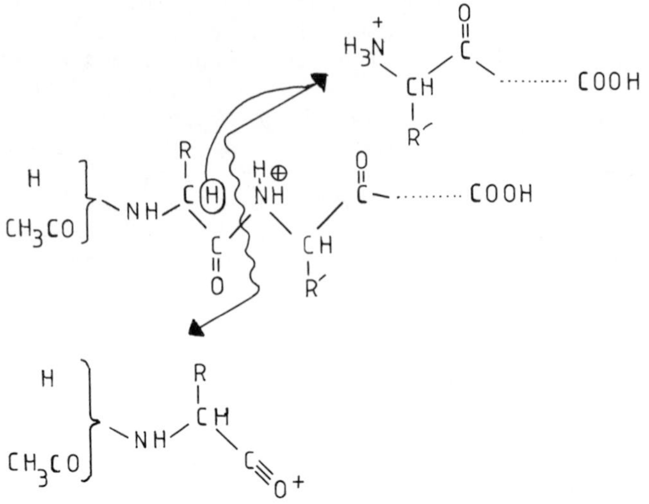

Fig. 5. Fragmentation of positive ions under FAB conditions.
N-and C-terminal fragment ions are produced.

These include blocked and free peptides, of known and unknown structure, containing examples of all the protein-
derived amino acids (9). Unlike the EI studies outlined
earlier, there is no need for derivatisation with FAB MS
analysis; each peptide gives molecular weight information
as the free material. Fragmentation, if it occurs, is via a
chemical ionisation type pathway (Fig. 5) to form both N-
and C-terminal sequence ions. The mass difference between
these ions generally corresponds to the mass of each residue
within the peptide (cf EI sequencing). In some cases, because
of complex or weak fragmentation, it can be difficult to
readily interpret the spectra. To ameliorate these problems
the N-(1:1-acetyl: ^{2}H-acetyl) or 1:1 methyl ester derivative
is prepared (10).   Ions, from the N- or C-terminus respectively then appear as 1:1 doublets, 3 mu apart (unless they
contain Lys/Arg or Asp/Glu which will also be isotopically
labelled). For example, the FAB spectrum of the 1:1 acetyl
d$_3$-acetyl derivative of Met-enkephalin is shown in Fig. 6;
N-terminal doublets at m/z 206, 263, 320 and 467 are readily
assigned to Ac Tyr-, Ac-Tyr-Gly-, Ac-Tyr-Gly-Gly- and Ac-Tyr-
Gly-Gly-Phe-. Note also the signal at m/z 178 - loss of CO-
common at the N-terminus. C-terminal fragments (singlets)

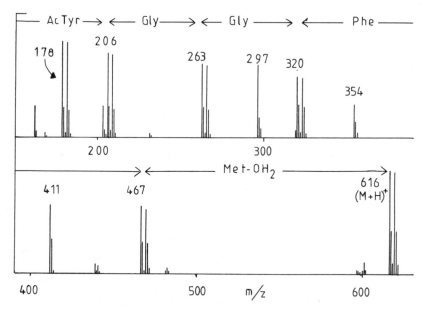

Fig. 6. FAB spectrum of the N-(1:1 acetyl:^{2}H-acetyl)
derivative of Met-enkephalin. N-terminal ions appear as
doublets 3 m.u. apart; C-terminal ions are singlets.

are seen at m/z 297, 354 and 411 (assignable to H$_2$Phe-Met-
OH,H$_2$ Gly-Phe-Met-OH and H$_2$-Gly-Gly-Phe-Met-OH respectively).
The intense doublet (m/z 616,M+H$^+$) confirms the molecular
weight and sequence (9). The negative spectra show a very
abundant (M-H)$^-$ ion at m/z 614 together with the N- and C-
terminal fragment signals. It is used to confirm the interpretation of the positive spectrum. The FAB spectrum of
c-endorphin is shown in Fig. 7. In a similar way, molecular
weight and sequence information are present.

FAB MS can play a valuable complementary role to classical protein sequence studies, an example of which was work
on approximately 10 nmoles of a peptide derived from bovine
ATPase (Dr J. Walker, Cambridge). Ions at m/z 360,459, 560,
661,732,847 in the FAB spectrum defined the partial sequence
...Val-Thr-Thr-Ala-Asp-, the N- and C-terminal sequences
were not observed in this experiment. The sequence although

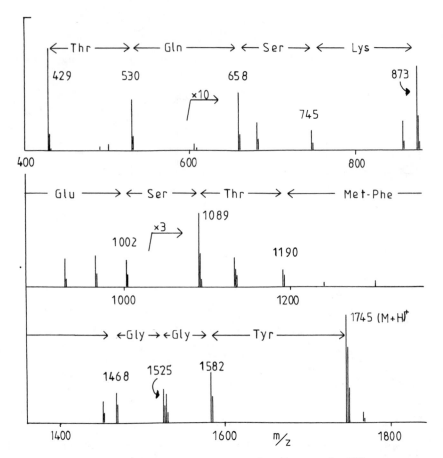

Fig. 7. FAB spectrum of α-endorphin above m/z 400; sequence
information is obtained from C-terminal ions.

unexpected from classical data has been confirmed
and extended further by classical studies (11). Other
examples of unknowns include an interesting new cardioact-
ive neuropeptide from Aplysia the sequencing of dynorphin-
related peptides from the adrenal - both studies carried
out at the 1-5 nmole level.

The great value of EIMS has been its mixture handling
capability. FAB MS of a peptide mixture (derived by tryptic
digestion of pepsin) clearly showed however, that selective

ionisation of components within the mixture can occur. Of
four peptides present only one Val-Gly-Leu-Ala-Pro-Val-Ala
gave both the molecular weight and sequence information;
Gln-Tyr-Tyr-Thr-Val-Phe-Asp-Arg and the corresponding
pGlu peptide,gave quasimolecular ions only, whilst the
fourth component, the tetrapeptide Ala-Asn-Asn-Lys, gave
no spectra at all except under extreme conditions (a weak
quasimolecular ion was obtained after total glycerol evaporation). As yet, no theory as to the processes involved in
FAB ionisation can account for such phenomena, although
presumably some form of glycerol-peptide interaction on the
matrix surface must be involved. It is clear however, that
care must be taken in interpreting the FAB spectra of mixtures. These data can illustrate the dangers of placing full
reliance on any one technique in structural analysis.

In summary, FAB MS provides a simple way of determining M.W., and often structure, at the 1-5 nanomole level
without the need for derivatisation. It is thus a very
powerful new adjunct to protein sequence methods. In other
studies we are using FAB in combination with the high field
magnet facility to study CNBr peptides up to mass 4000 (see
later)

Slow-Reacting Substance of Anaphlylaxis

Probably one of the most difficult problems solved by MS in
this laboratory was the structure elucidation of slow-react-
ing substance of anaphylaxis (SRS-A),   released in the
lung in picomole quantities during the asthma crisis. The
active material was purified from antigen-sensitised guinea
pig lungs by charcoal extraction, gel filtration, ether
extraction and two reverse phase HPLC steps (12). The pure
material posessed a characteristic triene chromophore
(λmax 280nm) (12). Both analytical protein chemical and EIMS
techniques were employed in the structure elucidation. An
essential aid to the interpretation of the spectrum was the
labelling of the active material (1-5µg) with acetic anhydride: d_6 acetic anhydride (1:1) in methanol. Signals in
this spectrum arising from the amino group in SRS-A (previously shown to be present from our classical studies) appeared as 1:1 doublets, 3 mu apart. This allowed identification
of the SRS-A spectrum from the impurities still present.
From the data obtained, SRS-A was defined as the novel pepti-
dolipid 5-(S)-hydroxy-6(R)-cysteinyl glycinyl-7,9-(trans)
-11, 14(cis) eicosa tetraenoic acid (13,14,15). This substance was later termed Leukotriene D_4(LTD$_4$). During the

structural investigation, attempts to define the molecular weight of SRS-A(LTD) by FD were not successful. Excellent quasimolecular ions on 1μg material have now been obtained with FAB (e.g. LTC_4M+H^+:626, $M-H^-$:624,LTD_4M+H^+:497,$M-H^-$: 495); little fragmentation occurs with these substances.

FAB will offer great advantages in the leukotriene/ prostaglandin field for the rapid and sensitive screening of new metabolites/derivatives. Had the technique been available 3-5 years ago, it would have greatly facilitated our original structural studies on SRS-A(LTD).

Neuropeptides and Related Substances

Neuropeptides play a major role in the control and regulation of many body processes. Indeed it is believed that an imbalance of these substances within the CNS may result in both mental and physical illnesses. MS has already played an important part in this field - perhaps the most striking example being the structure identification of the enkephalins (1). However, many present problems in the neuropeptide field are outside the range of present MS techniques. Sensitivity - the capability to specifically detect sub-picomole quantities - is perhaps the most pressing problem facing MS analysis today. We are currently exploiting FAB and negative chemical ionisation (NCI) in our attempts to compete with structural probes such as radioimmunassay. NCI analysis of fluorinated derivatives of peptides seems to be very promising. Using the N-(1:1 acetyl:d_3 acetyl) perfluorobenzyl ester of Met-enkephalin, we have obtained specific quantitation at the 10 pmole level. Selected ion monitoring of the major fragment ion ($M-C_7F_5H_2$ m/z 614) should reduce this level even further.

FAB will also be of great value in this field. We are currently obtaining peptide spectra at the 1-5 nanomole level with no difficulty. Recently, by a combined use of Xenon as bombarding species, and protonating the sample by loading in HCl/ glycerol, we have obtained quasimolecular ions at the picomole level,demonstrating an increase of a factor of 10 over the same experiment using Argon and without protonation in HCl. The excellent signal/noise ratio observed on 15 picomoles of the hexapeptide Lys-Phe-Ile-Gly-Leu-Met-NH_2 is shown in Fig.8. These handling techniques have also been applied to produce quasimolecular ions (but not necessarily fragmentation) on many other neuropeptides in the picomole range, including α-MSH ($M+H^+$:1664) α-endorphin ($M+H^+$:1745) and Lys-Vasopressin ($M+H^+$:1056).

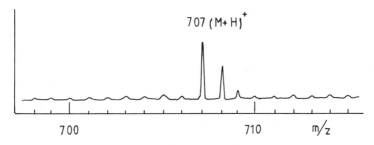

Fig. 8. The quasimolecular ion obtained from 15 picomoles of Lys-Phe-Ile-Gly-Leu-Met-NH$_2$ under FAB conditions (Xe, HCl/glycerol).

Neuropeptides exist as picomole components of biological fluids. To extract and characterise them, a high resolution reverse-phase HPLC system was developed in this laboratory using <u>volatile</u> buffers (n-propanol/acetic acid/water) - a prerequisite for MS analysis. Eighteen neuropeptides of general interest chromatograph in this system with distinct and invariant retention times (2). This system has found wide application in studies of neuropeptides in disease states, and our enkephalin precursor work (16,17). When a pituitary extract was chromatographed in this system and the fractions assayed for aldosterone release on the adrenals, three distinct peaks of UV/bioactive material were obtained (A,B,C). The first two peaks (A,B) were indistinguishable by amino-acid content, and gave identical EI spectra for the N-acetyl N,O permethyl derivatives. From the MS/protein chemical data the structure of peak A was determined as a known peptide Ac-Ser-Tyr-Ser-Met-Glu-His-Phe-Arg-Trp-Gly-Lys-Pro-Val-NH$_2$,-α-melanocyte-stimulating hormone (α-MSH)! (18). Peak B obviously had a similar, closely related structure. It could for example be readily converted to Peak A (α-MSH) on mild base treatment. The FAB spectrum of B showed an intense quasimolecular ion at m/z 1706, 42 mu higher than α-MSH (Fig. 9). It was concluded that peak B was the O-acetyl derivative of α-MSH. EI analysis of underivatised peptides derived from peak B located the acetyl group on the N-terminal Acetyl-serine (19). This was the first time that a regulatory role on adrenals had been demonstrated for α-MSH and derivatives. The third region of bioactivity "C" contained the peptide Val-Val-Tyr-Pro-Trp-Thr-Gln-Arg a haemoglobin fragment. We are currently investigating the biological properties of this material.

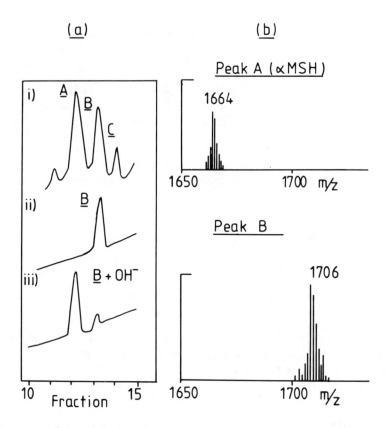

Fig. 9. (a) HPLC profile of a bovine pituitary extract
showing three regions of biological activity. Peak B is
converted to peak A on mild base treatment. (b) FAB spectra
of peak A (α-MSH) and peak B showing quasimolecular ions
(M+H)$^+$ at 1664 and 1706 respectively.

High Mass Studies

With the advent of soft ionisation techniques such as FAB
and FD, it is now possible to generate molecular species
from polar, thermally labile compounds. Many of the most
interesting problems in the area of biomolecular structure
determination require mass analysis in the 2000-4000 mu
range, well outside the capabilities of most mass spectrometers. To overcome these limitations, a High Field Magnet
(HFM) was commissioned for this laboratory, with a mass

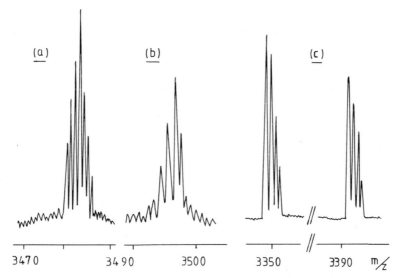

Fig. 10. High mass studies: molecular ion regions of (a) glucagon (FAB-magnet scan), (b) Insulin B chain (FAB - integrating scan) and (c) an acetylated human milk oligosaccharide (FD).

range of 3000 mu at full accelerating voltage (and thus full sensitivity).

Fig. 10 shows the molecular ion region of FAB spectra of some high mass peptides (Glucagon and Insulin B chain) and an oligosaccharide obtained using a High Field Magnet instrument. A full FAB spectrum of melittin, showing significant fragmentation information, has also been obtained. High mass quasimolecular ions are conveniently mass marked using reference compounds. Alkyl halides (e.g. KI, CsI) readily from cluster ions under FAB conditions and are eminently suitable as references. Indeed, we have obtained strong signals for $(CsI)n$ Cs^+ clusters from mass 392.71 (n=1) all the way up to mass 13,123.39 (n=50)!

For FD studies, the most suitable high mass reference compound available is Fomblin oil - a fluorinated polyether:

$$CF_3- \left[(O-CF-CF_2-O-CF-CF_2)_X-O-CF_2\right]_n-CF_3$$
$$\qquad\quad CF_3 \qquad\quad CF_3$$

We have used this reference (EIMS) to mass mark the FD
spectra of such compounds as Vitamin B_{12} and its biosynthet-
ic precursors, alkaloid tremogens, oligosaccharides and
glycopeptides (20,21). It is perhaps among the latter two
groups that the HFM facility will be of greatest value in
the future.

Oligosaccharides/Glycopeptides

MS offers many advantages in the analysis of oligosaccharides /glycopeptides as it is not restricted to compound
class. This was clearly demonstrated in the structure
elucidation of 'AF8' an antifreeze glycopeptide from

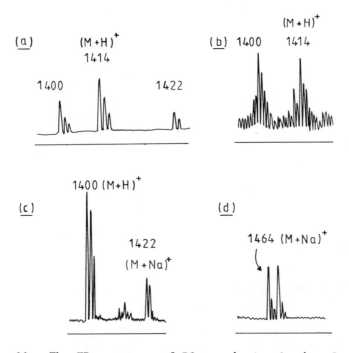

Fig. 11a. The FD spectrum of Bleomycin A_2 showing facile
demethylation from the quasimolecular ion (m/z 1414); the
FAB spectrum (b) is similar. (c) The FD spectrum of
Bleomycin demethyl A_2 ($M+H^+$:m/z 1400) and its N-(1:1 acetyl:
^2H-acetyl) derivative (d).

antarctic fish. The peptide sequence was determined by
EI/CI MS of the acetyl permethyl derivative. Signals were
also observed in the spectrum,corresponding to a dissacc-
haride:hexose -N-acetylhexosamine (m/z 219,260,464). In a
second MS experiment, using a trimethylsilyl derivative,
the carbohydrate was defined as galactosyl (1-3)-N-acetyl
galactosamine (22). In an extension of these techniques we
have partially sequenced a glycopeptide from Prothrombin
(23).

We are currently studying a variety of oligosaccharides
and related compounds using a combined EI/FD/FAB approach;
these include human milk suggars, blood group I glycosphin-
golipids, naturally methylated mycobacterial lipoplysacc-
harides and the Bleomycin family of glycopeptide antibiotics
(24,25). Some examples of the data obtained from the
Bleomycins are shown in Fig. 11.

CONCLUSION
Over the past 10-15 years MS has played an increasingly important part in biomolecular structure determination. Fast Atom
Bombardment MS is perhaps the most exciting recent development in this field and, as yet, has not been fully exploited.
The future developments of MS, probably in the areas of
sensitivity, high mass analysis ($>$ 5000mu) and on-line HPLC
coupling, should open the way for an even greater involvement of MS in studies of biologically important molecules.
Proteins have always proved the greatest challenge to mass
spectroscopists because of their inherent involatility,
high mass and complexity of structure. As this review hopefully will have shown, the challenge has been met both with
respect to the rapid generation of total protein sequence
and in high sensitivity studies on neuropeptides or posttranslationally modified substances. Most of the problems
were solved 5-10 years ago, in particular those relating to
the preparation of derivatives at the 20-100 nmole level and
to interpretation of complex mixture spectra. The application of MS techniques to protein sequencing has however
remained rather a specialist activity, and we now look forward to a new era in which the power and simplicity of the
newer MS approaches to sequencing are fully utilised by the
next generation of protein chemists.

REFERENCES

1. Morris, H.R. (1979) Phil.Trans R.Soc.Lond., A293, 39-51.
(and references cited therein).

2. Morris, H.R. (1980) Nature, 286, 447-452.

3. Hoffman, T. Kawakami,M., Hitchman, A.J.W.,Harrison, J.E.
and Darrington, K.I. (1979) Can.J. Biochem., 57, 737-740.

4. Fitton, J.E., Dell, A. and Shaw, W.V. (1980) FEBS Letters,
115, 209-212.

5. Nystrom, L.E., Lindberg, V., Keadrick-Jones, J. and Jakes,
R. (1979), FEBS Letters, 101, 161-165.

6. Morris, H.R. and Dell, A. (1975) Biochem.J., 149 754-755.

7. Dell, A., Williams, D.H., Morris, H.R., Smith, G.A.,
Feeney, J. and Roberts G.C.K. (1975) J.Am. Chem.Soc. 97,
2497-2502.

8. Barber, M., Bordoli, R.S., Sedgwick, R.D. and Tyler, A.N.
(1981) Chem.commun., 325-327.

9. Morris, H.R., Panico, M., Barber, M., Bordoli, R.S.,
Sedgwick, R.D. and Tyler, A.N. (1981) Biochem.Biophys.Res.
Comm., 101, 623-631.

10. Hunt, E. and Morris, H.R. (1973) Biochem.J. 135, 833-837.

11. Walker, J. and Brunswick, M. personal communication.

12. Morris, H.R., Taylor, G.W., Piper, P.J., Sirois, P. and
Tippins, J.R. (1978) FEBS Letters, 87, 203-206.

13. Morris, H.R., Taylor, G.W., Piper, P.J., Samhoun, M.N.
and Tippins, J.R. (1980) Prostaglandins, 19, 185-201.

14. Morris, H.R., Taylor, G.W., Piper, P.J. and Tippins,J.R.
(1980) Nature, 285, 104-106.

15. Morris, H.R., Taylor, G.W., Rokach, J., Girard, Y.,
Piper, P.J., Tippins, J.R. and Samhoun M.N. (1980)
Prostaglandins, 20, 601-607.

16. Dell A., Etienne, A.T., Morris, H.R., Beaumont, A.,
Burrell, R. and Hughes, J. (1979) in Molecular Endcrin-
ology ed. I. MacIntyre and M. Szelke, pp. 91-97. Elsevier/
North Holland Biochemical Press.

17. Morris, H.R., Dell, A. and Etienne, A.T. (1980) in The
Biochemistry of Schizophrenia and Addiction ed. G. Hemmings
pp. 77-84, MTP Press, Lancaster, UK.

18. Vinson, G.P., Whitehouse, B.J. Dell, A., Etienne, A.T.
and Morris, H.R. (1980) Nature, 284, 464-467.

19. Dell, A., Etienne, A.T., Panico, M., Morris, H.R.,
Vinson, G.P., Whitehouse, B.J., Barber, M., Bordoli, R.S.,
Sedgwick, R.D. and Tyler, A.N. submitted for publication.

20. Morris, H.R., Dell, A. and McDowell, R.A. (1981)
Biomed. Mass Spectrom., 8, 463-473.

21. Dell, A. in Soft Ionisation-Biological Mass Spectrometry ed. H.R. Morris, Heyden and Son, in press.

22. Morris, H.R., Thompson, M.R., Osago, D.T., Ahmed, A.I.,
Chan, S.M., Vandenheede, J. and Feeney, R. (1978) J.Biol.
Chem., 253, 5155-5162.

23. Morris, H.R., Taylor, G.W., Petersen, T.E. and
Magnusson, S. (1980) Adv.Mass Spectrom., 8, 1090-1096.

24. Dell, A., Morris, H.R., Hecht, S.M. and Levin, M.D.
(1980) Biochem.Biophys.Res.Comm., 97, 987-994.

25. Dell, A., Morris, H.R., Levin, M.D. and Hecht, S.M.
(1981) Biochem.Biophys.Res.Comm., 102, 710-738.

SEQUENCE ANALYSIS OF POLYPEPTIDES BY DIRECT

CHEMICAL IONIZATION MASS SPECTROMETRY

Steven A. Carr and Vernon N. Reinhold

Department of Biological Chemistry

Harvard Medical School, Boston, MA. 02115

INTRODUCTION

The identification and structural elucidation of small, biologically active peptides is a challenging analytical problem. These compounds are often not amenable to classical protein sequencing methods because of the presence of N-terminal blocking groups, carbohydrate side chains, or covalently modified amino acids. The low volatility and thermal lability of most polypeptides requires that they be derivatized prior to conventional electron impact (EI) or chemical ionization (CI) mass spectrometry for complete structural evaluation (1). Unfortunately, peptides of biological origin are often obtainable only in sub-nanomole amounts which makes sample consuming derivatization highly undesirable. Furthermore, chemical treatment may modify or destroy sensitive functional groups present in the peptide.

In recent years a number of new ionization techniques have been developed which enable mass spectral analysis of ionic, highly polar and thermally labile compounds with a minimum of decomposition. To date, the most widely utilized of these methods has been field desorption mass spectrometry (2). However, the lack of abundant, reliably recognizable sequence ions, and the relatively large sample requirements severely limit its usefulness for peptide sequence analysis. Recently, mass spectral techniques which utilize lasar induced desorption (3), low (KeV) energy ions (4),

263

high (MeV) energy fission fragments (5) or fast atoms (6) have been shown to produce ions characteristic of a peptide's molecular weight, and in some cases interpretable sequence ions.

Another recent addition to the mass spectroscopist's repertoire of soft ionization methods is direct chemical ionization (7,8), a sensitive and reproducible means of obtaining both molecular and sequence information on underivatized peptides, as well as many other involatile materials. Samples are coated on a wire attached to an extended direct insertion probe which when heated induces sample volatization and subsequent ionization by the charged reagent gas plasma.

We have explored the applicability of direct chemical ionization mass spectrometry (DCI) to the problem of oligopeptide sequence analysis at the subnanomole level. The DCI spectra of a series of synthetic underivatized and simply derivatized peptides up to ten amino acids in length have been obtained from 0.03 to 1 µg of material. Ions characteristic of the molecular weight and abundant, easily interpretable sequence ions are produced when ammonia is used as the reagent gas. Avoiding metal–sample interaction by desorbing peptides from polyimide coated wires was found to enhance volatility and decrease decomposition relative to the results obtained with bare wire filaments (9).

EXPERIMENTAL

The DCI spectra were obtained by loading dilute solutions of the sample (.01 to 1 µg/ul) onto a polyimide coated loop of nichrome wire (9). Current was applied to the wire using a programmable power supply. The mass spectrometer utilized in this study was a Finnigan-MAT 312 instrument with reverse Nier–Johnson geometry which is fitted with a combined CI/EI ion source.

RESULTS AND DISCUSSION

The most abundant sequence ions in the NH_3–DCI mass spectra of underivatized peptides (Scheme 1) are the N-terminal amide ions = $X-NH-CHR_n-CO-NH_3^+ + (NH_3)_m$, and C-terminal ammonium ions = $^+NH_3-CHR-CO-Y + (NH_3)_m$. These

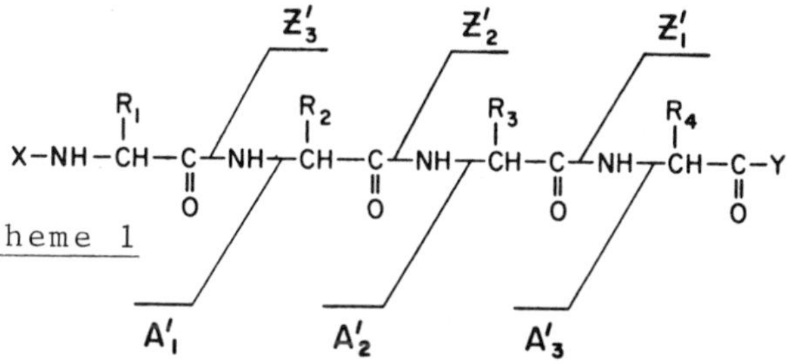

Scheme 1

structures are indicated on the mass spectra as $(A_n' + 2H)^+$ and $(Z_n' + 2H)^+$ for m = 0, and $\left|(Z_n' + H) + NH_4\right| +$ for M = 1; the X and Y groups denote the remainder of the peptide chain. These ions are not observed by conventional CI using methane or isobutane as the reagent gas (10) or tandem mass spectrometry using NH_4^+ as the ionizing species (11). For these latter techniques, the major N-terminal fragments are acylium ions of the type $(H_2N-CHR-CO^+)$ from which other abundant fragments are produced by neutral losses of CO, H_2O or NH_3.

In general, the ions of highest mass in the DCI spectra correspond to $(M + H)^+$ and/or $(M + NH_4)^+$ as well as ions arising via the loss of H_2O or NH_3 from the molecular ion. The molecular ion, together with the abundant A' and Z' series sequence ions are usually sufficient to establish the sequences. While not essential, the amino acid composition of the peptide is usually available and provides confirmatory evidence for the proposed structure.

The NH_3-DCI mass spectra of the underivatized peptides Tyr-Gly-Gly-Phe-Leu (Leu-Enkephalin), Phe-Asp-Ala-Ser-Val and Pro-Gln-Gln-Phe-Phe-Gly-Leu-Met-NH$_2$, (the C-terminal octapeptide of substance P), are typical of those obtained for a wide range of polar and non-polar linear peptides (Figs. 1a-c). The ion of highest mass in the spectrum of Leu-Enkephalin (Fig. 1a) corresponds to $(M + H)^+$ = 556. Ions arising by the loss of H_2O and/or NH_3 from the molecular ion occur at m/z 521 and 538. The abundant and intense fragment ions may be easily related to the sequence of the peptide. For example, the fragment at m/z 238 corresponds to $(A_2' + 2H)^+$, and the next most intense ion (m/z 353) is $[(Z_3' + H) + NH_4]^+$. Also present are the proton or ammonia cationized Z_1', Z_2', Z_4' and A_4' ion series which (together with the other ions discussed above) unambiguously allow the

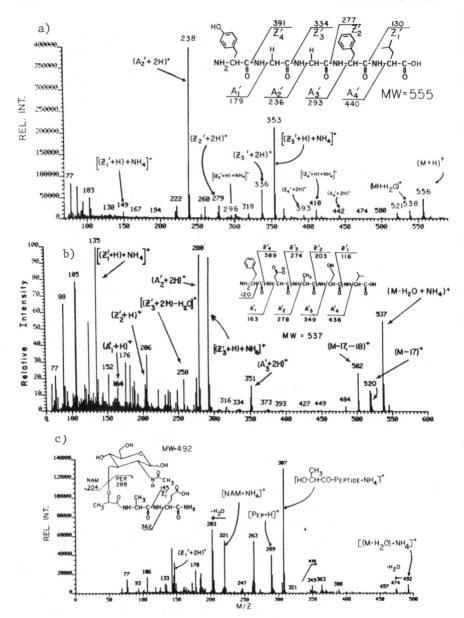

Figure 1.   Ammonia-DCI mass spectra of a)Tyr-Gly-Gly-Phe-Leu,
b)Phe-Asp-Ala-Ser-Val and, c)N-acetyl-muramyl-Ala-IsoGln.

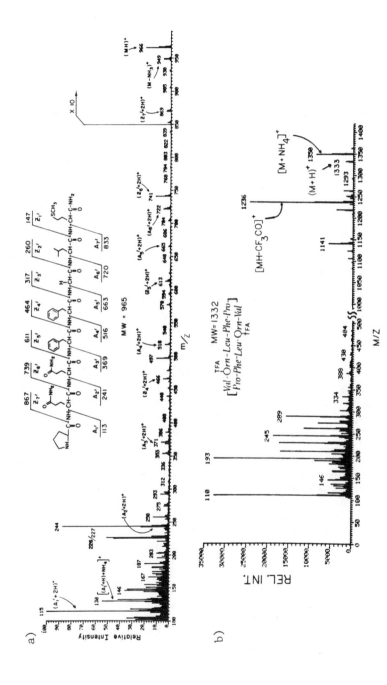

Figure 2. Ammonia–DCI mass spectra of a) Pro–Gln–Gln–Phe–Phe–Gly–Leu–Met–NH$_2$, and, b) ditrifluoroacyl–Gramicidin–S.

sequence of this pentapeptide to be determined.

The structures assigned to the sequence ions are based
not only upon correlation of the observed mass to the
expected mass, but on mass shifts of sequence ions which are
obtained when $^{15}NH_3$ or N^2H_3 are used as the reagent gas.
These procedures reduce the chance of misassigning fragment
structure and, therefore, amino acid sequence. For example,
m/z 537 in the mass spectrum of Phe-Asp-Ala-Ser-Val (Fig. 1b)
could be the molecular ion. However, when $^{15}NH_3$ is used as
the reagent gas m/z 537 shifts upwards to m/z 538 indicating
that this ion actually corresponds to $[(M-H_2O) + NH_4]^+$.
Abundant sequence ions are also present in the NH_3-DCI mass
spectrum including the protonated and/or ammonia cationized
A_1', A_2', A_3', Z_1', Z_2' and Z_3' ions which allow the sequence
of this pentapeptide to be deduced.

The sensitivity of peptide analysis by DCI depends upon
the amino acid composition and the length of the polypeptide
chain. In general, linear peptides up to 10 amino acids in
length which do not contain Arg, His or Trp (see below) can
be analyzed using less than 1.0 µg of material, and complete
mass spectra have been recorded on as little as 25 ng of
small non-polar peptides (e.g., tetravaline). The mass spectrum of 0.2 µg of Pro-Gln-Gln-Phe-Phe-Gly-Leu-Met-NH_3 (Fig. 2a),
the moderately polar C-terminal octapeptide of substance P,
still exhibits the expected molecular ion $((M + H)^+ = m/z\ 966)$
and a sequence ion series which is sufficient to establish
the structure. Small peptides up to four amino acids in
length which contain Arg, His or Trp may be analyzed successfully by NH_3-DCI. For example, the mass spectrum of tetra-
gastrin (Trp-Met-Asp-Phe-NH_2) exhibits both molecular and
sequence ions of the type described above. However, larger
peptides containing these amino acids (and linear peptides
of >10 amino acids) decompose prior to vaporization which
results in the loss of ions related to the intact peptide as
well as the formation of a multitude of pyrolysis fragments.
Larger peptides containing Arg may be analyzed if this residue
is converted to Orn by hydrozinolysis.

Glycopeptides and cyclic peptides are also ammenable to
the DCI mass spectral approach. The DCI mass spectrum of
N-acetyl-muramyl-Ala-Glu-NH_2 (adjuvant dipeptide, Fig. 1c)
exhibits ions characteristic of the molecular weight (m/z 492
$= [(M-H_2O) + NH_4]^+$) as well as fragment ions arising from
both the sugar and peptide portions of the molecule. The

cyclic decapeptide gramicidin-S (di-TFA derivative, Fig. 2b) yields abundant cationized molecular ions (m/z 1350 = $(M + NH_4)^+$), but no readily interpretable sequence ions.

Simple derivatization of peptides to the corresponding acetyl or perfluoroacyl methyl esters affects both molecular ion abundance and relative intensities of sequence ions. Masking of the amino terminus by acylation has the general effect of increasing the amount of ion current carried by the $(NH_4)^+$ cationized molecular ions and fragment ions. Introduction of an acetyl group does not cause significant change in the relative intensities of sequence ions, whereas perfluoroacyl groups cause a marked increase in the intensity of the $(A_n' + H)^+$ vs $(A_n' + 2H)^+$ ion series.

The effects obtained by modifying the desorption surface have also been examined (9). We have found that peptides (and many other compound classes) desorbed from DCI emitters coated with a thermally stabilized polyimide exhibit enhanced relative abundance and overall yield of molecular ions. and a decrease in the abundance of fragments that lack sequence information. For example, the intensity of m/z 875, the $(M + NH_4)^+$ ion for dipentafluoropropionyl-Phe-Asp⁻-Ala-Ser-Val-$(OCH_3)_2$ is approximately five times greater for equivalent amounts of peptide desorbed from coated wire versus bare metal wire. Furthermore, the number of spectra in which $(M + NH_4)^+$ is present are more than doubled for the sample desorbed from the coated wire. In addition, the spectrum of the sample desorbed from bare wire exhibits several new ions which correspond to loss of CF_3CF_2CO and $CF_3CF_2CONH_2$ from sequence ions.

The NH_3-DCI mass spectral approach is particularly well suited for determining the structure of minute amounts of individual oligopeptides, such as the many important physiologically active peptides and peptide antibiotics. It is not well suited for the complex mixtures of peptides encountered in protein sequencing because it would require extensive separation of these mixtures prior to mass spectrometric analysis.

ACKNOWLEDGEMENTS

We wish to thank Prof. K. Biemann, Dept. of Chemistry, M.I.T., Cambridge, MA. for many helpful discussions.

This work is supported by grant No. 5 P01 GM 26625 from the National Institute of General Medical Sciences, National Institutes of Health.

REFERENCES

1) Biemann, K. (1980) in Biochemical Applications of Mass
    Spectrometry, First Supplementary Volume (Waller, G.R.,
    and Dermer, O.C., eds) pp. 469-525, Wiley Interscience,
    New York.
2) Beckey, H.D. (1977)   Principles of Field Ionization
    and Field Desorption Mass Spectrometry, Pergamon
    Press, Oxford.
3) Mumma, R.O. and Vastola, F.J. (1972) Org. Mass Spectrom.
    6, 1373; Posthumus, M.A., Kistemaker, P.G., Meuzelaar,
    H.L.C. and Ten Noever de Brauw, M.C. (1978) Anal.
    Chem. 50, 985.
4) Grade, H. and Cooks, R.G. (1978) J. Am. Chem. Soc. 100,
    5615; Ens, W., Standing, K.G., Chait, B.T. and Field,
    F.H. (1981) Anal. Chem. 53, 1241.
5) Macfarlane, R.D. and Torgerson, D.F. (1976) Science
    191, 920; Macfarlane, R.D. (1980) in Biochemical
    Applications of Mass Spectrometry, First Supplementary
    Volume (Waller, G.R., and Dermer, O.C., eds) pp. 1209-
    1218, Wiley Interscience, New York.
6) Barber, M., Bordoli, R.S., Garner, G.V., Gordon, D.B.,
    Sedgwick, R.D., Tetler, L.W. and Tyler, A.N. (1981)
    Biochem. J. 197, 401.
7) Baldwin, M.A. and McLafferty, F.W. (1973) Org. Mass
    Spectrom. 7, 1353; Hunt, D.F., Shabanowitz, J., Botz,
    F.K. and Brent, D.A. (1977) Anal. Chem. 49, 1160.
8) Cotter, R.J. (1980) Anal. Chem. 52, 1589A.
9) Reinhold, V.N. and Carr, S.A. (1981), submitted to
    Anal. Chem.
10) Mudgett, M., Bowen, D.V., Field, F.H. and Kindt, T.J.
    (1976) Biomed. Mass Spectrom. 4, 159.
11) Beuhler, R.J., Flanigan, E., Greene, L.J. and Friedman,
    L. (1974) Biochemistry 13, 5060.

A NEW COMPUTER-AIDED METHOD FOR SEQUENCING A POLYPEPTIDE
FROM THE MASSES AND EDMAN-DEGRADATION OF ITS CONSTITUENT
PEPTIDE FRAGMENTS

YASUTSUGU SHIMONISHI

Institute for Protein Research, Osaka University

Suita, Osaka 565, Japan

INTRODUCTION
    Recently Edman degradation [1] has been greatly improved
and refined for determination of the amino acid sequences of
peptides and proteins in subnanomolar quantities, particularly by use of an improved automated sequencer [2]. However, the method generally requires the purification of
peptide mixtures generated from the polypeptide or protein
to be sequenced. During the past 3 years, we elaborated a
method [3,4] for sequencing peptide mixtures, that does not
require their purification, by a combination of field-
desorption mass spectrometry (FD-MS) and Edman degradation,
because the separation and purification of peptide mixtures
are still time consuming and tedious procedures.
    To our knowledge, FD-MS has not been widely used for
sequencing peptides, because the mass spectra give few of the
fragment ion peaks necessary for elucidating the structure of
peptides, although they give abundant molecular or quasimolecular ion peaks. However, we realized that FD-MS
provides useful information for structure determination in
combination with other techniques; that is, the N-terminal
amino acid residues of peptides in a mixture can be determined at a time by measurement by FD-MS of the mass values of
peptides before and after Edman degradation, and therefore
the sequences of peptides in a mixture can be determined by
repeating this procedure [3,4]. Furthermore, we realized

that when peptide mixtures are generated from a polypeptide
or protein by two or more kinds of specific cleavage methods,
the amino acid sequence of the polypeptide or protein can
be deduced from the molecular weights and N-terminal partial
sequences of the peptides in the mixtures, measured as
described above, using an appropriate computer program [5,
6], even if complete or almost complete sequences of peptide
fragments are not elucidated.

   In this paper, we describe a new method for sequencing
a polypeptide using Edman degradation, FD-MS and a computer
calculation. The method involves the following procedures:
1) measurement by FD-MS of the molecular weights of constituent peptides in mixtures generated from a polypeptide
by two or more kinds of specific cleavage methods, 2) Edman
degradation of the peptide mixtures and quantitation by
high-performance liquid chromatography (HPLC) of the phenylthiohydantoins (PTH's) released successively from the
peptide mixtures, and 3) computer calculation of candidate
amino acid sequences of the polypeptide from the data
obtained in 1) and 2).

METHODS
   FD mass spectra were measured with a Matsuda-type
second-order double-focusing mass spectrometer [7] with a
mono FD ion source. The conditions for measurement of mass
spectra were as described [4]. Edman degradation was performed manually as described [4] using a mixture of pyridine
and water (1/1, v/v) at pH 9.35 as buffer solution. The
resultant PTH's were subjected to HPLC, and the peptide
fractions to FD-MS. HPLC of PTH's was performed on a Zorbax
ODS column (4.6 x 250 mm) at 62°C using a mixture of acetonitrile and 0.01 M sodium acetate (pH 4.5) (42/58, v/v) [8].
Computer programs [5,6] were designed for reconstituting the
complete sequence or the major portions of the sequence of a
polypeptide or protein.

RESULTS and DISCUSSION
   First, for confirmation of the validity of the method,
the FD mass spectra of a single peptide degraded by the
Edman method and of peptide mixtures generated from a polypeptide by proteolytic digestion were measured. Figure 1

shows the mass spectra of a single peptide, bradykinin, and its degraded peptide fragments. The mass values were observed to be 1060, 904, 807, 710, 653, 506, 419, 322, and 175, in order of degradation. From these values the sequence was determined to be R(156)-P(97)-P(97)-G(57)-F (147)-S(87)-P(97)-F(147)-R-OH(175) simply by subtraction. Figure 2 shows the FD mass spectra of the chymotryptic and tryptic peptides of human β-endorphin as examples of peptide mixtures. Tryptic peptides gave five intense ion peaks, as expected from the sequence. Chymotryptic peptides also gave expected mass peaks, although mass peaks of some peptide fragments generated by nonspecific cleavage were also observed. Thus, these results demonstrate the possibility that a combination of FD-MS and Edman degradation is used for sequence determination of peptide mixtures.

Next, as typical examples of peptide mixtures, we measured the FD mass spectra of chymotryptic (Fig. 3) and tryptic peptides of glucagon and their degraded peptide fragments. The PTH's released were also measured by HPLC, as shown in Fig. 4. The mass spectra of chymotryptic peptides and HPLC of PTH's released showed that six peptides

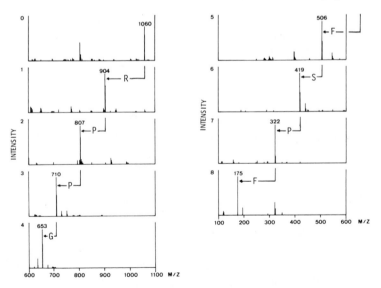

Fig. 1. FD mass spectra of bradykinin (0) and its degraded peptide fragments (1-8).

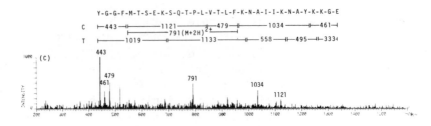

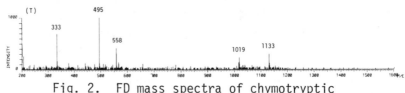

Fig. 2.   FD mass spectra of chymotryptic
(C) and tryptic (T) peptides of human β-endorphin

were present in the mixture; namely, L-D-S-R---(1107), H-S-
Q-G---(676), T-S-D-Y(485), L-M-N-T(478), V-Q-W(432), and
S-K-Y(397) (numbers in parentheses are mass values observed).
Since L has the same residual weight as I, it cannot be
distinguished from the latter by the mass spectra only, but
the PTH derivatives of the two are easily identified by HPLC.
Similarly, the sequences of tryptic peptides were determined
to be H-S-Q---(1357), A-Q-D---(1352), Y-L-D---(653), and R
(175) (data not shown).

Generally, almost all amino acid residues should be
addressed for reconstitution of the complete sequence of a
peptide or protein from the sequences of its constituted
peptide fragments. However, this is not necessary if the
molecular weights of the peptide fragments are precisely
determined. That is, the amino acid sequence of a peptide
or protein can be calculated from the molecular weights and
N-terminal amino acid residues (or partial sequences) of two
or more kinds of peptide fragments generated by two or more
kinds of specific cleavage methods and the PTH's released
from these fragments, as shown above with glucagon, since
the combinations (kind and number) of amino acid residues of
a peptide fragment with a given molecular weight can easily
be determined. We designed computer programs [5,6] for
calculation of the sequence of a peptide or protein from
such data as described above. Figure 5 shows the sequence

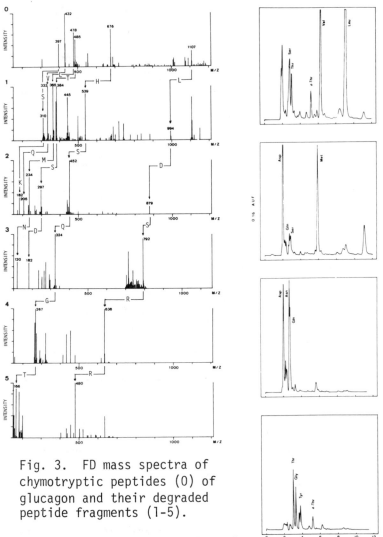

Fig. 3. FD mass spectra of chymotryptic peptides (0) of glucagon and their degraded peptide fragments (1-5).

Fig. 4. HPLC of PTH's released and extracted into the organic phase in order of degradation from chymotryptic peptides of glucagon.

of glucagon calculated using one of these programs. The results indicate that the entire sequence of glucagon could be determined from the molecular weights and N-terminal amino acid residues or partial sequences of the chymotryptic

and tryptic peptides and the PTH's released by Edman degradation.

The procedure was then applied to sequence determination of the N-terminal BrCN fragment of <u>Streptomyces</u>
<u>erythraeus</u> lysozyme [9], the structure of which was un-
known.   Figure 6 shows the FD mass spectra of the tryptic

INPUT DATA

AMINO ACID N-TERM C-TERM
COMPOSITION

MOL WEIGHTS,N-TERMINAL SEQUENCES AND PTH-AMINO ACIDS IN EACH CYCLE

CHYMOTRYPTIC PEPTIDES

TRYPTIC PEPTIDES

```
                                            3    4                                    2    1
  G   1                          396 SK     G   0  1,0                      174 P     C   0    0
  A   1                          431 VD     A   0    0                      652 Y     A 1,0    0
  U   0                          477 LM     U   0    0                     1351 A     U   0    0
  S   4                          484 TS     S 2,0    0                     1356 H*    S 1,0    0
  P   0                          675 HS     P   0    0                               P   0    0
  V   1                         1106 LD     V   0    0                               V   0    0
  T   3                                     T   0  1,0                               T   0    0
  L   2              T                      L   0    0                               L 1,0    0
  I   0                                     I   0    0                               I   0    0
  M+D 4                                     M 1,0    0                               M   0    0
  Q+E 3                                     D 1,0    0                               D   0  2,0
  K   1                                     Q 2,0    0                               Q 1,0  2,0
  H   1                                     K   0    0                               K   0    0
  H   1    H                                E   0    0                               E   0    0
  C   0                                     M   0    0                               M   0    0
  F   2                                     H   0    0                               H   0    0
  R   2                                     C   0    0                               C   0    0
  Y   2                                     F   0    0                               F   0    0
  W   1                                     R   0  1,0                               R   0    0
                                            Y 1,0  1,0                               Y   0    0
TOTAL 29                                    W 1,0    0                               W   0    0
```

CANDIDATE SEQUENCES

1   HSQGTFTSDYSKYLDSRRAQDFVQWLMNT

Fig. 5.   Computer output sequence of glucagon.

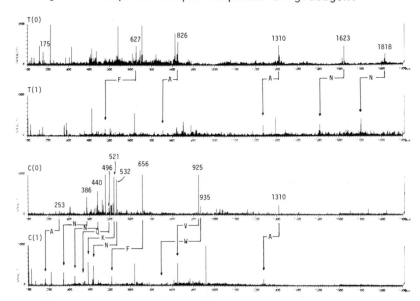

Fig. 6.   FD mass spectra of tryptic (T(0) and T(1)) and
chymotryptic (C(0) and C(1)) peptides of N-terminal BrCN
fragment of <u>Streptomyces</u> <u>erythraeus</u> lysozyme.

and chymotryptic peptides and their peptide fragments obtained after one-cycle of Edman degradation. The spectra showed six tryptic and ten chymotryptic peptides and their molecular weights and N-terminal amino acid residues were determined. Figure 7 shows the results of HPLC of PTH's released from the tryptic and chymotryptic peptides. The PTH's released at each cycle of Edman degradation were quantitated and their kind and ratio were selected as integer values within the limits of their possible maxima and minima, as input data in Fig. 8. The amino acid sequences

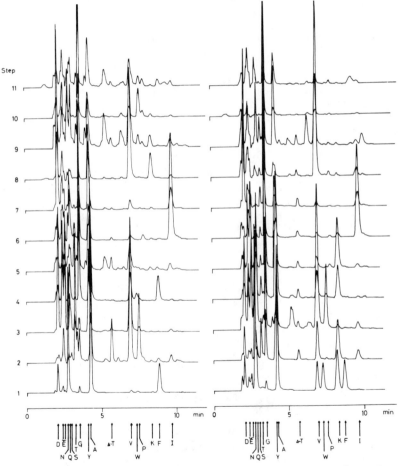

Fig. 7. HPLC of PTH's released from tryptic (left) and chymotryptic (right) peptides.

INPUT DATA

AMINO ACID COMPOSITION N-TERM C-TERM

MASS VALUES, N-TERMINAL SEQUENCES AND PTH-AMINO ACIDS IN EACH CYCLE

CHYMOTRYPTIC PEPTIDES

	2	3	4	5	6	7	8	9	10	11
G	1,0	1,0	1,0	1,0	1,0	0	1,0	0	1,0	0
A	2,0	1,0	2,0	1,0	0	0	0	0	0	0
S	0	1,0	0	0	0	1,0	0	1,0	1,0	0
P	0	1,0	0	0	0	0	0	0	0	0
V	1,0	2,0	0	0	0	1,0	1,0	0	0	0
T	2,0	0	1,0	0	1,0	1,0	0	0	0	0
C	0	0	0	0	0	0	0	0	0	0
L	0	0	0	0	0	0	0	0	0	0
I	1,0	0	0	0	1,0	0	0	0	0	0
N	0	0	0	0	0	0	0	0	0	0
D	0	1,0	0	0	0	1,0	0	0	0	0
Q	1,0	2,0	1,0	0	0	0	0	0	0	0
K	1,0	0	1,0	1,0	0	0	0	0	0	0
E	1,0	1,0	1,0	1,0	1,0	1,0	0	0	0	0
H	0	0	0	0	0	0	0	0	0	0
M	0	0	0	0	0	0	0	0	0	1,0
F	0	0	0	0	1,0	0	0	0	0	0
Y	2,0	1,0	2,0	2,0	0	1,0	1,0	1,0	1,0	1,0
W	0	1,0	1,0	1,0	1,0	1,0	1,0	1,0	1,0	1,0
U	1,0	1,0	1,0	1,0	1,0	1,0	1,0	1,0	1,0	1,0

TRYPTIC PEPTIDES

	2	3	4	5	6	7	8	9	10	11
G	0	0	1,0	2,0	2,0	1,0	0	1,0	2,0	2,0
A	1,0	0	1,0	1,0	0	0	0	0	0	0
S	1,0	0	0	0	0	0	0	2,0	1,0	2,0
P	1,0	0	0	0	0	0	0	0	1,0	0
V	1,0	1,0	1,0	0	0	0	1,0	1,0	0	1,0
T	2,0	0	0	1,0	1,0	0	0	0	0	1,0
C	0	0	0	0	0	0	0	0	0	0
L	0	0	0	0	0	0	0	0	0	0
I	0	0	0	0	2,0	0	0	0	0	0
D	0	1,0	0	0	0	2,0	0	0	0	1,0
Q	0	1,0	1,0	1,0	1,0	1,0	0	0	1,0	0
K	0	0	0	1,0	0	1,0	1,0	0	0	0
E	0	2,0	1,0	1,0	1,0	0	0	0	0	0
H	0	0	0	0	0	0	0	0	0	0
M	0	0	0	0	0	0	0	0	0	1,0
F	0	1,0	0	0	0	0	0	0	0	0
R	0	0	0	0	0	0	0	0	0	0
Y	0	2,0	0	1,0	1,0	1,0	1,0	1,0	1,0	1,0
W	0	0	1,0	1,0	1,0	1,0	1,0	1,0	1,0	1,0
U	0	0	0	1,0	1,0	1,0	1,0	1,0	1,0	1,0

CANDIDATE SEQUENCES

1   ATVAGIDVSGHQRNVDWQYWWNQGKRFAYVKATEGTGYKNPYFAQQYNGSYNIGU

Fig. 8. Computer output sequence of N-terminal BrCN fragment of Streptomyces erythraeus lysozyme

were sought using a computer program [6] and there was only one output sequence as shown in Fig. 8.

Acknowledgments

The author expresses his sincere thanks to Prof. Y.Izumi for valuable advice and to Prof. H.Matsuda, Drs. T.Matsuo and I.Katakuse, Prof. T.Ikenaka, Drs. S.Hara and S.Aimoto, Mr. Y.-M.Hong, Miss T.Kitagishi, and Mr. T.Takao (Osaka University) for collaboration. He is also grateful to Yamada Scientific Foundation for financial support.

REFERENCES
[1] Edman,P. and Henschen,A.(1975) in Protein Sequence Determination, Needlemann,S.B. ed., 2nd edn., Springer-Verlag, Berlin, pp. 232-279.   [2] Hunkapiller,M.W. and Hood,L.E. (1980) Science 207, 523.   [3] Shimonishi,Y., Hong, Y.-M., Matsuo,T., Katakuse,I. and Matsuda,H.(1979) Chem.Lett. 1369. [4] Shimonishi,Y., Hong,Y.-M., Kitagishi,T.,Matsuo,T., Matsuda,H. and Katakuse,I.(1980) Eur.J.Biochem.112, 251. [5] Kitagishi,T., Hong,Y.-M. and Shimonishi,Y.(1980) Int.J. Pept.Protein Res.17, 436.   [6] Kitagishi,T., Hong,Y.-M., Takao,T., Aimoto,S. and Shimonishi,Y. Bull.Chem.Soc.Jpn. to be submitted.   [7] Matsuda,H.(1976) Atomic Masses Fundam. Constants 5, 185.   [8] Zimmerman,C.L., Appella,E. and Pisano, J.J.(1977) Anal.Biochem. 77, 569.   [9] Morita,T., Hara,S. and Matsushima,Y.(1978) J.Biochem.(Tokyo) 83, 893.

THE COMPLEMENTARITY OF MASS SPECTROMETRY TO

EDMAN DEGRADATION OR DNA BASED PROTEIN SEQUENCING

K. Biemann

Department of Chemistry
Massachusetts Institute of Technology
Cambridge, MA   02139

MASS SPECTROMETRIC MULTIPLE PHASE CHECK FOR
LONG PROTEIN SEQUENCES DEDUCED FROM DNA SEQUENCES

Over the past few years the advances in the methodologies for the rapid sequencing of long DNA chains have advanced to the point that it has become feasible to deduce the amino acid sequence of a protein from the base sequence of the structural gene responsible for the synthesis of the protein, using the genetic code. In principle, it is only necessary to determine a short segment of the NH_2-terminal and COOH-terminal sequences of the protein and the remaining amino acid sequence should fall into place.

Although DNA sequencing can be accomplished quite accurately, and the confidence improved by sequencing both strands of the DNA it has to be kept in mind that the omission of a single base at one point and erroneous insertion of another one later on will lead to a derived protein structure which is correct at both ends but may have an entirely incorrect segment in the center. While the occurrence of a stop codon would alert one to this fact, there may well be extended segments within the structural gene where no such codon appears in any of the three reading phases.

Clearly, establishing the correctness of the translation of the DNA sequence by spot-checking the protein sequence randomly along the entire protein, rather than only at the NH_2- and COOH-termini greatly enhances the

confidence in the correctness of the amino acid sequence
of the protein.

Unfortunately, conventional sequencing of even parts
of a large protein is tedious, time-consuming and requires
a relatively large amount of protein. For this reason we
have developed a strategy which provides a significant
number of such randomly spaced peptide sequences long enough
to be unique and at a speed that is comparable to that of
the base sequencing of a gene. In order to cut the time
requirement it is clearly necessary to eliminate the need
for the preparation and separation of complex enzymatic
digests of the protein, i.e. one has to use the entire protein or segment thereof.

The gas chromatographic mass spectrometric (GCMS)
technique for protein sequencing which we have developed in
the past (1,2) lends itself well to this task because it
permits the selective identification of peptides of the
required size (tetra-and pentapeptides) in the very complex
mixtures produced by the digestion of the intact protein
or a large segment with appropriate enzymes. These mixtures
are converted to analogous mixtures of derivatives (N-tri-
fluorodideuteroethyl-polyaminoalcohol-O-trimethylsilyl
ethers) as outlined in Figure 1. The resulting complex
mixture of derivatives is then injected into a gas chromatograph coupled to a fast scanning mass spectrometer which
records the mass spectra of all components with a computer-
based data acquisition system. The spectra are interpreted
in terms of the amino acid sequence of the individual
peptides with the aid of a computer program (3). In the
earlier stages of this process the emphasis is placed chiefly
on the identification of tetra- and pentapeptides because
the chance that they appear twice or in more than one
reading frame is very remote.

Having in hand such a set of peptide sequences derived
from all areas of the protein at a time frame faster than
the determination of the entire DNA sequence of the structural gene makes it possible to match these peptides with
each set of base sequences as they become available. Much
time is saved because this can be done even before the base
sequences have been rigorously checked by repetition of the
experiment or sequencing the complementary strand.

The availability of these random peptide sequences
allows one to 1) locate the structural gene along the DNA,
at least approximately, because the NH_2- and COOH-termini

Protein

| Specific enzymatic
| or chemical cleavage
↓

Complex mixture of di-to hexapeptides
| 3N HCl/MeOH or CH_2N_2
↓

| $CF_3CO_2CH_3$ + TEA
↓

| B_2D_6/THF
↓

| TMSDEA/Pyridine
↓

Complex mixture of O-TMS polyamino alcohols

↓

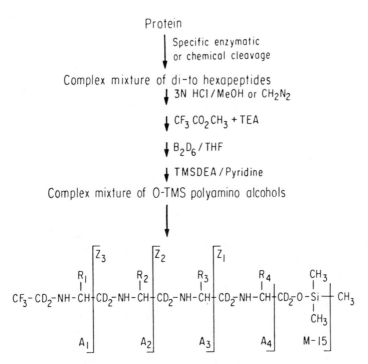

Figure 1. Conversion of a protein into a mixture of O-TMS amino alcohols (structure indicates major cleavage upon electron impact and fragment nomenclature).

must be within the size of the protein (deduced from its approximate molecular weight) from the points where such peptides match the DNA sequence; 2) the direction of transcription can be deduced from the first few peptides that are found to match one of the two DNA strands; 3) the reading phase of the sense strand is established at each region where an identified peptide sequence matches a base sequence; 4) an error such as the omission or erroneous insertion of a base manifests itself by finding a match of one peptide in a given reading phase while another peptide matches another reading phase of the same DNA segment, pinpointing the error to the region between the two peptides. This error can then be found by scrutinizing that particular gel or repeating the electrophoresis experiment of that DNA segment or of a related one. In this way it is possible to concentrate the effort on those sections of the base sequence which seem to be doubtful without unnecessarily

double-checking sequences that were shown to fit the peptides identified and must therefore be correct.

The Amino Acid Sequence of Alanine-tRNA synthetase

This strategy has been developed and evaluated in the
course of the determination of the amino acid sequence of
alanine-tRNA synthetase, an enzyme that turned out to be
a tetramer of a single chain of 875 amino acids. The DNA
sequence of the structural gene was determined by Professor
Paul R. Schimmel and Dr. Scott D. Putney of the Department
of Biology, M.I.T., using the new method of Maxim and
Gilbert (4). This sequence is shown in Figure 2 along with
the structure of the enzyme derived therefrom (5). Short
peptide sequences were determined by gas chromatographic
mass spectrometry using an NH_2-terminal segment of about
380 amino acids in length obtained by limited tryptic
cleavage of the AlaRS (6). In a second set of experiments
peptide sequences from the remainder of the protein were
obtained. All these peptides are underlined in Figure 2,
which shows that about 25% of the peptide bonds have been
confirmed by GCMS sequencing (the 11 NH_2-terminal amino
acids were determined by an Edman degradation) and we believe
that this level of coverage suffices to have high confidence
in the DNA-derived protein structure.

USE OF FAST ATOM BOMBARDMENT MASS SPECTROMETRY (FABMS)
FOR THE DETERMINATION OF PROTEIN STRUCTURE

The GCMS technique is well suited for the elaboration
of partial peptide sequences even to the point that they
can be assembled to an entire protein sequence but it turns
out that it is more efficiently combined with other methods,
such as the one described in the foregoing or the Edman
degradation. An example of the latter, where some of the
unique features of the GCMS technique were particularly
useful was the determination of the structure of bacteriorhodopsin (7,8). A disadvantage is the need for the multi-
step chemical reaction sequence (see Fig. 1) necessary to
convert the peptide mixture to the derivatives suitable
for GCMS.

Figure 2.   Base sequence of the structural gene of AlaRs
and the amino acid sequence of the protein.   Identified
peptide sequences are underlined (5).

Recently a new ionization technique was developed by Barber (9) and termed "fast atom bombardment" mass spectrometry (FABMS) which lends itself particularly well to the ionization and mass analysis of very polar, involatile and large molecules, particularly peptides (10). It simply involves the bombardment of the sample (less than a microgram dissolved in a small drop of glycerol suffices) with a beam of argon or xenon atoms in the ion source of the mass spectrometer.

The Sequence of a Tryptic Fragment of Macromomycin

An example of our current work on the amino acid sequence of macromomycin (MCR) which was carried out in collaboration with Drs. T.S.A. Samy and K.S. Hahm of the Sidney Farber Cancer Institute (SFCI), Boston, illustrates one of the uses of this approach, in this case finalizing a peptide sequence chiefly deduced from Edman and GCMS data.

The first 45 amino acids of the NH_2-terminal tryptic peptide had been determined by Sawyer et al. (11) and identified as follows:

Ala Pro Gly Val Thr Val Thr Pro Ala Thr Gly Leu Ser Asn Gly[15]

Glu Thr Val Thr Val Ser Ala Thr Gly Leu Thr Pro Gly Thr Val[30]

Tyr His Gly Gln Ser Ala Val Ala Glu Pro Gly Val Ile Gly Pro[45]

On the basis of two Edman degradations (carried out at SFCI) one covering the NH_2-terminus to His-32 and the other extending from Pro to the COOH-terminus and a set of GCMS experiments the following sequence was derived for this tryptic peptide T-1:

Ala Pro Gly Val Thr Val Thr Pro Ala Thr Gly Leu Ser Asp Gly[15]

Gln Thr Val Thr Val Ser Ala Thr Gly Leu Thr Pro Gly Thr Val[30]

Tyr His Gly Gln Cys Ala Val Val Glu Pro Gly Val Ile Gly Cys[45]

Asp Ala Thr Thr Ser Thr Asp Val Thr Ala Asp Ala Ala Gly Lys[60]

In order to verify the sequence immediately after His-32 where we had at this point neither Edman nor GCMS information, T-1 was digested with α-chymotrypsin and *S. aureus* V8 protease and the resulting mixture was then subjected to FABMS. One would expect a peptide His-Gly-Gln-Cys-Ala-Val-Val-Glu of MW 899 (Cys is carboxymethylated) but there

was no peak for $(MH)^+$ at m/z 900. However, as Table I
shows, there were signals at m/z 999, 1021 and 1043, indicating a peptide of MW 998. This difference of 99 mass
units corresponds to valine which must be placed after
His-32.

Table I

Peptide	MH^+	MNa^+	MNa_2^+
Gly-24........Tyr-31	807	822	
His-32........Glu-40	999	1021	1043
Ala-1........Leu-12	1083	1105	
Pro-41........Ala-56	1565		
Pro-41........Lys-61	2007	2024	2051

Peptides derived from T-I by digestion with S. aureus V8 protease and
α-chymotrypsin, and the m/z values of their molecular ion clusters as
determined by FABMS.

A FAB mass spectrum of a thermolysin digest of T-1
revealed ten peptides and these are listed in Table II.
Their molecular weights all agree with the insertion of the
previously missing valine. The GCMS data combined with
the specificity of thermolysin place it at position 33.
The complete sequece of T-1 is shown in Figure 3. The
remaining three tryptic peptides of MCR have been sequenced
using the same combination of techniques and the structure
of this protein, which is 112 amino acids long will be
published elsewhere.

Table II

Peptide	MH^+	MNa^+	MNa_2^+	Peptide	MH^+	MNa^+	MNa_2^+
Val-20........Gly-24	434	456		Leu-12........Thr-17	620	642	664
Leu-25........Thr-29	488	510	532	Val-54........Lys-61	732	754	
Val-38........Gly-42 orVal-39......Val-43	500	522		Val-4........Gly-11	745	767	789
				Leu-25........His-32	887	909	931
Val-6........Gly-11	545	567		Val-33........Gly-42	1016	1038	1060
Ala-37........Gly-42	571	593	615				

Peptides derived from T-I by digestion with thermolysin and the m/z
values of their molecular ion clusters as determined by FABMS.

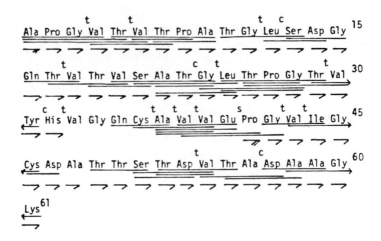

Figure 2. Sequence of the tryptic peptide T-I from macromomycin. Arrows: Edman degradation; underlining: peptides
   identified by GCMS; c,s,t: cleavage sites of α chymotrypsin, S. Aureus V8 protease and thermolysin, respectively.

FABMS Peptide Mapping for Phase Checking of DNA Sequences

Inspection of the data summarized in Tables I and II
indicate that these two experiments cover 110 of the 112
amino acids of MCR in 15 individual fragments. The information obtained from a FAB mass spectrum of an enzyme digest
of even a much larger protein should therefore provide data
which could be used to correlate the base sequence of the
gene with the amino acid sequence of its product in much the
same way as we had used the GCMS technique on AlaRS.

In order to test this hypothesis we have calculated the
composition and exact mass of all peptides that would be
generated by tryptic digestion of AlaRS. These are 72 peptides containing more than three amino acids, and only two
pairs of these are of the same nominal mass (but they could
be differentiated in a high resolution mass spectrum because
they differ by 0.0364 and 0.1453 amu, respectively).

When comparing the peptides expected from the translation of the correct reading frame with the tryptic peptides
corresponding to either of the two incorrect reading frames,
it is found that only two peptides (larger than tripeptides)

of the correct sequence have the same nominal mass as two peptides from either incorrect reading frame. In all four instances the mass of the peptides differ by more than 0.025 amu.

It thus should be possible to use a FAB mass spectrum of a tryptic digest with no or minimal separation to carry out a phase check of a DNA sequence. Clearly this approach would be even simpler and less time and material consuming than the GCMS method we used in the case of alanine-tRNA-synthetase.

While the data obtained with MRC demonstrate that for a protein of slightly more than 100 amino acids almost complete coverage can be achieved, it remains to be seen whether the much more complex mixtures obtained from larger proteins can be analyzed to a sufficient degree. As a test case we have used bovine serum albumin which is 582 amino acids long. In a very preliminary experiment we have subjected a tryptic digest of carboxymethylated protein to FABMS without any further separation. The data contained 10 peaks which are clearly due to $(MH)^+$ ions of peptides and these cover about 11% of the sequence. Under different conditions one should be able to obtain tryptic peptides from the more resistant regions of the protein and thus extend the coverage to the extent (about 25%) which we had found to be sufficient with the GCMS methodology.

Acknowledgement

The author is pleased to acknowledge the close association and collaboration of S.A. Carr, C.E. Costello, B. Gibson, W.C. Herlihy, and S. Martin who carried out the GCMS and FABMS experiments; to T.S.A. Sami and K.-S. Hahm of the Sidney Farber Cancer Institute, Harvard Medical School, who provided the tryptic peptides of macromomycin and neocarzinostatin and carried out the Edman degradations, and to H. Umezawa, of the Institute of Microbial Chemistry, Tokyo for the macromomycin. This work was supported by grants GM-05472 and RR-00317 from the National Institutes of Health. The funds for the acquisition of the Ion Tech Neutral atom gun were provided through a research contract (N00014-78-C-0421) from the Office of Naval Research.

References

1. K. Biemann in Biochemical Applications of Mass Spectrometry, Supplement I, G.R. Waller, Ed., John Wiley and
   Sons, 1980.
2. S.A. Carr, W.C. Herlihy, and K. Biemann, Biomed. Mass
   Spectrom., 8, 51 (1981).
3. W.C. Herlihy and K. Biemann, Biomed. Mass Spectrom. 8,
   70 (1981).
4. A.M. Maxam and W. Gilbert, Proc. Natl. Acad. Sci. USA,
   74, 560 (1977).
5. S.D. Putney, N.J. Royal, H. Neuman de Vegvar, W.C. Herlihy, K. Biemann and P. Schimmel, Science, 213, in press
   (September 1981).
6. W.C. Herlihy, N.J. Royal, K. Biemann, S.D. Putney and
   P.R. Schimmel, Proc. Natl. Acad. Sci. USA 77, 6531-6535
   (1980).
7. H.G. Khorana, G.E. Gerber, W.C. Herlihy, C.P. Gray, R.J.
   Anderegg, K. Nihei and K. Biemann, Proc. Natl. Acad.
   Sci. USA 76, 5046-5050 (1979).
8. W.C. Herlihy, R.J. Anderegg and K. Biemann, Biomed. Mass
   Spectrom. 8, 62 (1981).
9. M. Barber, R.S. Bordoli, R.D. Sedgwick, and A.N. Tyler,
   J. Chem. Soc. Chem. Comm., 1981, 325.
10. M. Barber, R.S. Bordoli, G.V. Garner, D.B. Gordon, R.D.
    Sedgwick, L.W. Tetler and A.N. Tyler, Biochem. J., 197,
    401 (1981).
11. T.H. Sawyer, K. Guetzow, M.O.J. Olson, H. Busch, A.
    Prestakyo andS.T. Crooke, Biochem. Biophys. Res. Commun.
    86, 1133-1138 (1979).

Cleavage of Proteins

ENZYMIC CLEAVAGE OF PROTEINS

KEIL B.

Protein Chemistry, Institut Pasteur

28, rue du Dr. Roux, PARIS 75015, France

It would be an impossible task to survey in the limited space all the problems of the use of proteolytic enzymes, after all the excellent reviews which appear every year. In the following chapters will therefore be discussed only a very limited number of questions on new development in this field.

Before discussing the specificity of proteolytic enzymes, let's review some recent changes in their list. The classical and generally accepted separation of endo-peptidases according to the active site in four groups (serine, cysteine, aspartyl and metalloproteinases) is presumably only temporary : there are indications that other structures of active sites exist. A recent example is the *Astacus* protease composed of only hundred residues and with the primary structure different from all known pro-teases. This enzyme cannot be included in any of the existing four groups, because it is not inhibited by any of the natural or synthetic proteinase inhibitors known, included DFP, iodoacetic acid, EDTA and phenantrolin [1]. It does not cleave the usual synthetic peptide or ester subs-trates of any other proteinase, and it has a characteristic specificity of its own, as we will see later.

A clear subgrouping exists already even inside the four actual groups of endopeptidases. Among the serine proteinases are the trypsin family on one side, the subtilisin family on the other. It seems that the same situation is in the

cysteine proteases, where clostripain differs sharply from
the papain family : its partial sequence already known has
no homology with any other protein [2] and it is the only
cysteine proteinase which is not inhibited by E64 (L-trans-
epoxysuccinyl-leucyl-agmatine) [3].

The list of newly characterized proteinases is growing
steadily and the well established views on the structural
basis of proteolysis change accordingly. Side by side with
the classical globular molecules of trypsin or elastase,
reinforced by internal disulfide bridges, we encounter
proteinases of the complement system in which the specific
proteolytic activity is created by formation of non-covalent
complexes of polypeptides originated from different parent
molecules, like in C3 or C5 convertase. The existence of a
protease with two chains bound non-covalently was proposed
in the case of cathepsin D [4] and recently for the highly
active α-clostripain [5].

Since 1973 sixty new proteolytic enzymes were retained
by the Commission on Enzyme Nomenclature. Out of every ten
or twenty new proteases hardly one is of some interest as a
tool in sequence work, but their growing list gives us a
more comfortable choice.

FRAGMENTATION OF NATIVE PROTEINS IN DOMAINS

Two approaches using proteolytic enzymes are currently
used in the protein structure work. One is the mild partial
hydrolysis of a native protein to obtain selective cleavage
of particularly vulnerable bond, the second is a systematic
degradation of the unfolded polypeptide. The difference
between the two approaches is that in the first case, both
bond specificity of the enzyme and the tertiary structure
of the native substrate intervene, in the second case the
result depends exclusively on the broader or narrower bond
specificity of the proteinase.

Many high molecular weight proteins are composed of two
or several domains with exposed joining segments that can
be easily cleaved by proteinases. Currently, the domains
with intact specific function can be obtained using proteinases of very different specificity : the single chain
molecule of aspartokinase-homoserine dehydrogenase can be
cleaved in two domains by proteinases of such widely

different specificity as α-chymotrypsin, trypsin, subtilisin
or papain [6]. The Fab and Fc domains can be obtained from
immunoglobulins by splitting the labile segments with
papain, pepsin as well as with bacterial proteinases from
S. sanguis or N. gonorrhoe [7,8].

Native collagens contain in their compact triple-helical
structure a "weak" segment (in $\frac{3}{4}$ of the molecule from the
N-terminal) which is susceptible to cleavage by several
proteinases accordingly to their different specificity :
thus tadpole, V2 ascites carcinoma and synovial collagenases
split in type I collagen one bond Gly$\downarrow$Ile-Ala in α1 chain
and one Gly$\downarrow$Ile-Leu in α2 chain [9,10], Achromobacter collagenase the bond X$\downarrow$Gly-Pro [11] and Hypoderma collagenase
the bond X$\downarrow$Ala... [12]. Trypsin splits specifically an
Arg$\downarrow$Gly bond in the analogous segment of type III collagen
[13], thrombin in type V collagen [14]; in the latter case
trypsin, chymotrypsin, elastase and vertebrate collagenase
do not act.

The conversion of procollagen in collagen by procollagen
peptidase is another good example, that in a native protein
the distinct three-dimensional structure of the cleavage
site is indispensable for the action of the enzyme : this
proteinase splits one Pro$\downarrow$Glu bond 1000 fold rapidly in
native than in denatured procollagen and it does not split
at all the other Pro-Glu bonds in collagen [15].

Numerous other recent examples of cleavage of long
polypeptides into distinct domains exist, like in the work
on ceruloplasmin [16], transferrin, haemocyanin by plasmin
[17], etc... . From the point of view of general strategy
of sequence work, a preliminary cleavage of high molecular
weight proteins in domains can be the first indispensable
step to the success.

ENZYMIC CLEAVAGE OF DENATURED POLYPEPTIDES AT CHARGED
RESIDUES

Let's now consider the relative value of classical and
new enzymatic tools for a specific breakdown of unfolded
proteins. The relative occurence of amino acid residues in
the pool of known proteins when confronted with the specificity of proteinases indicates that the next to best choice
to obtain large fragments is either trypsin after the

blocking of lysine residues or clostripain. The best choice,
however would not be enzymes which split next to a given
amino acid residue, but those which need for their action a
specific dipeptidic or even longer sequence. So far, however,
with all the other problems ahead most workers in the field
make a conservative option.

Trypsin remains the most used proteinase in sequence
studies, directly or after chemical modifications of the
protein substrate. In many papers the authors comment atypical "chymotrypsin-like" cleavages, even if trypsin was
treated previously by TPCK, diphenylcarbamylchloride [18]
or octylisocyanate [19]. These "atypical" cleavages, however,
statistically do not correspond to the specificity of
α-chymotrypsin. They occur mostly on the carboxyl side of
tyrosines and phenylalanines, but also of dicarboxylic
acid amides, methionine etc... May I remind that it was
shown ten years ago, on glucagon (Fig. 1), that pseudo-
trypsin which is generated slowly by autolysis of β- and
α-trypsin has a similar specificity [20]. Crystalline

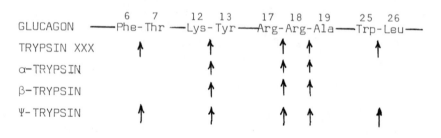

Fig. 1

Digestion of glucagon by commercial trypsin and pure
α-,β- and Ψ-trypsin [20]

trypsin is always composed of β- and α-trypsin as major
components and of a minor amount of Ψ-trypsin, and by a
simple chromatography on SP-Sephadex all three components
are readily separated. In view of the efforts spent usually
to purify a protein to homogeneity before tryptic cleavage,
it seems that a one-column operation to purify the β-tryp-
sin for the cleavage would be justified.

Previous observations that trypsin cleaves poorly the bond Arg-Pro or Lys-Asp [21] cannot be considered as strict rules : two out of four Arg-Pro bonds (Arg$_{94}$ and Arg$_{145}$) were cleaved neatly in fibrinogen [22,23] ; on the contrary the bond Lys$_{63}$-Pro remained resistant. Arg-Pro bonds were also cleaved by trypsin *i.e.* in chicken erythrocyte histone 45 [24] and in carbonic anhydrase [25]. The Lys-Asp bond was readily cleaved in the lectin favin [26].

An interesting successful assay was reported to use Sepharose-linked trypsin for a limited cleavage : a fragment of colipase containing 94 residues was split at one single arginine [27,28].

The limited cleavage by trypsin after chemical modification of different residues were not included in this review ; the discussion will be restrained to the existing ways how to obtain the same results by an enzyme without chemical substitution.

There are many proteinases splitting specifically at the carboxyl side of arginines. Among them the use of clostripain in sequence studies is increasing in the last few years and this for three reasons : as far, a non-specific cleavage was never observed, secondly, the C-terminal arginine improves the properties on the sequenator [29], and third, clostripain remains active in 6M urea for 20 hrs, which allows good degradation of otherwise insoluble proteins [30]. We have obtained with highly active α-clostripain cleavages exclusively at arginines in ribonuclease and a specific cleavage at one single arginine in parvalbumin, which otherwise contains 14 lysines [5] ; the same specific cleavages were observed and used in sequence determination of the glycoprotein-binding receptor [31]. On shorter time of hydrolysis, 3 out of 10 arginines were cleaved in allophycocyanine [30] and 3 out of 14 in the retinol-binding protein [32]. In both cases all lysine bonds remained intact. Commercial clostripain is now readily available.

Thrombin seems to be another arginine specific enzyme, under conditions, that it was carefully purified [23]. In fibrinogen it splits most rapidly an Arg-Gly bond [23] and in a CNBr fragment of H-2K^{b} major histocompatibility complex alloantigen it cleaved two out of four arginines and no lysines [33].

Recent studies show that the submaxillary gland proteinase, now commercially available, is much more specific
in its action than was supposed [34] : it cleaved in phos-
phatidylcholine-exchange protein only one Arg-Glu bond [35]
and in fibrinogen one Arg-Arg bond [36].

As regards specific cleavage at lysine bonds, the choice
of proteinases is much more restricted. Plasmin splitted in
fibrinogen apart of lysine bonds also arginine bonds [23].
Armillaria mellea proteinase which preferentially cleaves
at the amino group of lysine [37] is restricted in this
action by Glu, Asp or Asn on its carboxyl side. This enzyme
is not devoid of activity on arginine. In aspartate amino
transferase the enzyme cleaved apart of lysine residues
three Arg-Leu (Ile) bonds [38]. Myxobacter Al-1 proteinase
has also affinity for lysine residues [37]. In contrast to
the previously observed specificity for bonds on amino
groups of lysine, the recent work on the sequence of *p*-hy-
droxybenzoate hydroxylase [39] seems to indicate that the
enzyme cleaves bonds on the carboxyl of lysine, with minor
cleavages at the Tyr, Ala and Gln residues. A very selective cleavage at one single Lys-Leu bond was obtained in
human fibrinogen using a thrombin-like enzyme from *Agkis-
trodon contortrix* [36].

The choice in search for an enzyme which would cleave
at dicarboxylic acid residues is simple : most people use
only one, the *Staphylococcus aureus* protease which cleaves
preferentially at the carboxyl of glutamic acid [40].
Everybody agrees that it cleaves that way, but for some
reasons which are not yet clear, some proteins were cleaved
exclusively at glutamic residues whereas in some others was
also obtained important number of fragments with C-terminal
aspartic acid :

Cleavages at Glu- only :

        Ribosomal protein S7 [57], L23 [58], EL12 [59],
        Retinol binding protein [32], calcium binding protein
        [60], α1 (III) collagen [61], β2 microglobulin [62].

Additional cleavages at Asp- :

   1 X   Ribosomal S3 protein [59], phosphatidylcholine-exch.
         protein [35]

   2 X   Fibrinogen α chain [63], elongation factor TU [64],
         alligator myoglobin [52]

3 X   Immunoglobulin α1 chain [65], alcohol dehydrogenase
        [66]

4 X   Yeast enolase [19]

7 X   Lens lectin [41]

    In lens lectin the enzyme cleaved even more at the as-
partyl then at the glutamyl residue [41]. It is difficult
to the observer to find out, how far this is due to the
different sequences or to different experimental conditions.
Only in rare cases atypical cleavages were determined, like
the bond Tyr- in aubergine trypsin inhibitor [42], Tyr- and
Asn- in phosphatidylcholine exchange protein [35], and a
Ser- bond in fibrinogen [22]. This unusual cleavage is explained that possibly this serine is phosphorylated. Some
glutamic acid bonds seem to be particularly resistant to
the Staphylococcal proteinase, like Glu-Asn [42] or Glu-Pro
[35] ; as regards bonds GluGlu and GluAsp, observations
differ from one protein to other [35,43,44].

ENZYMIC CLEAVAGE AT NEUTRAL RESIDUES

    The specificity of enzymes which are usually used to
cleaved at neutral amino acid residues is in average much
broader than of enzymes splitting next to charged residues.
Their action can be better compared on basis of fragments
which were obtained from individual proteins than on basis
of kinetic studies with synthetic substrates. Fig. 2 represents a comparison of the average values of relative
rates of cleavage by chymotrypsin, thermolysin and pepsin,
computed from results of sequence studies published in last
two years. Taking in account the relative occurence of
different amino acid residues in proteins, it can be concluded that the action of chymotrypsin is slighthy more specific than that of thermolysin, whereas the action of
pepsin is definitely less advantageous. This of course can
vary from one individual protein to another. Bonds involving proline are in average more resistant ; sometimes are
observed on contrary neat cleavages at less usual residues
like the bond Glu-CMCys in ferredoxin by thermolysin [45].

    Only sporadically are used in sequence work other pro-
teinases of less pronounced specificity. Some examples

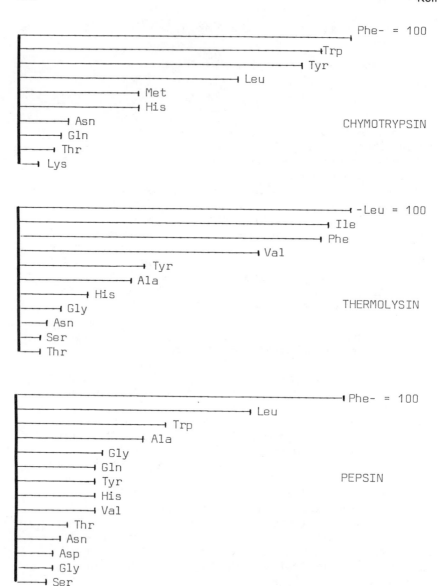

Fig. 2

Average values of relative rates of
cleavage by chymotrypsin, thermolysin and pepsin

will be mentioned here : cathepsin D degraded preferentially a Phe-Phe bond in myeline basic protein [46], papain
splitted a Gly-Leu bond in hypothalamic substance P [47],
elastase degraded histone H5 [24] at residues containing
short lateral chains. Proteinase K splitted in the nucleolar nonhistone protein C 23 preferentially bonds Met-Glu and
Ala-Ser P [48]. An enzyme from porcine pancreas, close to
chymotrypsin, proteinase E cleaved insulin B chain and ribonuclease at carboxyles of Val, Ala, Ser and Thr [49].

Quite interesting new kind of specificity was observed
with the new low molecular weight protease from *Astacus,*
which as mentioned above differs markedly from all other
groups of proteinases : it cleaves preferentially at the
amino group of -Gly, -Ala, -Ser and -Thr. In chain A of
oxidized insulin only one bond was cleaved [1], and in tubulin the enzyme acted preferentially on ⁺Ala [50].

The observation, that certain cathepsins act also as
dipeptidyl-peptidases, can be of interest as complementary
in sequence work : cathepsin B was found to liberate
C-terminal dipeptides sequentially from glucagon and aldo-
lase [51], a preparation of cathepsin C was used to degrade
to dipeptides fragments of alligator myoglobin [52].

                   TOWARDS A HIGHER SPECIFICITY

The search for a higher specificity can progress it two
ways : either by discovery of proteinases which cleave next
to a residue poorly represented in proteins or by better
knowledge of the specificity of proteinases with extended
substrate-binding site. The average per cent of arginine in
proteins is 4,9, whereas that of histidine only 2,0, of
methionine 1,7 and of tryptophan 1,3. The popularity of
chemical cleavages next to these residues (CNBr, skatol) is
well founded. However, corresponding enzymatic means are
lacking.

Many proteinases with extended substrate binding site
exist, but only very few of them were systematically assayed
on polypeptides except their natural substrate. With the
proteinases of the blood clotting cascade, snake proteinases,
kallikrein, enterokinase, acrosin, urokinase, post-proline
endopeptidase, chymosin etc... agreeable surprises are still

possible. Cathepsin D is a protease with extended substrate
binding cite which orients the enzyme to cleave only a
restricted number of bonds in certain proteins. Thus in
denatured collagen chain α1 (I) it cleaves only at one
Leu-Ser bond, in chain α2 (I) only at one Phe-Leu bond
(Fig. 3) [67]. In both adjacent sequences is a leucine in
position P3' ; however a sequence very close to the latter
one is resistant. Brain cathepsin D cleaves in myeline
basic protein also one bond with a leucine and another with
an isoleucine in position P3', but still more rapidly a
segment which does not follows this rule [46].

The influence of amino acids adjacent to the susceptible
bond is much more clear in the action of streptococcal
proteinase [68] : the chief requirement for hydrolysis is
the presence of a bulky side chain on the amino acid adjacent
on the amino-terminal side to the cleaved dipeptide structure
(Fig. 3).

CATHEPSIN D

COLLAGEN α1 (I)	-Tyr-Asp-Leu▼Ser-Phe-LEU- +
COLLAGEN α2 (I)	-Hyp-Gly-Phe-Leu-Gly-LEU- +
COLLAGEN α1 (I)	-Hyp-Gly-Glu-Ala-Gly-LEU- -
MYELINE PROTEIN	-Gly-Ile-Leu-Asp-Ser-LEU- +
	-Phe-Lys-Asn-ILE- +
	-Gly-Arg-Phe-Phe-Gly-Ser- ++
	-Phe-Lys-Asn-Ile-Val-Thr- +

STREPTOCOCCAL PROTEINASE

INSULIN	-VAL-Asn▼Gln-His- ++
B chain	-VAL-Glu-Ala-Leu- ++
	-TYR-Leu-Val-CmC- ++
	-PHE-Phe-Tyr-Thr- ++

Fig. 3

Specificity of cathepsin D and of streptococcal proteinase

Collagenases will serve as example of enzymes whose specificity has been studied exclusively on their natural protein substrate-collagen. As a matter of fact, collagenases were defined as enzymes 1. Cleaving native collagen in its helicoidal structure and 2. without action on any other protein. Achromobacter collagenase falls well within the first part of the definition. It cleaves the same segment of native collagen [11] as vertebrate "true" collagenases do [9] with bond specificity $\downarrow$Gly-Pro-. However, a negligible, but still detectable residual caseinolytic activity remained in the purest sample of the enzyme. Because in the structure of casein are present several short sequences resembling to those in collagen, the sequential study was undertaken on the action of Achromobacter collagenase on pure β-casein. It revealed that this protein is cleaved by the enzyme at four sites of common character X$\downarrow$Gly-Pro and X$\downarrow$Ala-Pro [53] . The same bonds were cleaved during the autolysis of the enzyme [54]. A very limited cleavage was also observed in the λ and κ chains of Bence-Jones protein [55] and in a fragment of 59 residues from fibrinogen, where only one bond X$\downarrow$Gly-Thr was cleaved [36].

The sequence Pro-X$\downarrow$Gly-Pro which is susceptible to be cleaved by Achromobacter collagenase was found only in few protein structures as far determined. The enzyme can therefore not only be useful to cleave some proteins into large fragments but also in genetic engineering : the sequence Pro-X-Gly-Pro mounted in the fused polypeptide between the bacterial moiety and the desired productive sequence could allow their separation by a highly selective splitting with Achromobacter collagenase at this site [56].

REFERENCES

1. Zwilling, R., Dörsam, H., Torff, H.-J. and Rödl, J. (1981) FEBS Letters 127, 75-78.
2. Gilles, A.-M. and Keil, B. prepared for publication.
3. Barrett, A.J. (1981) private communication.
4. Sapolsky, A.J. and Wolssner, Jr. J.F. (1972) J. Biol. Chem. 247, 2069-2076.
5. Gilles, A.-M., Imhoff, J.-M. and Keil, B. (1979) J. Biol. Chem. 254, 1462-1468.
6. Véron, M., Falcoz-Kelly, F. and Cohen, G.N. (1972) Eur. J. Biochem. 28, 520-527.
7. Putnam, F.W., Yu-Sheng, V.L. and Low, T.L.K. (1979) J. Biol. Chem. 254, 2865-2874.

8. Plant, A.G., Gilbert, J.V., Artenstein, M.S. and Capra, J.D. (1975) Science 190, 1103-1105.
9. Harper, E., Bloch, K.J. and Gross, J. (1971) Biochemistry 10, 3035-3041.
10. Dixit, S.N., Mainardi, C.L., Seyer, J.M. and Kang, A.H. (1979) Biochemistry 18, 5416-5422.
11. Lecroisey, A. and Keil, B. (1979) Biochem. J. 179, 53-58.
12. Lecroisey, A. and Keil, B. manuscript in preparation.
13. Miller, E.J., Finch, J.E., Chung, E. and Butler, W.T. (1976) Arch. Bioch. Bioph. 173, 631-637.
14. Sage, H., Pritzl, P. and Bornstein, P. (1981) Biochemistry 20, 3778-3784.
15. Horlein, D., Fietzek, P.P., Wachter, E., Lapiere, C.M. and Kuhn, K. (1979) Eur. J. Biochem. 99, 31-38.
16. Dwulet, F.E. and Putnam, F.W. (1981) Proc. Nat. Acad. Sci. USA 78, 790-794.
17. Gullick, W.J., Head, E.J. and Wood, E.J. (1981) Biochem. J. 197, 23-29.
18. Chung Wong, R.S., Hofmann, T. and Bennick, A. (1979) J. Biol. Chem. 254, 4800-4808.
19. Chin, C.C., Brewer, J.M., Eckard, E. and Wold, F. (1981) J. Biol. Chem. 256, 1370-1376.
20. Keil-Dlouha, V., Zylber, N., Imhoff, J.-M., Tong, N.T. and Keil, B. (1971) FEBS Letters 16, 291-295.
21. Cunningham, B.A., Wang, J.L., Berggard, I. and Peterson, P. (1973) Biochemistry 12, 4811-4821.
22. Strong, D.D., Watt, K.W.K., Cottrell, B.A. and Doolittle, R.F. (1979) Biochemistry 18, 5399-5404.
23. Hessel, B., Makino, M., Iwanaga, S. and Blombeck, B. (1979) Eur. J. Biochem. 98, 521-534.
24. Briand, G., Kniecik, D., Sautière, P., Wouters, D., Borie-Loy, O., Biserte, G., Mazen, A. and Champagne, M. (1980) FEBS Letters 112, 147-151.
25. Nyman, P.O., Strid, L. and Vestermark, G. (1966) Biochim. Biophys. Acta 122, 554-556.
26. Hemperly, J.J., Hopp, T.P., Becker, J.W. and Cunningham, B.A. (1979) J. Biol. Chem. 254, 6803-6810.
27. Bonicel, J., Couchoud, P., Fogglizzo, E., Desnuelle, P. and Chapus, C. (1981) Biochim. Biophys. Acta 669, 39-45.
28. Chapus, C., Desnuelle, P. and Fogglizzo, E. (1981) Eur. J. Biochem. 115, 99-105.
29. Pfletschinger, J. and Braunitzer, G. (1980) Z. Physiol. Chem. 301, 925-931.
30. Sidler, W., Gysi, J., Isker, E. and Zuber, H. (1981) Z. Physiol. Chem. 362, 611-628.

31. Dirckamer, K. (1981) J. Biol. Chem. 256, 5827-5839.
32. Rask, L., Anundi, H. and Peterson, P.A. (1979)FEBS
    Letters 104, 558.
33. Uehara, H., Ewenstein, B.M., Martinko, J.M., Nathenson,
    S.G., Coligan, J.E. and Kindt, T.J. (1980) Biochemistry
    19, 306-315.
34. Schenkein, I., Levy, M., Franklin, E.C. and Frangione, B.
    (1977) Arch. Biochem. Biophys. 182, 64-70.
35. Moonen, P., Akeroyd, R., Westerman, J., Puijk, W.C.,
    Smits, P. and Wirtz, K.W.A. (1980) Eur. J. Biochem. 106,
    279-290.
36. Henschen, A., Lottspeich, F. and Hessel, B. (1979) Z.
    Physiol. Chem. 360, 1951-1956.
37. Wingard, M., Matsueda, G. and Wolfe, R.S. (1972) J.
    Bacteriol. 112, 940-949.
38. Doonan, S., Doonan, J.H., Hanford, R., Vernon, C.A.,
    Walker, J.M., Airoldi, L.P., Bossa, F., Barra, D.,
    Carloni, M., Fasella, P. and Riva, F. (1975) Biochem. J.
    149, 497-506.
39. Weijer, W.J., Jekel, P.A.and Beintema, J.J. (1981)
    private communication.
40. Drapeau, G.R., Boily, Y. and Houmard, J. (1972) J. Biol.
    Chem. 247, 6720-6726.
41. Foriers, A., Lebrun, E., Rapenbusch, R.V., de Neve, R.
    and Strossberg, A.D. (1981) J. Biol. Chem. 256, 5550-5560.
42. Richardson, M. (1979) FEBS Letters 104, 322-326.
43. Bianchetta, J.D., Bidaut, J., Guidoni, A.A., Bonicel, J.J.
    and Rovery, M. (1979) Eur. J. Biochem. 97, 395-405.
44. Wooton, J.C., Barron, A.J. and Fincham, J.R.S. (1975)
    Biochem. J. 149, 749-755.
45. Takruri, J., and Boulter, D. (1980) Biochem. J. 185,
    239-243.
46. Whitaker, J.N. and Seyer, J.M. (1979) J. Biol. Chem. 254,
    6956-6963.
47. Carraway, R., Leeman, S.E. (1979) J. Biol. Chem. 254,
    2944-2945.
48. Mamrack, M.D., Olson, M.O.J. and Busch, H. (1979)
    Biochemistry, 18, 3381-3386.
49. Kobayashi, R., Kobayashi, Y. and Hirs C.N.W. (1981)
    J. Biol. Chem. 256, 2460-2465.
50. Ponstingl, H. et al. (1981) Proc. Nat. Acad. Sci. USA in
    press.
51. Aronson, N.N., Jr., and Barrett, A.J. (1978) Biochem. J.
    171, 759-765.
52. Dene, H., Sazy, J., Goodman, M. and Romero-Herrera, A.E.
    (1980) Biochim. Biophys. Acta 624, 397-408.

53. Gilles, A.M. and Keil, B. (1976) FEBS Letters 65,369-372.
54. Keil-Dlouha, V. (1976) Biochim. Biophys. Acta 429, 239-251.
55. Coletti-Previero, M.-A. , private communication.
56. Töpert, M., private communication.
57. Reinbolt, J., Tritsch, D., and Wittmann-Liebold, B. (1979) Biochimie, 61, 501-522.
58. Wittmann-Liebold, B. and Grener, B. (1979) FEBS Letters 108, 69-74.
59. Amons, R., Pluijms, W. and Moller, W. (1979) FEBS Letters 104, 85-89.
60. Hofmann, T., Kawakami, M., Hitchman, A.J.W., Harrison, J.E. and Dorrington, K.J. (1979) Canad. J. Biochem. 57, 737-748.
61. Dewes, H., Pfietzek, P.P. and Kühn, K. (1979) Z. Physiol. Chem. 260, 821-832.
62. Gates, F.T. III, Coligan, J.E. and Kindt, T.J. (1979) Biochemistry 18, 2267-2272.
63. Watt, K.W.K., Cottrell, B.A., Strong, D.D. and Doolittle, R.F. (1979) Biochemistry 18, 5410-5416.
64. Jones, M.D., Petersen, T.E., Nielsen, K.M., Magnusson,S., Sottrup-Jensen, L., Gausing, K. and Clark, B.F.C. (1980) Eur. J. Biochem. 108, 507-526.
65. Infante, A.J. and Putnam, F.W. (1979) J. Biol. Chem. 254, 9006-9016.
66. Thatcher, D.R. (1980) Biochem. J. 187, 875-883.
67. Scott, P.G. and Pearson, H. (1981) Eur. J. Biochem. 114, 59-62.
68. Gerwin, B.I., Stein, W.H. and Moore, S. (1966) J. Biol. Chem. 241, 3331-3334.

UNUSUAL ENZYMATIC CLEAVAGE AND PREPARATIVE PEPTIDE SEPARATION BY HIGH PRESSURE LIQUID CHROMATOGRAPHY IN THE ESTABLISHMENT OF THE AMINO ACID SEQUENCE OF TUBULIN

Herwig Ponstingl, Erika Krauhs, Melvyn Little,

Wolfgang Ade and Roland Weber

Institute of Cell and Tumor Biology, German Cancer Research Center, D-6900 Heidelberg, Federal Republic of Germany.

In establishing the complete amino acid sequences of α-tubulin and ß-tubulin (1,2), both of M_r 50,000, we employed two proteases which, to our knowledge, have never before been used to generate peptides for the elucidation of primary structures. One is an endopeptidase from the digestive fluid of the crayfish Astacus fluviatilis (E.C. 3.4.99.6). It has a single polypeptide chain with an apparent M_r of only 11,000, whose NH_2-terminal sequence bears no resemblance to that of any other known protease (3). As it is not blocked by known inhibitors, it seems that this enzyme belongs to none of the established families of proteases. It preferentially cleaves at the amino side of small uncharged residues (4).

The other enzyme was recently isolated by Drapeau from a mutant of Pseudomonas fragi. It is a metallo-protease specific for the peptide bonds at the amino side of aspartic and cysteic acid (5).

The peptides resulting from digestion of tubulin subunits with these and several commonly used enzymes were purified by reversed phase high pressure liquid chromatography. Apart from its high speed and powerful resolution for fragments of a M_r up to 40,000, the method practically eliminated the soluble aggregates that required 8M urea in all previous separation steps.

MATERIALS AND METHODS

Enzymatic digestion of the tubulin chains with the
Astacus protease, donated by R. Zwilling, Heidelberg, was
done at pH 8.0 (0.1 M ammonium bicarbonate adjusted with
dilute ammonium hydroxide) and 20°C for 4 hr with 50 mg
protein at a concentration of 1 mg/ml and an enzyme : substrate ratio of 1:50 (w/w). The protease from a mutant of
Pseudomonas fragi was a gift from G. Drapeau, Montreal.
It was used to digest α-tubulin in 0.01 M ammonium bi-
carbonate/2 M urea for 24 hr at 37°C and an enzyme : substrate ratio of 1:100 (w/w). Overlapping fragments were
obtained by digestion with trypsin, thrombin, chymotrypsin,
cyanogen bromide and with proteases from Staphylococcus
aureus and mouse submaxillary glands.

The digests were fractionated by gel filtration in 8 M
urea containing 0.1 M ammonium bicarbonate on Sephadex G50
superfine and desalted in 0.1 M ammonium bicarbonate on
Sephadex G10.

Peptide separation by reversed phase high pressure
liquid chromatography. For fractions that had entered
Sephadex G50 a Zorbax C-8 column, 250 x 4.6 mm, was used
in a DuPont 850 liquid chromatograph equipped with a
temperature controlled column compartment set at 40°C, a
septumless injection system with a 1.5 ml sample loop and
a UV-spectrophotometer set at 220 nm. Buffer A was 0.05 M
ammonium bicarbonate, adjusted to pH 7.5 with acetic acid,
buffer B contained 40 % A and 60 % acetonitrile (v/v). The
buffer reservoirs were continuously sparged with a stream
of dispersed helium bubbles to remove dissolved air and
carbon dioxide. The buffer flow rate was 1.5 ml/min.

Fractions that had been excluded from Sephadex G50
were chromatographed under the same conditions, except that
Buffer B contained 60 % n-propanol, the flow rate was 0.7
ml/min and the peptides were detected at 220 nm or 280 nm.
Fragments with a M_r > 10,000 were chromatographed under the
latter conditions on a Zorbax TMS column, 250 x 4.6 mm.

Sequence determination. Automated sequencing was performed on a Beckman 890C-sequencer using 0.1 M quadrol in
a program with single cleavage. To reduce losses, 3 mg Polybrene was degraded together with 200 μg glycylglycine for
3 cycles before adding 10 to 50 nanomoles of the peptide.
PTH-norleucine was added to each fraction as internal
standard. Thiazolinone derivatives were converted to the

corresponding phenylthiohydantoins in 1 M HCl and identified quantitatively by reverse phase high pressure liquid
chromatography with a linear gradient of 16 to 44 % of
acetonitrile in 0.01 M sodium acetate pH 4.50 at 62°C on a
DuPont Zorbax ODS column 250 x 4.6 mm (6).

RESULTS AND DISCUSSION

The amounts of tubulin subunits available for
sequence studies were somewhat small for secondary
cleavage of larger fragments. As we wanted to generate new
sets of primary fragments with unusual cleavage sites, we
checked the patterns of tubulin digests resulting from
various proteases by gel electrophoresis and by end group
determination. Two enzymes proved particularly valuable
and were used preparatively.

The low molecular mass protease from Astacus did not
attack native tubulin. From reduced, alkylated, and
separated α- and ß-tubulin, however, we isolated a total
of 96 peptides, all soluble at pH 8.0 in buffers of low
ionic strength. They partially overlapped and from the
completed sequences it was evident that they had been
derived from 71 cleavage sites in a total of 895 residues
in α and ß tubulin. These fragments varied from 6 to 61
residues with an average length of about 20 residues, the
sequences derived from them cover 514 of the 895 residues.

The major aminoterminal residues were Ala, Thr, Ser,
Gly and Val. From table 1, however, it can be seen that
these residues at the cleaved bonds comprise only a
fraction - in summary one fifth - of their total number in
tubulin. In addition, 10 % of the peptides had other amino
termini. The residues preceding the cleavage sites did not
appear to have any influence on digestion.

Most of the cleavage pattern may be explained by
assuming that the cleft at the active site of the Astacus
protease sterically restricts the residue on the amino side
of the hydrolyzed bond to the short uncharged side chains,
Val being the largest one that can be accomodated.

Furthermore, prediction of the secondary structure of
tubulin according to the method of Chou and Fasman (7) indicated either a reverse turn conformation or an intrinsic
reverse turn potential masked by larger helices or ß-sheets
around 90 % of the cleavage sites. We have speculated (4)

Table 1   Aminoterminal residues in peptides from α- and
           ß-tubulin generated by the Astacus protease

	Number of residues in α- and ß-tubulin	Number of times this residue has been found amino-terminal in the generated peptides	Percentage of this residue found at cleavage sites
CM-Cys	20	0	0
Asp	53	0	0
Asn	39	1	3
Thr	59	14	24
Ser	50	11	22
Glu	73	0	0
Gln	38	1	3
Pro	40	0	0
Gly	72	11	15
Ala	65	21	32
Val	65	7	11
Met	28	1	4
Ile	45	1	2
Leu	63	2	3
Tyr	34	0	0
Phe	43	0	0
His	23	1	4
Lys	34	0	0
Trp	8	0	0
Arg	43	0	0
Total	895	71	

that a narrow cleft at the active site may be responsible
for the limitations in size of side chains and conformations
of the peptide backbone on the sites for cleavage. This is
supported by the observation that only denatured tubulin can
be digested with the Astacus protease.

Digestion of α-tubulin with the Pseudomonas fragi
protease which is highly specific for aspartic and cysteic
acid required denaturation of the polypeptide chain in 2 M
urea. As we had alkylated all cysteines for the separation
of subunits, hydrolysis was restricted to aspartyl residues.
In spite of incubation for 24 hrs, only 15 out of 27
aspartic acid residues gave rise to new peptides. It was

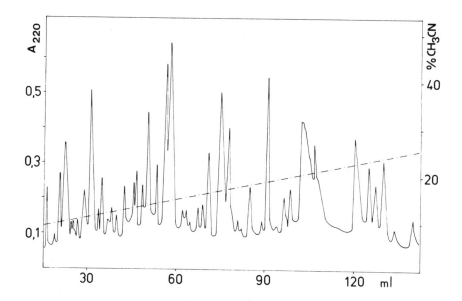

Fig. 1. Separation of peptides 15-20 residues long, derived
from α-tubulin by digestion with Astacus protease, on a
Zorbax C-8 column in 0.05M ammonium bicarbonate pH 7.5
with an acetonitrile gradient.

not evident from the sequence patterns around expected and
observed cleavage sites, why a considerable number of bonds
had been resistant to proteolytic attack. However, as a
clearcut pattern was achieved only in 2 M urea, we assume
features of the secondary structure to be responsible.

Of the resulting peptides, all soluble in 0.1 M
ammonium bicarbonate, 13 were purified by HPLC, spanning
90 % of the sequence and ranging from four to 88 residues
in length. Their unusual starting points greatly facilitated
overlapping of peptide sets obtained with other proteases.

In our initial attempts to fractionate tubulin digests,
soluble peptide aggregates presented a major problem. The
addition of 8 M urea in most purification steps eliminated
this difficulty, but caused partial blockage of amino groups
by cyanate from decomposing urea. The application of high
pressure liquid chromatography techniques, however, made
further use of urea unnecessary, as the organic solvent

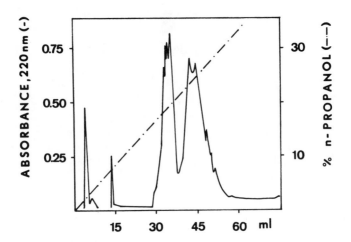

Fig. 2. Separation of two large thrombic peptides from ß-
tubulin on a Zorbax C-8 column in 0.05M ammonium bicarbonate pH 7.5 with a linear gradient of n-propanol, flow
rate 0.7 ml/min. The first main peak has a M_r of 7,000, the
second 12,000, heterogeneity is due to sequence variants
and to partially oxidized Trp and Met residues.

effectively reduced ionic interactions, and the high resolution enabled larger amounts of pure fragments to be
obtained in fewer steps.

     Furthermore, the high resolution helped to establish
sequence microheterogeneity. We had previously found heterogeneous PTH's in a single Edman degradation step. Only with
HPLC, however, was it possible to support these findings
by amino acid analysis of separated variants of the same
peptide and to link the exchanges by sequence determination
with homogeneous material.

     As our tubulin sequences represent the first members
of a new protein class with no known homologs, it was not
possible to predict which peptides to expect and which
methods would be most successful. To reduce the amount and
number of peptides to be separated in a single run, we retained one initial fractionation step in 8M urea on
Sephadex G50 superfine. After desalting, each dried
fraction was taken up in 1 to 1.5 ml of 0.05 M ammonium bi-

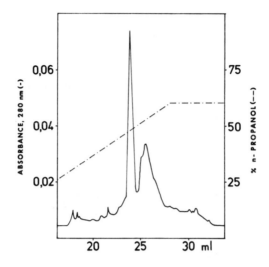

Fig. 3. Separation of two homologous fragments of tubulin, each having a M_r of 40,000, on a Zorbax-TMS column in 0.05 M ammonium bicarbonate pH 7.5 with a gradient of n-propanol.

carbonate and injected via a large sample loop. To avoid solubility problems, no attempt was made to inject more concentrated samples in smaller volumes, which might have increased resolution. It was, therefore, sometimes necessary to make repetitive runs with small volumes of the same sample.

Fig. 1 gives an example of a separation of peptides generated from 100 nanomoles of α-tubulin by the Astacus protease. pH 7.5 was selected because most tubulin peptides are only soluble in alkaline media. On the other hand, the column matrix gradually breaks down at a pH of more than 7.5. In general, shallow gradients proved to be best for the separation of small peptides and steeper gradients for large fragments.

Peptides eluted in the exclusion volume of Sephadex G50 usually required n-propanol as a more potent eluant. Its higher viscosity implied lower flow rates. The two large thrombic peptides (M_r 7,000 and 12,000) from α-tubulin, shown in the chromatogram of Fig. 2, were separated under these conditions.

Taking advantage of the low hydrophobicity of TMS-columns, still larger polypeptides may be separated. Fig. 3 shows the separation of two homologous tubulin fragments of equal size (M_r 40,000) and very similar charge, which were not separated by either gel filtration or ion exchange chromatography.

In total, we have separated more than 200 fragments of α and ß-tubulin. As this is a new family of proteins, we were not able to optimize our procedure by analogy with methods used for other proteins. The digests with various enzymes have been fractionated in quantities up to 700 nanomoles and molecular masses up to 40,000. Thus, the method appears to be as generally applicable to mixtures of unknown peptides as any of the previously used separation procedures. Its main advantages are separation on the basis of side chain hydrophobicity, high resolution, speed and reduction of aggregates.

Acknowledgements

We wish to thank Mr. Jürgen Kretschmer and Miss Herta Scherer for their skilful technical assistance and J.D. Capra, Dallas, for suggestions in the initial steps of HPLC. This work was supported by the Deutsche Forschungsgemeinschaft.

References

1. Ponstingl, H., Krauhs, E., Little, M. & Kempf, T. (1981) Proc.Natl.Acad.Sci. USA 78, 2757-2761.
2. Krauhs, E., Little, M., Kempf. T., Hofer-Warbinek, R., Ade, W. & Ponstingl, H. (1981) Proc.Natl.Acad.Sci. USA 78, 4156-4160.
3. Zwilling, R., Dörsam, H., Torff, H.-J. & Rödl, J. (1981) FEBS Lett. 127, 75-78.
4. Krauhs, E., Dörsam, H., Little, M., Zwilling, R. & Ponstingl, H., Anal.Biochem., in press.
5. Drapeau, G.R. (1980) J.Biol. Chem. 255, 839-840.
6. Zimmerman, C.L., Appella, E. & Pisano, J.J. (1977) Anal.Biochem. 77, 569-573.
7. Chou, P.Y. & Fasman, G.D. (1978) Ann.Rev.Biochem. 47 251-276.

CHEMICAL CLEAVAGE OF PROTEINS

Dr. Mark A. Hermodson

Purdue University

West Lafayette, Indiana 47907

I have been charged with introducing the topic of chemical cleavage of proteins and also with "presenting new data." Since virtually all of the new developments in this area in the last five years concern techniques for cleaving peptides at tryptophanyl residues, I will briefly review the methods available for cleaving at sites other than tryptophan and then discuss tryptophanyl bond cleavage in more detail. My evaluations of the various techniques will be from the distinctly biased perspective of the researcher interested in producing fragments of proteins suitable for sequence analysis by modern automated sequencing equipment, and given the title of this conference, I suspect those biases are widely shared in this audience.

The ideal method for cleaving proteins to generate fragments for automated sequence analysis meets four criteria. First, quantitative cleavage should occur at the specificity sites. Any method which gives much less than quantitative cleavage yields very complex mixtures of peptides having overlapping amino acid sequences, which, due to similarities in size and chemical characteristics, may be impossible to separate. Second, the method should exhibit very high or absolute specificity, both in terms of where peptide cleavage occurs and also with respect to which amino acid residues are modified by the reagent. Low specificity causes too frequent cleavage of the chain and also low yields of cleavage at certain sites, and extensive modification of residues other than those cleaved causes problems in identification of the modified residues during

313

sequence analysis. Third, the amino-termini of the
peptides produced must be susceptible to Edman degradation.
Fourth, the specificity sites should be infrequent enough
to produce peptides 25 to 100 residues long, a practical
range for automated sequence analysis.

CLEAVAGE AT METHIONYL RESIDUES

It is instructive to note in these days of recombinant
DNA technology, when many of our colleagues are suggesting
that protein sequence analysis is in its final stages of
senescence, that the first practical method for chemically
cleaving polypeptides, the cyanogen bromide reaction, was
developed only 20 years ago (1,2). In many ways it is
still the most useful cleavage method for sequence
analysis. The main precaution that needs to be observed
for its use (besides the obvious safety problems) is that
only colorless reagent should be employed. Most preparations of cyanogen bromide develop yellow and then orange
coloration on prolonged storage, and these colored impurities can cause destruction of tyrosyl and tryptophanyl
residues. With good reagent in 70% formic acid (room
temperature, 20-24 hours), most methionyl bonds are cleaved
in near quantitative yield without modification of any
other residue provided cysteine residues are suitably
alkylated. In instances of extreme insolubility of the
polypeptide other acidic solvents may be used, and
guanidine·HCl or sodium dodecyl sulfate may be added to
the mixture as well. Strongly acidic solvents like 75%
trifluoroacetic acid may cause low yields of cleavage at
Asp-Pro bonds which complicates the mixture (3).
Occasional methionyl bonds give less than quantitative
cleavage yields, particularly where the Met is followed by
serine or threonine, but this method generally comes as
close to meeting the four criteria above as anything in our
chemical arsenal.

CLEAVAGE AT CYSTEINYL RESIDUES

Very high yield cleavage at cysteinyl residues can be
accomplished with 2-nitro-5-thiocyanobenzoic acid (4,5,6).
The reagent is highly selective for cysteinyl residues, and
the peptides produced are often of ideal size. Unfortunately, the peptides are blocked by the iminothiazolidine

carboxylyl group and are not susceptible to Edman
degradation. Treatment of the blocked peptides with a
reduced nickel catalyst converts the blocking group in
reasonable yield to alanine and the methionyl residues to
aminobutyrate (7). However, the reagent causes further
cleavage of the peptide chain in many cases, and other side
reactions such as deamidation of Asn and Gln or reduction
of other residues have not been fully characterized. Due
to these problems this method has not been employed to any
extent in sequence analyses except to "crack" a large
polypeptide into two or three smaller pieces and thus
simplify isolation of the cyanogen bromide fragments (for
example).

Reaction of S-methylcysteine residues with cyanogen
bromide to yield O-serine peptides which can be cleaved
with hydroxylamine (8) has not been used to any extent in
peptide chemistry, probably due to the propensity of the
intermediate to undergo β-elimination and produce a blocked
pyruvyl peptide. Thus, even though a method to cleave
peptides at cysteinyl residues would be very useful, no
method for doing so is suitable for sequence analyses.

CLEAVAGE AT PROLYL RESIDUES

Sodium metal in liquid ammonia has been used to cleave
proteins on the amino side of prolyl residues (9,10). The
amino acid preceding the prolyl residue is reduced to a
certain extent in the reaction, but under controlled conditions (-33°, 210 sec., 600 fold excess of sodium) few side
reactions, if any, are observed. There is evidence,
however, that the time, temperature, and sodium concentrations are all critical variables for the success of this
reaction (10). Rigorous exclusion of water from the
reactants is obviously necessary, and the mixture is potentially explosive. Due to these difficulties this method
has found little use in protein sequence studies, but when
suitably desperate, the protein chemist should be aware of
its existence.

CLEAVAGE AT TYROSYL AND HISTIDINYL RESIDUES

N-Bromosuccinimide and similar active halogen compounds oxidize and cleave peptide chains at tryptophanyl,

tyrosyl, and histidinyl bonds, in the order of decreasing
susceptibility (11). The yields of cleavage at any of the
bonds is generally terrible, and specificity is nonexis-
tent. Tryptophan, tyrosine, histidine, methionine, and
cysteine and its derivatives are all modified by these
reagents, most to the extent where they cannot be recognized in subsequent sequence analyses. Even amino groups
can be oxidized by these reagents. The use of this
procedure is not recommended except in situations of
extreme duress.

CLEAVAGE AT ASPARAGINYL-GLYCYL BONDS

The asparaginyl side chain can cyclize on the amino
group of an adjacent glycyl residue with loss of ammonia to
produce a cyclic imide which is susceptible to cleavage by
hydroxylamine (12). Formation of the cyclic imide is
certainly catalyzed by acid and may also be catalyzed by
base. The latter seems likely, since prolonged incubation
of proteins in the 2 $\underline{M}$ hydroxylamine solution at pH 9
results in further cleavage of the chain at asparaginyl
residues followed by amino acids other than glycine. The
yields of cleavage at Asn-Gly bonds tend to be in the
50-80% range, which, if there are a number of cleavage
sites, are low enough to cause real problems in purifying
the mixture of fragments and "overlap" fragments. Fortun-
ately, this bond is rather infrequent in proteins, occurring no more than once in 150 peptide bonds on the average,
so there are generally no more than two such sites in a
given polypeptide. The specificity of the reaction is
excellent; no other residue is affected, and with four hour
incubations the yields of cleavage at sites other than
Asn-Gly are very low while giving optimal cleavage at the
desired site.

This has proven to be a very useful method for amino
acid sequence analysis both for generating large fragments
of proteins and for providing a way around the worst
impediment to automated Edman degradation found within a
polypeptide. The cyclic imide is not susceptible to Edman
degradation, and the hot acid treatment during each cycle
in the sequenator apparently catalyzes imide formation,
since yield drops of 50-75% or more are frequently observed
at Asn-Gly sequences (13). This often makes sequence
analysis for more than five or six residues beyond an

Asn—Gly sequence impossible unless the bond is cleaved with hydroxylamine and a new peptide generated.

CLEAVAGE AT ASPARTYL-PROLYL BONDS

The most acid-sensitive bond in proteins is the aspartyl-prolyl bond. Low to moderate yields of cleavage at that site are frequently observed as a side reaction of the cyanogen bromide or tryptophanyl-bond cleavage procedures, since those reactions are performed in acidic solutions (14). In addition the frequent use of 1-2 $\underline{M}$ formic or acetic acid solutions as gel filtration solvents for peptide separations exposes the peptides to acidic conditions for prolonged periods and can lead to further fragmentation at Asp-Pro bonds. A host of protocols have been devised to exploit the acid susceptibility of this bond (14). The conditions range from four day incubations at 40°C in 10% acetic acid adjusted to pH 2.5, to 70-90% formic acid solutions with high concentrations of guanidine. The wide range of conditions used reflects the fact that the susceptibility of a given Asp-Pro bond to acid cleavage varies dramatically, and the yields of cleavage at particularly resistant sites can be improved somewhat by use of denaturants and more vigorous conditions. Cleavage yields range from less than 30% to 100%. As with cleavage of Asn-Gly bonds, the Asp-Pro bond is rather infrequent in proteins, so it is rare to have more than two sites in a polypeptide, thus making this a practical method in spite of the yield problems. The specificity of the reaction is very good unless quite vigorous conditions are employed. No other residues are modified, but strongly acidic conditions and higher temperatures can lead to low yields of cleavage elsewhere in the peptide. The next most acid-sensitive bonds are those on either side of aspartyl residues, but the conditions needed to obtain high yields of cleavage at Asp are so severe that nothing remotely resembling specificity can be accomplished (15).

CLEAVAGE AT TRYPTOPHANYL BONDS

Prior to 1970 most attempts to cleave polypeptides at tryptophanyl bonds employed N-bromosuccinimide or a similar strong oxidant, and the results invariably showed all the

problems with that reagent (above). M. Z. Atassi developed
a periodate oxidation method in 1967 which cleaved the
tryptophanyl bonds in sperm whale myoglobin in 93% and 20%
yields at tryptophans 7 and 14, respectively (16). The
reagent oxidized the methionyl residues to the sulfoxide
and modified one residue of tyrosine out of the three
total, so the side reactions were minimal. However, the
wide variation in cleavage yields made this method of
little use to sequence analyses.

The most heavily used method for tryptophanyl bond
cleavage during the past ten years employs the method of
Omenn, Fontana, and Anfinsen (17). The reagent,
2-(2-nitrophenylsulfenyl)-3-methyl-3-bromoindolenine, or
BNPS-skatole, is used in several-fold excess over the molar
amount of sensitive residues in the peptide in aqueous
acetic acid solutions. With fresh reagent of high quality,
side reactions are limited to oxidation of methionine and
alkylated cysteine derivatives to the sulfoxides and oxidation of cysteine to cystine. Phenolic compounds are
sometimes added to scavenge reactive contaminants which
can oxidize tyrosyl residues. In addition to the far
higher specificity of BNPS-skatole for tryptophan residues
than that observed for N-bromosuccinimide, the cleavage
yields at most tryptophanyl residues is better, generally
in the 50-70% range. While that cleavage yield is
perfectly acceptable for cleaving one or two sites in a
polypeptide, multiple sites result in a very complex
mixture of peptides with overlapping sequences which are
often difficult to separate. An additional disadvantage to
this reagent is its instability to storage. Free bromine
is apparently generated slowly, the reagent darkens from
light yellow to orange color, and side reactions, particularly at tyrosyl residues, become a problem with use of old
reagent.

Exposure of polypeptides to very large amounts of
cyanogen bromide in a one to one mixture of heptafluorobutyric acid and 88% formic acid cleaves both methionyl and
tryptophanyl bonds in high yields (18). Tyrosyl residues
are apparently brominated, but not cleaved. Cysteine
residues and its derivatives are presumably oxidized, too,
but careful documentation of this is lacking to my knowledge. Cleavage at methionyl residues can be prevented by
mild oxidation to the sulfoxide making the method selective
for tryptophan. Cleavage yields at tryptophan are reported

to be around 80%, but we have evidence that this is quite
variable (W. J. Ray and M. A. Hermodson, unpublished observations). The method has been used to advantage a number
of times, but is not in wide use in sequencing studies,
probably due to its lack of ready specificity.

Mild oxidation of polypeptides in acetic acid/12 N
HCl/DMSO mixtures followed by more vigorous oxidation by
addition of 48% HBr causes cleavage of tryptophanyl bonds
in roughly 60% yields with oxidative side reactions
confined to methionyl and cysteinyl residues (19). The
reaction conditions are strongly acidic, and acid hydrolysis of bonds around aspartyl residues could well occur in
low yields. In addition, the yields of cleavage are lower
than those for the BNPS-skatole or CNBr digestions, so the
method has not been used to any extent in sequence
studies.

In 1979, Walter Mahoney and I discovered that o-iodosobenzoic acid in aqueous acetic acid containing guanidine
hydrochloride gave us very high yields of cleavage at most
tryptophanyl bonds (20). We selected human and bovine
serum albumins and α, β, and γ-chains of human hemoglobin
for test proteins because they have ideally placed tryptophan residues for determining cleavage yields. This turned
out to be fortunate in some respects and unfortunate in
others. Tryptophanyl residues preceding glycyl, threonyl,
glutaminyl, seryl, alanyl, and arginyl residues were found
to cleave in 90% yield or more. Trp-Val and Trp-Ile bonds
consistently gave around 70% cleavage yields, and a single
instance of a bond preceding a pyridylethylcysteine residue
cleaved in 80% yield. Optimal yields were obtained with
5-10 mg/ml of protein and 2 mg reagent per mg protein in
80% acetic acid made 4 M in guanidine hydrochloride. More
than 20 hours in the dark at room temperature were required
for complete cleavage. Under those conditions using our
test proteins, we observed no cleavage at other residues,
and amino acid compositions of treated and untreated
proteins were identical. We did observe, however, that
cyanogen bromide no longer cleaved the methionyl residues,
which, taken with the observation that HCl hydrolysis gave
quantitative recovery of methionine, suggested that methionine was converted to the sulfoxide. Very high levels of
the reagent (50 times optimal) produced methionine sulfone
and essentially destroyed tyrosyl residues, but still did

not cleave our test proteins anywhere except at tryptophan.

We sent a pre-print of our manuscript to Prof. Walsh
at Seattle and received a phone call from Lowell Ericsson
in his research group about two days later. He reported
that he had tried the reagent on sperm whale myoglobin and
that he had obtained complete cleavage at the tryptophanyl
residues and also about 30% cleavage between Tyr_{103}
and Leu_{104}, but no cleavage at the other two tyrosyl
residues. We confirmed his results immediately. Further
experimentation clearly showed that some tyrosyl residues,
but by no means all, were susceptible to oxidation and
cleavage by the reagent under optimal conditions. Bovine
serum albumin has 18 tyrosyl residues including Tyr-Gly,
Tyr-Leu, and Tyr-Phe bonds which cleaved in other proteins,
but none cleaved in albumin. Examination of a number of
preparations of o-iodosobenzoic acid showed highly variable
results with respect to tyrosine modification and cleavage,
so we established a collaboration with Dr. Paul Smith from
Pierce Chemical Co. to produce high quality reagent free of
the contaminant which modified tyrosine (21). The main
result of this investigation was that careful oxidation of
iodobenzoic acid with 90% nitric acid at low temperature
(<15°C) produced a product which gave a very low yield of
cleavage (<10%) at Tyr_{103} in myoglobin while still
giving excellent yields at the tryptophans. Inclusion of
10-20 mole percent of p-cresol in the reaction mixture
completely eliminated tyrosine modification and cleavage
without affecting the reaction at tryptophan. The temperature during the synthetic procedures was absolutely critical for production of good reagent. The best preparation
produced at low temperature was an amorphous powder giving
8% cleavage at Tyr_{103} in myoglobin. An attempt to
purify it further by recrystallization from a large volume
of hot (95°C) water produced beautiful crystals which
looked nicer than any preparation we had ever had, and
which gave a 35% yield of cleavage at Tyr_{103}. Clearly
heating produced a reactive contaminant which modified
tyrosine. We suggested that o-iodoxybenzoic acid was a
likely candidate for such a contaminant for three reasons.
First, we observed that mass spectrometry of our best
preparation of o-iodosobenzoic acid showed only
o-iodobenzoic acid and o-iodoxybenzoic acid in about equi-
molar quantities. This was undoubtedly due to the disproportionation of the o-iodosobenzoic acid on the hot probe

(225°C) of the mass spectrometer. Second, synthetic
o-iodoxybenzoic acid cleaved myoglobin in 70% yield at the
tryptophan residues and more than 50% at Tyr_{103}.
Third, mixing o-iodoxybenzoic acid into a good preparation
of o-iodosobenzoic acid caused increasing tyrosine cleavage
with increasing levels of o-iodoxybenzoic acid. This
effect was obvious at rather low levels of o-iodoxybenzoic
acid.

Some conclusions may be drawn from the above observations. First, very high reagent to protein ratios are
required for good cleavage yields. Two milligrams of
reagent per milligram of protein means that there is
typically a 20 fold excess of reagent present over the
molar quantities of sensitive residues. This means that
trace contaminants can be a problem since they can easily
be present in stoichiometric amounts and still be hard to
detect. It also requires that the medium used be a good
solvent for both the reagent and the protein. The o-iodosobenzoic acid is very sparingly soluble in aqueous solutions of acetic acid. Experiments in which only 80% acetic
acid is used as the solvent are probably meaningless, since
most of the reagent fails to dissolve in that medium. The
inclusion of 4 M guanidine hydrochloride was originally
intended to accomplish solubilization of the reagent and
protein and to denature the protein to make the tryptophan
residues accessible. We did notice some time ago that urea
was not an effective substitute but failed to recognize the
significance of that fact (21). Very recent evidence from
Dr. Angelo Fontana's laboratory neatly explains it by
demonstrating the participation of chloride in the reaction
itself (A. Fontana, personal communication).

Second, the effectiveness of 0.1-0.2 molar equivalents
of p-cresol in preventing tyrosine modification while still
allowing the reaction at tryptophan to proceed in good
yield absolutely demonstrates that the side reaction at
tyrosines is due to a contaminant which is either present
in the reagent preparation or is generated during the reaction in situ. This is so because elementary chemical
kinetic theory requires that if compound A and compound B
react at a certain rate at given concentrations of the
reactants, the reaction rate is entirely independent of any
other constituent in the solution. Thus, if o-iodosobenzoic acid (or an in situ derivative of it which is the
oxidant active on tryptophan residues) were the compound

which also modified tyrosine residues, nothing could be
added to the solution which would effectively prevent the
side reaction without simultaneously preventing the reaction of tryptophan. While the exact chemical nature of the
tyrosine-reactive contaminant is still not rigorously
proven, for the reasons stated above, we still suggest that
o-iodoxybenzoic acid is one likely candidate, particularly
since pre-incubation of the reagent solution with p-cresol
appears to be more effective than adding all the reactants
at once, implying a pre-existing contaminant in the
reagent.

The o-iodosobenzoic acid reagent has gained widespread
use in sequence studies since its introduction two years
ago, and due to the very high cleavage yields and preventable side reactions, it now appears to be the reagent of
choice for cleaving polypeptides at tryptophanyl residues.

REFERENCES

1. Gross, E., and Witkop, B. (1961). J. Am. Chem. Soc. 83, 1510.
2. Gross, E., and Witkop, B. (1962). J. Biol. Chem. 237, 1856.
3. Titani, K., Hermodson, M. A., Ericsson, L. H., Walsh, K. A., and Neurath, H. (1972). Biochemistry 11, 2427.
4. Jacobson, G. R., Schaffer, M. H., Stark, G. R., and Vanaman, T. C. (1973). J. Biol. Chem. 248, 6583.
5. Degani, Y., and Patchornik, A. (1974). Biochemistry 13, 1.
6. Stark, G. R. (1977). Meth. Enzymol. 47, 129.
7. Schaffer, M. H., and Stark, G. R. (1976). Biochem. Biophys. Res. Commun. 71, 1040.
8. Gross, E., and Morell, J. L. (1974). Biochem. Biophys. Res. Commun. 59, 1145.
9. Benisek, W. F., Raftery, M. A., and Cole, R. D. (1967). Biochemistry 6, 3780.
10. Atassi, M. Z., and Singhal, R. P. (1970). Biochemistry 9, 3854.
11. Ramachandran, L. K., and Witkop, B. (1967). Meth. Enzymol. 11, 283.
12. Bornstein, P., and Balian, G. (1977). Meth. Enzymol. 47, 132.
13. Hermodson, M. A., Ericsson, L. H., Titani, K., Neurath, H., and Walsh, K. A. (1972). Biochemistry 11, 4493.
14. Landon, M. (1977). Meth. Enzymol. 47, 145.
15. Schultz, J. (1967). Meth. Enzymol. 11, 283.
16. Atassi, M. Z. (1967). Arch. Biochem. Biophys. 120, 56.
17. Omenn, G. S., Fontana, A., and Anfinsen, C. B. (1970). J. Biol. Chem. 245, 1895.
18. Ozols, J., Gerard, C., and Stachelek, C. (1977). J. Biol. Chem. 252, 5986.
19. Savige, W. E., and Fontana, A. (1977). Meth. Enzymol. 47, 459.
20. Mahoney, W. C., and Hermodson, M. A. (1979). Biochemistry 18, 3810.
21. Mahoney, W. C., Smith, P. K., and Hermodson, M. A. (1981). Biochemistry 20, 443.

PROTEIN FRAGMENTATION WITH o-IODOSOBENZOIC ACID

A REINVESTIGATION

A. Fontana, D. Dalzoppo, C. Grandi and M. Zambonin

Institute of Organic Chemistry, Biopolymer Research Centre, C.N.R., University of Padova, Via Marzolo 1, I-35100 Padova, Italy.

The reagent o-iodosobenzoic acid (IBA)[+] has been reported to cleave specifically tryptophanyl peptide bonds in proteins with 70-100% yields (1). However, the results obtained in a number of laboratories (2-4) indicated that the reagent is not as selective as originally reported, since cleavage occurs also at tyrosine in moderate to high yields. Subsequently, Mahoney et al. (5,6) proposed that o-iodoxybenzoic acid thought to be present as a contaminant in the commercially available reagent is responsible for the observed cleavage at tyrosine. The results of the studies reported herein rule out this proposal and show that IBA, under the experimental conditions of protein fragmentation (80% aqueous acetic acid containing 4M Gdn.HCl), mediates the oxidative chlorination of both the indole nucleus of tryptophan and the phenol ring of tyrosine with concomitant peptide bond cleavage[++].

[+] Abbreviations used: IBA, o-iodosobenzoic acid; BNPS-skatole, 2-(2-nitrophenylsulfenyl)-3-methyl-3-bromoindolenine; NBS, N-bromosuccinimide; DMSO, dimethyl sulfoxide; Gdn.HCl, guanidine hydrochloride.

[++] A more comprehensive report on the fragmentation of peptides and proteins using IBA will be published elsewhere (7).

Purity of o-Iodosobenzoic Acid. The controversial issue
of the purity of IBA (1,5,6) prompted us to develop suitable
analytical methods in order to check unambigously the quality of the reagent employed. The homogeneity of the reagent
commercially available (Pierce Chem. Co.) or synthesized in
our laboratory according to literature (8,9) was established on the basis of several criteria, including elemental
analysis (Anal. Calcd. for $C_7H_5O_3I$: C, 31.8; H, 1.9; I, 48.1.
Found: C, 32.08; H, 1.9; I, 47.6) and determination of its
oxidizing power by iodometry, which is the routine procedure used to standardize the reagent (10). This method of analysis, based on the reaction Ar-IO + $2H^+$ + $2I^-$ $\longrightarrow$ Ar-I +
+ H_2O + I_2, would permit unambigously to detect components
of higher oxidation state than IBA such as o-iodoxybenzoic
acid, if present. Iodometry indicated 99-100% purity of the
samples of IBA used throughout this study.
     Thin layer and column chromatography were also employed.
TLC on a cellulose plate with methanol:acetic acid:water
(6:3:1, v/v) as eluent gave a single spot for IBA (R_F 0.82),
well separated from o-iodoxybenzoic acid (R_F 0.54). Ion exchange chromatography on a DEAE-cellulose column eluted with
a gradient (0.01-0.5M) of Tris-acetate buffer, pH 9.0, allows
complete separation of IBA and o-iodoxybenzoic acid. As seen
in Fig. 1, IBA is not contaminated by the iodoxy-derivative,
whereas this last compound synthesized according to the method reported in literature (11) is slightly contaminated by
IBA. A homogeneous sample of o-iodoxybenzoic acid was prepared by preparative chromatography on a Sephadex G-10 column
eluted with 10% acetic acid, which also allows complete separation of IBA from the iodoxy-derivative.

Modification of Amino Acids with IBA. It has been previously reported by us (2) that the reaction of free tryptophan with IBA in 80% aqueous acetic acid, in the absence of
Gdn.HCl, occurred very slowly, if occurred at all. On the
other hand, in presence of halides (Gdn.HCl, HCl, NaCl),tryp
tophan disappeared rapidly from the reaction mixture and
oxindolylalanine (2-hydroxytryptophan) and dioxyndolylalani-
ne diastereoisomers were identified in the reaction mixture
by amino acid analysis (2). The participation of halide
ions in the reaction of tryptophan with IBA was clearly demonstrated by spectrophotometric monitoring of the reaction.

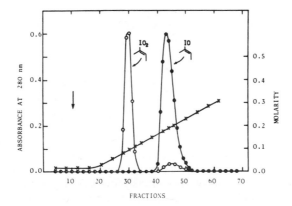

FIGURE 1: Ion exchange chromatography of IBA (●) and o-io-doxybenzoic acid (O). A DEAE-cellulose (DE-52, Whatman) column (2x5 cm) was equilibrated with 0.01M Tris-acetate buffer, pH 9.0, and eluted with a linear gradient (0.01-0.5M) (x) of the same buffer. Gradient elution was started at the 11th fraction (arrow). Fractions of 2 ml were collected at a flow rate of 20 ml/h and analyzed spectrophotometrically at 280 nm. The compounds ($\sim$2 mg) were applied to the column dissolved in 2 ml of Tris buffer.

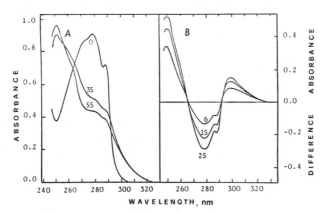

FIGURE 2: Spectrophotometric monitoring of the reaction of tryptophan with IBA. (A) Absorpton spectra of the reaction mixture of tryptophan (5.5×10^{-5}M) treated with IBA (11.5×10^{-5}M) in 80% acetic acid in presence of NaCl (8.5×10^{-5}M) at 22°C. (B) Difference spectra obtained during the course of the reaction. Numbers near the curves indicate time of incubation.

When tryptophan was reacted with IBA under the conditions
reported in the legend to Fig. 2, without NaCl added, the
ultraviolet spectrum of the reaction mixture was unchanged
after 30 min. After addition of NaCl, a relatively fast de
crease in absorption near 280 nm and an increase near 250 nm
(corresponding to oxindole formation) (12) was observed
(Fig. 2). Over a 22-h reaction time, even in the absence of
added halides, some tryptophan (10-20%) was modified, as de
termined from its recovery on the analyzer. This could be
considered as an indication of an intrinsic slow reactivity
of IBA towards the indole nucleus. However, it should be
considered that oxidative halogenation of the indole nucleus
involves halogenation/dehalogenation steps (2,13-15), so
that catalytic amounts of halides present in the solvents
used would be sufficient to cause some modification of tryp
tophan to occur.

    Tyrosine was recovered unchanged after reaction with
6-20 fold molar excess of IBA in 80% acetic acid for 22 h
at room temperature. On the other hand, when halides (Gdn.HCl,
HCl, NaCl, HBr, KBr) were included in the reaction mixture,
3-halo- and 3,5-dihalotyrosine derivatives were identified
as the products of the reaction. The data shown in Table I
indicate that the formation of chlorinated derivatives of
tyrosine by the IBA/Gdn.HCl reagent accounts for the destruc
tion of tyrosine at a relatively low molar ratio of reagent
to amino acid. With higher amounts of reagent it appears
that 3,5-dichlorotyrosine reacts further to give compounds
which are either ninhydrin-negative or are not eluted from
the column of the amino acid analyzer under standard elution
conditions. In fact, it has been previously shown that the
reaction pathway of tyrosine halogenation with NBS involves
two discrete steps, i.e., (i) rapid consumption of two moles
of halogenating agent to give ortho halogenation of the phenolic moiety of tyrosine and (ii) slower consumption of a
third mole of reagent to afford a spirodienolactone via an
intermediate iminolactone (16-18).

    An amino acid mixture as used for calibration of the
amino acid analyzer, containing all common amino acids except for tryptophan and cysteine, was allowed to react with
1-5 μmol of IBA in 80% acetic acid containing 1 μmol of HCl
for 22 h at room temperature. As seen in Table II, even in
the presence of such a low amount of HCl, IBA causes the mo
dification of methionine, cystine, tyrosine and histidine.
Methionine is partly recovered on the analyzer as the sulfo-
xide- and sulfone-derivative, while cystine as cysteic acid.

TABLE I: Reaction of Tyrosine with IBA/Gdn.HCl.[a]

IBA (equiv.)	Recovery (%) [b]		
	Tyr	Cl-Tyr	Cl_2-Tyr
0	100	0	0
0.5	63	38	0
1	24	74	0
2	0	64	31
5	0	2	46
10	0	0	7

[a] Tyrosine (0.5 µmol) was reacted with the indicated amount of IBA in 400 µl of 80% acetic acid containing 4M Gdn.HCl for 22 h at room temperature. Alanine (0.5 µmol) was added to the reaction mixture as an internal standard.

[b] The per cent recovery on the analyzer (short column) was calculated assuming an identical color factor (ninhydrin reaction) for tyrosine and its chloro-derivatives.

It is also seen (Table II) that tyrosine modification can be completely prevented by conducting the reaction in the presence of an excess of p-cresol. Under more drastic reaction conditions than those described in Table II (much higher concentrations of IBA) partial destruction of amino acids to products that no longer produce a color with ninhydrin was observed. This parallels the NBS-degradation of free amino acids described previously (19,20).

In summary, the IBA/Gdn.HCl reagent shows a reactivity towards the amino acid side chains similar to that previously observed with the very reactive reagent NBS (16-20). This reactivity contrasts with that of other milder reagents which have been proposed, subsequently to NBS, for protein fragmentation at tryptophan (BNPS-skatole, tribromocresol, N-chlorosuccinimide, DMSO/haloacid) (2, and references cited therein).

Cleavage of Tryptophanyl and Tyrosyl Peptides. The IBA--mediated peptide bond fission was studied with several

TABLE II: Per Cent Recovery of Oxidation and/or Halogenation Sensitive Amino Acids after Reaction with IBA [a].

Amino acid	1 μmol IBA		5 μmol IBA	
	[b]	[c]	[b]	[c]
Methionine	0	0	0	0
Meth. sulfoxide	9	50	0	traces
Meth. sulfone [d]	41	22	40	30
Half-cystine	traces	62	0	0
Cysteic acid	27	13	43	38
Tyrosine	99	99	0 [e]	98
Histidine	100	99	95	100
Alanine [f]	100	100	100	100

[a] A mixture of all common amino acids (0.25 μmol each), with the exception of tryptophan and cysteine, was incubated in 80% aqueous acetic acid (725 μl) containing 1 μmol HCl at room temperatuture for 22 h with the indicated amount of IBA. Some reactions were carried out in presence of p-cresol. Yields of recovery (%) of amino acids were determined by automatic amino acid analysis.

[b] Without p-cresol added to the reaction mixture.

[c] With 50 μmol of p-cresol added to the reaction mixture.

[d] Since methionine sulfone coelutes with aspartic acid (Jeol amino acid analyzer, model JLC-6AH, single column procedure), its % recovery was calculated by difference, assuming quantitative recovery of aspartic acid.

[e] Peaks of 3-chloro- and 3,5-dichlorotyrosine appeared in the amino acid chromatogram.

[f] All other amino acids were recovered quantitatively.

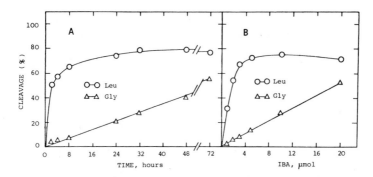

FIGURE 3: Cleavage of tryptophanyl (o) and tyrosyl (Δ) peptide bonds as function of time and addition of IBA. (A) Cleavage was effected by incubating a mixture of Gly-Pro- -Trp-Leu and Gly-Tyr-Gly (2 μmol each) with IBA (30 μmol) in 1 ml of 80% acetic acid/4M Gdn.HCl at room temperature. (B) A mixture of the two peptides (0.4 μmol each) was incubated with increasing amounts (1-20 μmol) of IBA in 500 μl of ace tic acid/Gdn.HCl for 22 h at room temperature. Yields of cleavage were determined from the recovery on the analyzer of leucine (cleavage at tryptophan) and glycine (cleavage at tyrosine).

---

tryptophanyl (Gly-Trp-Gly, Leu-Trp-Leu, Gly-Pro-Trp-Leu, Phe- -Val-Gln-Trp-Leu) and tyrosyl (Gly-Tyr-Gly, Val-Tyr-Val, Leu- -Tyr-Leu, Lys-Tyr-Lys) peptides. It was found that IBA (5-15 equiv.) in 80% acetic acid containing 4M Gdn.HCl cleaved both series of peptides in 25-83% yields, cleavage at tryptophan being generally higher (7). It was also found that NaCl can be used as a source of halide alternative to Gdn.HCl, with com parable yields of peptide bond fission. When the cleavage reactions were conducted without added halides and over a 22 h reaction time,variable, but low (0-19%), yields of clea vage at tryptophan were observed, whereas cleavage at tyro- sine did not occurred.
     Fig. 3 shows the effect of time and equivalents of IBA on the extent of cleavage of the tryptophanyl and tyrosyl pep- tide bond using Gly-Pro-Trp-Leu and Gly-Tyr-Gly as model pep tides. Cleavage of the tryptophanyl peptide bond occurs fas- ter and needs much less equiv. of reagent than that of the tyrosyl peptide bond. In addition, from the data of Fig. 3, it appears that approximately 50% yield of cleavage at trypto- phan occurs quite rapidly in few hours, whereas additional cleavage up to about 80% occurs at a much slower rate.

As expected from the results on the reactivity of IBA
with free amino acids (Table II) and in agreement with the
results reported by Mahoney et al. (6), it was found that
excess p-cresol added to the reaction mixture acts as an efficient scavenger for tyrosine modification and cleavage,
allowing quite selective cleavage at tryptophan to occur. The
scavenger effect of p-cresol was demonstrated also with NBS.
As expected from previous studies (16-18), NBS (6 μmol) in
80% aqueous acetic acid cleaved both Gly-Pro-Trp-Leu and Lys-
-Tyr-Lys (0.5 μmol each) at tryptophan and tyrosine in 40 and
43% yield, respectively. In presence of an excess of p-cresol
cleavage occurred selectively at tryptophan in 38% yield.

Conclusions. The results of this study confirm our earlier proposal that IBA might cleave tryptophanyl peptide bonds
via oxidative halogenation (2). The participation of halide in
the IBA-mediated modification of the indole nucleus was demonstrated in a number of ways. The fact that over a longer reaction time some modification of tryptophan and variable (but
low) cleavage of the tryptophanyl peptide bond occurs even in
the absence of added halide, can well be explained in terms
of presence of catalytic amounts of halide in the solvent mixture employed (cf. above). That the IBA/Gdn.HCl reagent acts
as a halogenating agent was clearly demonstrated by the isolation on the amino acid analyzer of 3-chloro- and 3,5-dichlo-
rotyrosine derivatives. With tyrosine peptides, not a trace of
peptide bond cleavage was observed in the absence of added halide. This clearly shows that cleavage at tyrosine is an inherent property of the IBA/Gdn.HCl reagent and not due to contaminant o-iodoxybenzoic acid (5,6). In fact, the homogeneity of
IBA was established on the basis of a number of analytical criteria, which showed unambigously the purity of the reagent employed.

To explain the mechanism of the IBA-mediated halogenation,
we have previously proposed (2) that halogenating species (including iodobenzene dichloride and chlorine) are formed when
IBA is incubated in aqueous acid solution in presence of halides, as a result of a series of equilibrium reactions. That halides are oxidized also to halogen by iodoso-compounds is well
known and is the basis of their standard analysis by iodometry (10). There is evidence that IBA (1) might be in equilibrium with the cyclic structure 1-hydroxy-1,2-benziodoxolin-
-3-one (2) (23). In fact, the properties and reactions of IBA
are quite different from those of its meta or para-isomers.In
particular, IBA is rather stable to hot water, which converts

normal iodoso-compounds into iodo- and iodoxy-derivatives, as reported in the literature (cf. 23, and references cited therein) and checked by us. This contrasts with the proposal of Mahoney et al. (6) that heating IBA in aqueous solution causes its disproportionation into iodo- and iodoxy-compound.

The IBA/Gdn.HCl reagent shows a reactivity towards amino acid side chains similar to that previously reported for NBS (16-20) and, similarly, cleaves at both tryptophan and tyrosine. The addition of p-cresol as a scavenger for tyrosine modification and cleavage permits the achievement of quite selective and efficient cleavage at tryptophan, with yields of cleavage comparable (up to 83%) to those obtained with other reagents presently often employed for protein fragmentation at tryptophan (e.g., BNPS-skatole).

In summary, the results of this study clarify the mechanisms of the IBA-mediated peptide bond fissions at tryptophan (1) and tyrosine(4), as well as the nature of side reactions at the level of the side chain functions of other amino acid residues. Despite the limitations above discussed, the IBA/Gdn. .HCl/p-cresol reagent appears to be of practical utility in protein fragmentation studies, as an useful alternative to BNPS-skatole (19,24) or DMSO/haloacid (22).

References

1. Mahoney, W.C. and Hermodson, M.A. (1979) Biochemistry 18, 3810-3814.

2. Fontana, A., Savige, W.A. and Zambonin, M.(1980) in Methods in Peptide and Protein Sequence Analysis (Birr, Chr., ed.) Elsevier/North-Holland Biomedical Press, Amsterdam, 309--322.

3. Wachter, E. and Werhahn, R.(1980) in Methods in Peptide and Protein Sequence Analysis (Birr, Chr., ed.),Elsevier/North--Holland Biomedical Press, Amsterdam, 323-328.

4. Johnson, P. and Stockmal, V.B.(1980) Biochem. Biophys. Res. Commun. 94, 697-703.

5. Mahoney,W.C. and Hermodson,M.A.(1980) in Methods in Peptide and Protein Sequence Analysis (Birr, Chr.,ed.) Elsevier
   North Holland Biomedical Press, Amsterdam, 323-328.

6. Mahoney,W.C., Smith,P.K. and Hermodson,M.A.(1981) Biochemistry 20, 443-448.

7. Fontana,A., Dalzoppo,D., Grandi,C. and Zambonin,M.(1981)
   Biochemistry, in press.

8. Meyer,V. and Wachter,W.(1892)Ber. 25, 2632-2635.

9. Askenasy,P. and Meyer,V.(1893)Ber. 26, 1354-1370.

10. Hellerman,L., Chinard,F.P. and Ramsdell,P.A.(1941) J.Amer.
    Chem. Soc. 63, 2551-2553.

11. Hartmann,C. and Meyer,V.(1893) Ber. 26, 1727-1732.

12. Spande,T.F. and Witkop,B.(1967) Methods Enzymol.11, 498-
    506.

13. Patchornik,A. Lawson,W.E., Gross,E.and Witkop,B.(1960) J.
    Amer. Chem. Soc. 82, 5923-5927.

14. Fontana,A. and Toniolo,C.(1976) Progr.Chem.Org.Natl. Com-
    pounds 33, 309-409.

15. Savige,W.E. and Fontana,A.(1980) Int. J. Peptide Protein
    Res. 15, 285-297.

16. Witkop,B.(1961) Adv. Protein Chem. 16, 221-321.

17. Wilson,I.G. and Cohen,L.A.(1963) J. Amer. Chem. Soc. 85,
    560-564.

18. Ramachandran,L.K. and Witkop,B.(1967)Methods Enzymol. 11,
    283-308.

19. Omenn,G.S., Fontana,A. and Anfinsen,C.B.(1970)J.Biol.Chem.
    245, 1895-1902.

20. Burstein,Y. and Patchornik,A.(1972) Biochemistry 11, 4641-
    4650.

21. Fontana,A., Vita,C. and Toniolo,C.(1973) FEBS Lett. 32,
    139-142.

22. Savige,W.E. and Fontana,A.(1977) Methods Enzymol. 47, 459-
    469.

23. Baker,G.P., Mann,F.G., Sheppard,N. and Tetlow,A.J.(1965)
    J. Chem. Soc. 3721-3728.

24. Fontana,A.(1972) Methods Enzymol. 25, 419-423.

Coordination of Protein and DNA Sequencing

ANALYSIS OF THE <u>ESCHERICHIA COLI</u> ATP-SYNTHASE COMPLEX BY DNA AND PROTEIN SEQUENCING

by

John E. Walker, Alex Eberle, Nicholas J. Gay, Peter Hanisch,
Matti Saraste and Michael J. Runswick
Laboratory of Molecular Biology, The MRC Centre,
Hills Road, Cambridge CB2 2QH, England

ABSTRACT

DNA cloning and sequencing offer an alternative and often more rapid approach to protein sequence analysis. We have used such a strategy to sequence the eight constituent proteins of the <u>Escherichia coli</u> ATP-synthase complex which is bound to the inner membrane of the bacterium. The genes for these proteins are organised in the <u>unc</u> operon. We have cloned this region of the <u>E. coli</u> chromosome, determined the DNA sequence by the dideoxy chain termination procedure coupled with cloning into bacteriophage M13. The α, β, γ and ε subunits have been identified by fractionation of the enzyme complex by polyacrylamide electrophoresis and solid phase sequencing of the Coomassie Blue stained bands. Protein c is identified from a published protein sequence; the identities of genes for 8, a and b are inferred from various properties of the proteins.

INTRODUCTION

Direct protein sequence analysis is no longer the most efficient way to determine the total sequence of many

---

Abbreviations: H^+-ATPase, H^+ translocating ATPase compled
F_1, extrinsic ATPase sector of H^+-ATPase
F , intrinsic membrane sector of H^+-ATPase

proteins. It is often quicker to isolate the gene and
determine its DNA sequence with the rapid DNA sequencing
techniques now available [1-3]. This is particularly true
when gene isolation is based on genetic techniques
(mutation, selection, complementation) applicable both to
many prokaryotes and viruses, and to lower eukaryotes such
as yeast, fungi and nematodes. The power of this approach
can be exemplified by the sequence analysis of proteins
involved in protein synthesis in Escherichia coli [4-7] of
mitochondrial cytochrome oxidase polypeptides [8-10] and of
nematode myosin [J. Karn, unpublished work]. In higher
eukaryotes mixed oligonucleotide probes, synthesised on the
basis of partial protein sequences, have been introduced as
a means of finding genes in complemetary DNA libraries [11]
and potentially also in genomic libraries, thereby
extending the range of possibilities.

    A modern strategy for protein sequence analysis
employing both protein sequencing and DNA sequencing of
recombinant DNA is shown in Fig. 1. We have employed it to
determine the sequences of the eight polypeptides which
constitute the E. coli ATP-synthase complex.

    The structure of this membrane-bound enzyme is very
similar to those found in mitochondria and chloroplasts [12,
13]. It has an intrinsic membrane fraction, F_0, and an
extrinsic portion, F_1, which can be solubilised intact. F_1
contains five different polypeptides designated α, β, γ, δ
and ϵ for which a stoicheiometry of 3:3:1:1:1 has been
proposed [14], although this is not universally accepted
[15]. The catalytic site of synthesis of ATP from ADP is
found within F_1, probably in the β subunit [12,13].
Bacterial F_0 contains three polypeptides a, b and c [16] and
forms a proton channel coupling the vectorial movement of
protons by energy transducing mem branes to ATP synthesis
[12]. The entire bacterial complex is encoded by the atp
[17] (or unc [13]) operon located at about minute 83 on the
E. coli chromosomal map [18] close to the origin of replication, oriC. Genetic analyses suggested that the F_0 polypeptides were clustered at the promoter proximal portion of
the operon and are followed by the genes encoding F_1 subunits [13,17,19], but an unambiguous order of genes has not
been established [17,19,20]. This region of the E. coli
chromosome has been cloned into the transducing phage λAsn5
[20] (Fig. 2a) and a restriction map (Fig. 2b) established
[20]. As described below, by taking advantage of this

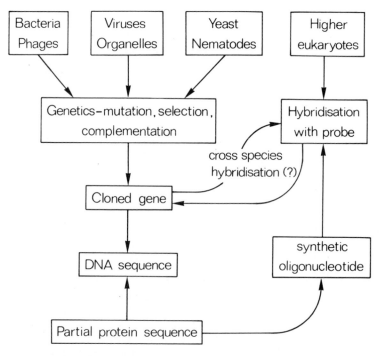

<u>Fig. 1.</u> Strategies of protein sequencing by gene cloning and sequencing of recombinant DNA.

information we cloned the appropriate regions into a plasmid and into the filamentous phage M13 and determined the DNA sequence of the <u>unc</u> operon (and adjacent regions of the <u>E. coli</u> chromosome). By protein sequence analysis of <u>E. coli</u> F_1 polypeptides in the amino-terminal regions and of peptides isolated from homologous bovine proteins, four genes, corresponding to those of the α, β, γ and ϵ subunit, were identified. An F_0 protein, c, was identified from a published total protein sequence [21]. A variety of criteria were invoked to identify genes for other ATP-synthase polypeptides.

MATERIALS AND METHODS

Enzymes, Proteins and Peptides

Proton translocating ATPase was prepared from beef

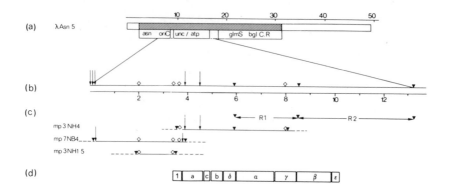

Fig. 2. (a) The extent of the E. coli chromosome (hatched) in λAsn5 showing genetic markers. The scale is in kilobases. (b) A restriction map with nucleotide numbers (kilobases) of part of region in (a). Arrows denote restriction sites for BamHI, ▼, EcoRI and ◊, HindIII. (c) Alignment of cloned EcoRI fragments R1 and R2 and the primary fragments used for sequence analysis in clones mp3.NH4, mp7.NB4 and mp3.NH1.5. mp3.NH4 and mp3.NH1.5 contain 4.0 and 1.5 kilobase HindIII fragments and mp7.NB4 a 4.0 kilobase BamHI fragment. (d) Alignment of the nine genes of the operon with restriction map.

heart mitochondria by gel filtration through Sephacryl S300, following solubilisation with 1.5% cholic acid, 3% ammonium sulphate, 10 mM Tris HCl pH 7.5, 0.1 mM MgSO$_4$, 0.5 mM EDTA, 50 mM sucrose, 0.02% sodium azide [M.J.R., M.A. Naughton, J.E.W., unpublished work]. E. coli membranes were made as described by Cox et al. [22]. Bovine F$_1$-ATPase was isolated from sub-mitochondrial particles and E. coli F$_1$ from E. coli membranes by the chloroform extraction method [23] followed by gel filtration through Ultrogel AcA34 [M.S., unpublished work].

The β subunit of bovine ATPase was purified from H$^+$-translocating ATPase by a procedure [M.J.R. and J.E.W., unpublished work] similar to that starting from F$_1$-ATPase described by Knowles and Penefsky [24], or from urea-solubilised bovine F$_1$ by gel filtration on Sepharose 6B and Sepharose 4B to remove the γ, δ and ε subunits, and ion

exchange chromatography of the fraction containing the α
and β proteins on DEAE-cellulose in 8 M urea, 10 mM Tris
HCl (pH 7.2), 1 mM β-mercaptoethanol and 0.1% phenylmethyl
sulphonyl fluoride with a NaCl gradient of 0-0.15 M. The
α subunit was not retained by the column under these conditions whilst the β subunit eluted at 0.06 M to 0.12 M NaCl.

    Tryptic arginine peptides were isolated from a tryptic
digest of succinyl β protein (250 nmol) by fractionation
first by gel filtration on Sephadex G75 (superfine) in 0.5%
ammonium bicarbonate.   Pooled fractions were subjected to
high pressure liquid chromatography on a radially
compressed column (Waters C18 Radialpak) in 10 mM sodium
acetate pH 4.5 with linear acetonitrile gradients up to 60
or 100% using an Altex chromatography system. The pumping
rate was 2 ml/min.

    The α, β, γ and ε subunits of E. coli ATPase (Fig. 3)
were isolated from E. coli F₁ by preparative gel electrophoresis in gradient gels (10-25% acrylamide) in 0.1% SDS
[25]. Proteins were visualised by staining the gel for 5
min with 0.2% Coomassie Blue in 50% methanol, 7% acetic
acid and destained in the same solution without dye, for 20
min. The subunits were excised and eluted and dialysed by
a published procedure [26].

DNA Sequencing

    DNA from bacteriophage λAsn5 containing about 26 kilo-
bases of the E. coli chromosome including unc was prepared
as described previously [27] The details of cloning of
fragments containing the unc locus into plasmid pACYC184
and into bacteriophage M13 are given elsewhere [27].
Recombinant M13 phages were sequenced by the dideoxy chain
termination method [1] using a synthetic primer 17 nucleotides in length complementary to a region of M13mp7
immediately adjacent to linker sequence [28]. Computer
programs for compilation and analysis of sequences are
described by Staden [29-31].

Protein Sequence Analysis

    Freeze dried subunits of E. coli F₁-ATPase recovered
from polyacrylamide gels were redissolved in 0.1 M sodium

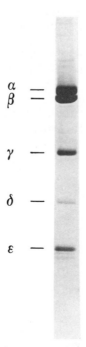

α —
β —

γ —

δ —

ε —

Fig. 3. The E. coli F₁-ATPase complex. The apparent
molecular weights of subunits are as follows:- α, 57; β, 52;
γ, 32; δ, 20-22; and ε, 13 k daltons. For protein
sequencing about 5 mg of F₁ protein were fractionated.

bicarbonate containing 10% propan-1-ol, added to DITC-glass
(25 mg) [32] and the level adjusted by addition of more
buffer until liquid just covered the beads. Attachment was
complete after 12 h, whereupon the beads were introduced
into the column of a solid phase microsequencer [unpublished
work], washed with methanol and degraded using a 34 min
programme. Conversion was effected by drying off the trifluoracetic acid at 60° in vacuo. The resulting phenylthiohydantoins were identified by reverse phase chromatography on
Zorbax-ODS as described before [33]. Manual sequence
analysis of peptides was performed by a scaled down version
of the method described by Chang [34]. The first coupling
of the peptide was carried out at 55° for 20 min with 10 μl
of dimethylaminoazobenzene-4'-isothiocyanate (2.82 mg/ml in
pyridine) and 20 μl 50% pyridine. The second coupling with
phenylisothiocyanate (2 μl) for the same duration also at

55° and then extraction twice with 200 μl heptane:ethyl
acetate 2:1 (v/v) saturated with 67% pyridine. Cleavage
with trifluoracetic acid (20 μl) was performed by reaction
for 10 min at 55° and followed by extraction with 25 μl H_2O
and 100 μl butyl acetate. Thiazolinones were converted at
80° for 15 min with 50% trifluoracetic acid. With this
procedure 2-10 nmol of peptide could be sequenced up to the
twentieth residue.

RESULTS

N-Terminal Sequences of E. coli F_1 Polypeptides

The α, β, γ and ε polypeptides were isolated by preparative gel electrophoresis of E. coli F_1-ATPase. The
preparation of F_1-ATPase employed was deficient in the δ
subunit, a feature common to F_1 preparations involving
chloroform extractions [23]. The sequences derived by the
solid phase technique were used to identify the four
corresponding genes (see below and Fig. 4.). It is interesting to note that in the cases of the α, β and ε subunits
the formylmethionine is removed by post-translational
cleavage, whereas the formyl group alone is removed from
the α subunit. The sequence in the N-terminal region of
the α subunit was also determined independently by Dunn
[35].

Cloning of the unc Operon and DNA Sequence

EcoRI restriction fragments R1 and R2 were amplified
in plasmid pACYC184 and then recloned into M13 [27]. By
sequence analysis of resulting recombinant clones it was
immediately apparent that EcoRI sites were to be found in
regions corresponding to the amino-terminal regions of α in
R1 [27] and of β in R2 [36] thereby establishing the orien-
tation of R1 and R2. The sequences of fragments R1 and R2
were completed by a random strategy employing restriction
enzymes with four or six base recognition sites as described
by Sanger et al. [2] and were found to contain the genes
for α, γ, β and ε (Fig. 2d) [27,36]. The remainder of the
operon was cloned directly into M13 (Fig. 2c) and sequenced
as described in detail elsewhere, by a directed rather than
a random approach [37]. The sequence with the deduced
amino acid sequences for the ATPase subunits is shown in

α subunit
Protein   M   Q   L   N   S   T   E   V   X   E   L
DNA   G G G G A C T G G A G C ATG CAA CTG AAT TCC ACC GAA ATC AGC GAA CTG

β subunit
Protein   A   T   G   K   I   V   Q   V   I   G   A   V   V   D   V   E   F   P   Q   D
DNA   T A G A G G A T T T A A G ATG GCT ACT GGA AAG ATT GTC CAG GTA ATC GGC GCC GTA GTT GAC GTC GAA TTC CCT CAG GAT

γ subunit
Protein   A   G   A   K   E   I   X   S   K   I   A   S   V   Q   N
DNA   T G A G G A G A A G C T C ATG GCC GGC GCA AAA GAG ATA CGT AGT AAG ATC GCA AGC GTC CAG AAC

ε subunit
Protein   A   M   T   Y   X   L   D   V   V   S   A   E
DNA   A A T C G G A G G G T G A T ATG GCA ATG ACT TAC CAC CTG GAC GTC GTC AGC GCA GAG CAA CAA

Fig. 4. Protein sequences determined in the amino-terminal regions of subunits of E. coli F_1-ATPase. Protein sequences were derived from material eluted from bands from 20 slots of a gel. ATG codons in boxes are proposed initiation codons and are preceded by boxed sequences complementary to the 3' end of ribosomal RNA.

Fig. 5 and the evidence for the identification of subunits is summarised in Table 1. Thus the unequivocal gene order in the unc operon shown in Fig. 2(d) was established.

DISCUSSION

The sequence analysis of the eight constituent proteins of the H+-ATPase complex was accomplished in less than nine months providing a further illustration of the power of the combined DNA:protein sequencing approach. An important facet of the work was the development of a rapid procedure for isolation of F_1 polypeptides by gel electrophoresis and their partial sequence analysis with a solid phase micro-sequencer [38] thereby enabling us to identify gene starts. This procedure has also been employed for identification of five genes from capsid proteins of bacteriophage lambda [38].

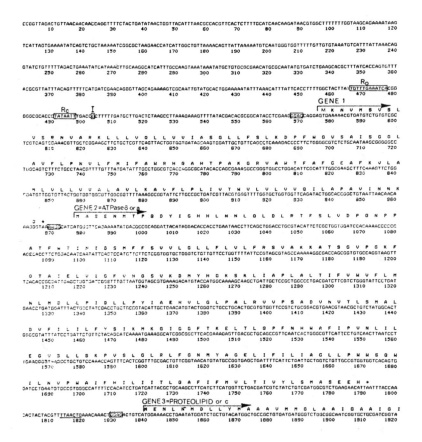

Fig. 5, Part (i)

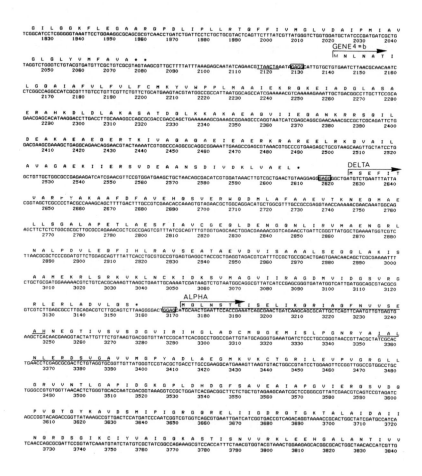

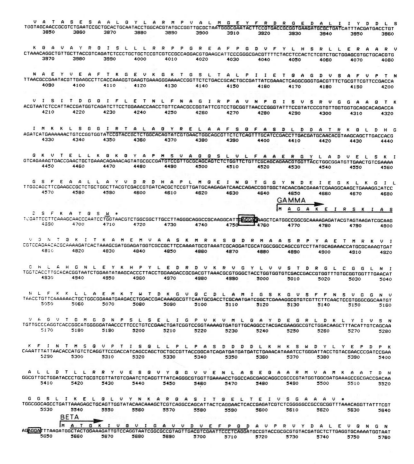

Fig. 5, Part (iii)

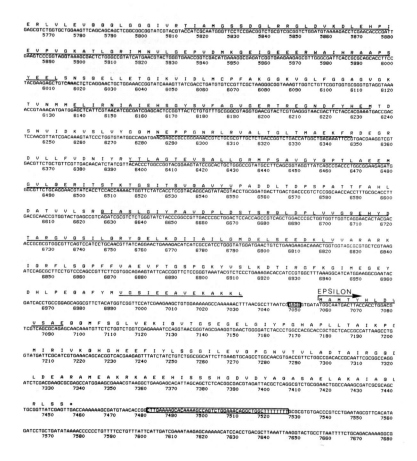

Fig. 5, Part (iv)

Table 1. Criteria employed for gene identification

Gene	Criteria
1	Unidentified; possibly pilot for assembly
a	Homology mt-ATPase 6
c	Total protein sequence
b	MW 17,212 kD, amphiphilic
δ	MW 19,310 kD, charge -7.5
α	N-terminal sequence; internal sequences from bovine mt-α
γ	N-terminal sequence
β	N-terminal sequence; internal sequences from bovine mt-β
ε	N-terminal sequence

A number of important deductions relating to both the function and assembly of the enzyme can be made from the sequences of the H^+-ATPase complex. Comparison of the sequences of the α and β subunits shows that they are weakly homologous [39]. This homology persists throughout the polypeptide chain but is strongest in the N-terminal region.   It is presumably related to the common property of the two proteins of binding adenine nucleotides [12,13]. An important consequence is that the proteins

Fig. 5. DNA sequence in the region of the unc operon. The genes coding for proteins are translated into one-letter code. Sequences boxed between residues 460-510 constitute the promoter, R_σ and R_C being recognition sequences for σ and core polymerase and I the site of initiation of transscription. The box before each gene contains the proposed ribosome binding site. Underlined DNA sequences before and after gene 3 are similar to each other and to the R_C sequence of trp. Underlined protein sequences are E. coli N-terminal sequences or homologous internal sequences in bovine β and α proteins. The boxed sequence following ε is the proposed transcription termination signal.

(a) Beef

E. coli

Thermophile

E. coli α

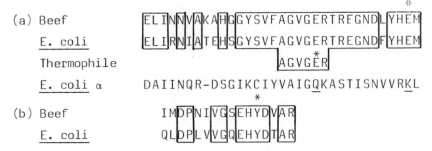

DAIINQR-DSGIKCIYVAIGQKASTISNVVRKL

(b) Beef

E. coli

Fig. 6. Comparison of sequences in F_1-ATPase β subunits
containing residues involved in catalysis. (a) Glutamic
acid residues involved in Mg^{2+} binding in different
species, marked with *. The residues were identified by
the covalent labelling with dicyclohexylcarbodiimide [43,
44]. Corresponding residues in E. coli α are underlined.
(b) An essential tyrosine covalently labelled in beef β by
p-fluorosulphonyl [^{14}C]-benzoyl-5'-adenosine [42]. In the
weakly homologous E. coli α sequence the residue corresponding to tyrosine is lysine.

probably have similar folds, possibly related to the widespread nucleotide binding fold [40,41]. Another interesting
facet of this homology is that amino acids in the β subunit
shown by Allison and colleagues [42-44] to be important for
catalysis (summarised in Fig. 6) are not conserved in the α
subunit. This is in accord with the widely held view that
the catalytic sites of the enzyme are in the β subunit [12,
13], whereas α, which binds nucleotides irreversibly [45],
may have a regulatory role.

     The δ and ε subunits which are required for binding F_1
to F_0 [46] are highly charged and in addition δ is
predicted to have an extremely α-helical secondary structure
[unpublished results]. Thus, the α-helical structure
strongly predicted from the sequence of membrane-bound subunit b, with a hydrophobic N-terminal domain followed by a
highly charged helical region (Fig. 7), is very striking.
It suggests that it may well be involved in making important ionic bonds with δ, ε and possibly other F_1 proteins.
Other charged residues could be involved in linking the
proton channel of F_0 to the catalytic sites in F_1.

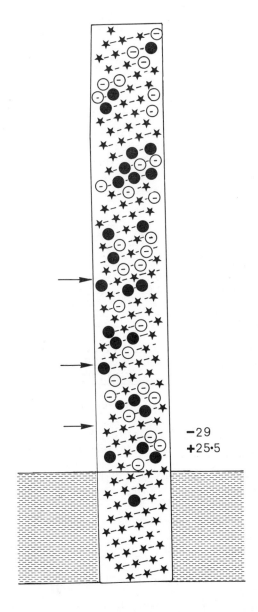

Fig. 7. Proposed secondary structure of membrane subunit b. The sequence is plotted on a helix surface and reads upwards, left to right. ⊖ , Asp or Glu,●, Lys, Arg or His,★, other. Arrows denote possible breaks in the α-helix.

−29
+25·5

We have proposed elsewhere that gene 1 protein may be important in serving as a pilot for assembly [37] by binding the messenger RNA to the membrane [unpublished work].

Given the striking order of the genes in the operon,
viz. intrinsic membrane genes a and c: amphiphilic protein
b: F_1 genes δ, α, γ, β, ε, this would provide a spatial
constraint which would influence assembly of the complex.

ACKNOWLEDGEMENTS

We thank S. Brenner for bacteriophage λAsn5, a gift to
him originally from Dr. Yura, and F. Sanger for his
encouragement. N.J.G. is in receipt of an MRC research
studentship, A.E. of support from the Swiss National Science
Foundation and M.S. of an EMBO fellowship.

REFERENCES

1. Sanger, F., Nicklen, S. and Coulson, A.R. (1977) Proc.
   Nat. Acad. Sci. USA 74, 5463-5467.
2. Sanger, F., Coulson, A.R., Barrell, B.G., Smith, A.J.H.
   and Roe, B.A. (1980) J. Mol. Biol. 143, 161-178.
3. Maxam, A.M. and Gilbert, W. (1977) Proc. Nat. Acad. Sci.
   USA 74, 4401-4405.
4. Post, L.E., Strycharz, G.D., Nomura, M., Lewis, H. and
   Dennis, P.P. (1979) J. Biol. Chem. 76, 1697-1701.
5. Post, L.E. and Nomura, M. (1980) J. Biol. Chem. 255,
   4660-4666.
6. Yokota, T., Sugisaki, H., Takanami, M. and Kaziro, Y.
   (1980) Gene 12, 25-31.
7. An, G. and Friesen, J.D. (1980) Gene 12, 33-39.
8. Fox, T.D. (1979) Proc. Nat. Acad. Sci. USA 76, 6534-
   6538.
9. Bonitz, S.G., Coruzzi, G. Thalenfeld, B.E., Tzagoloff, A.
   and Macino, G. (1980) J. Biol. Chem. 255, 11927-11941.
10. Anderson, S. et al. (1981) Nature 290, 457-465.
11. Agarwal, K.L., Brunstedt, J. and Noyes, B.E. (1981)
    J. Biol. Chem. 256, 1023-1028.
12. Futai, M. and Kanazawa, H. (1980) Curr. Topics Bionerg.
    10, 181-215.
13. Downie, J.A., Gibson, F. and Cox, G.B. (1979) Ann. Rev.
    Biochem. 48, 103-131.
14. Bragg, P.D. and Hou, C. (1975) Arch. Biochem. Biophys.
    167, 311-321.
15. Vogel, G. and Steinhardt, R. (1976) Biochemistry 15,
    208-216.

17. von Meyenberg, K. and Hansen, F.G. (1980) In: "Mechanistic Studies of DNA Replication and Genetic Recombination" ICN-UCLA Symp. Vol. XIX, Alberts, B. and Fox, C.F., eds., Academic Press, New York, in the press.
18. Bachman, B.J. and Low, B.K. (1980) Microbiol. Rev. 44, 1-56.
19. Downie, J.A., Langmans, L., Cox, G.B., Yanofsky, C. and Gibson, F. (1980) J. Bacteriol. 143, 8-17.
20. Kanazawa, H., Tamura, F., Mabuchi, K., Miki, T. and Futai, M. (1980) Proc. Nat. Acad. Sci. USA 77, 7005-7009.
21. Hoppe, J., Schairer H.U. and Sebald, W. (1980) Eur. J. Biochem. 112, 17-24.
22. Cox, G.B., Downie, J.A., Fayle, D.R.H., Gibson, F. and Radik, J. (1978) J. Bacteriol. 133, 287-292.
23. Beechey, R.B., Hubbard, S.A., Linnett, P.E., Mitchell, A.D. and Munn, E.A. (1975) Biochem. J. 148, 533-537.
24. Knowles, A.F. and Penefsky, H.S. (1972) J. Biol. Chem. 247, 6624-6630.
25. Laemmli, U.K. (1970) Nature 227, 680-685.
26. Walker, J.E., Auffret, A.D., Carne, A.F., Naughton, M.A. and Runswick, M.J. (1980) In: "Methods in Peptide and Protein Sequence Analysis", Birr, C., ed., Elsevier/North-Holland, pp. 257-265.
27. Gay, N.J. and Walker, J.E. (1981) Nucleic Acids Research 9, 2187-2194.
28. Duckworth, M.L., Gait, M.J., Goelet, P., Hong, G.F., Singh, M. and Titmas, R.C. (1981) Nucleic Acids Research 9, 1691-1706.
29. Staden, R. (1977) Nucleic Acids Research 4, 4037-4051.
30. Staden, R. (1978) Nucleic Acids Research 5, 1013-1015.
31. Staden, R. (1979) Nucleic Acids Research 6, 2601-2610.
32. Wachter, E., Machleidt, W., Hofner, H.G. and Ho, J. (1973) FEBS Lett. 35, 97-102.
33. Brock, C.J. and Walker, J.E. (1980) Biochemistry 19, 2873-2882.
34. Chang, J.Y., Brauer, D. and Wittmann-Liebold, B. (1978) FEBS Lett. 93, 205-214.
35. Dunn, S.D. (1980) J. Biol. Chem. 255, 11857-11860.
36. Saraste, M., Gay, N.J., Eberle, A., Runswick, M.J. and Walker, J.E. (1981) Nucleic Acids Research, submitted for publication.

37. Gay, N.J. and Walker, J.E. (1981) <u>Nucleic Acids Research</u> <u>9</u>, 3919-3926.
38. Walker, J.E., Auffret, A.D., Carne, A., Gurnett, A., Hanisch, P., Hill, D. and Saraste, M. (1981) <u>Eur. J. Biochem.</u>, submitted for publication.
39. Walker, J.E., Gay, N.J., Runswick, M.J. and Saraste, M. (1981) <u>Nature</u>, submitted for publication.
40. Rossman, M.G., Liljas, A., Bränden, C.I. and Banasak, L.K. (1975) In: "The Enzymes" Vol. XI, ed., Boyer, P.D., Academic Press, London and New York, pp. 62-102.
41. Risler, J.L., Zelwer, C. and Brunie, S. (1981) <u>Nature</u> <u>292</u>, 384-386.
42. Esch, F.S. and Allison, W.S. (1978) <u>J. Biol. Chem.</u> <u>253</u>,
43. Yoshida, M., Poser, J.W., Allison, W.S. and Esch, F.S. (1981) <u>J. Biol. Chem.</u> <u>256</u>, 148-153.
44. Yoshida, M., Allison, W.S. and Esch, F.S. (1981) <u>Fed. Proc.</u> <u>40</u>, 1734.
45. Harris, D.A. (1978) <u>Biochim. Biophys. Acta</u> <u>463</u>, 245-273.
46. Sternweis, P.C. (1978) <u>J. Biol. Chem.</u> <u>253</u>, 3123-3128.

THE PRIMARY STRUCTURE OF *ESCHERICHIA COLI* RNA POLYMERASE.

NUCLEOTIDE SEQUENCES OF THE *rpoB* AND *rpoC* GENES AND AMINO ACID SEQUENCES OF THE β AND β' SUBUNITS.

V.M.Lipkin, E.D.Sverdlov, G.S.Monastyrskaya,Yu.A.Ovchinnikov

Shemyakin Institute of Bioorganic Chemistry,

USSR Academy of Sciences, Moscow

INTRODUCTION

Elucidation of the transcription mechanism requires detailed knowledge of the active centers organization of RNA polymerase at the various stages of the RNA synthesis. This, in turn, can be obtained only after determining the primary and spatial structure of the enzyme.

Earlier we had established the amino acid sequence of the α subunit of *E.coli* DNA-dependent RNA polymerase by resorting solely to the ordinary methods of protein chemistry[1]. In the case of the β and β' subunits with their much higher molecular weights (I50000 and I55000, respectively), such an approach no longer suffice.

The progress in DNA sequencing methods allowed to realize the possibility of using the genetic code to obtain information on the primary protein structure from the nucleotide sequences. However here there are many pitfalls requiring considerable caution to avoid possible sources of error. In the first place the mRNA can undergo processing, leading to erroneous deduction of the protein structure. This holds particularly for eukaryotic cells, wherein "splicing" has been noted. Secondly, the protein itself can be processed. Thirdly, it is often difficult to recognize in the overall DNA structure the beginning of a structural gene. Moreover one has to bear in mind that a single error (deletion or insertion) in the DNA sequence could lead to a completely erroneous amino acid sequence of the protein.

Thus, primary structure determination of DNA cannot serve as a substitute for the direct sequencing of the protein. In view of this, we decided to utilize the methods of both protein and nucleotide chemistries, performing the parallel sequencing of the structural genes $rpoB$ (β subunit) and $rpoC$ (β' subunit) and of the corresponding proteins. Knowledge of the nucleotide sequence of the pertinent DNA segments would permit aligning of the peptide fragments from the protein analysis into an uninterrupted polypeptide chain. Such an approach provides the key to the most complicated problem in the primary structure analysis of high molecular proteins.

RESULTS

In fig.I restriction endonucleases cleavage map of $E.coli$ DNA region containing the structural genes of the β and β' subunits of the RNA polymerase ($rpoB$ and $rpoC$ correspondingly) is given. We determined the total sequence of the EcoRI-C, EcoRI-F, EcoRI-D and EcoRI-A - $Hind$III fragments and partial sequence of the EcoRI-G fragment carrying the beginning of the $rpoB$ gene[2-5]. These fragments were obtained from DNA of λrif^d47 and λrif^d18 transducing phages, containing the $E.coli$ $rpoBC$ operon[6,7], or corresponding plasmids by EcoRI restriction endonuclease digestion. In the case of EcoRI-A - $Hind$III fragment EcoRI and $Hind$III digestions were used.

The fragments were consecutively digested with one of the restriction endonucleases (Sau3AI, $Hinf$I, HpaII, and TaqI) cleaving the DNA into relatively small blocks. The resulting subfragments were phosphorylated by means of $|\gamma-^{32}P|$-ATP and phage T4 polynucleotide kinase and the mixture was separated by electrophoresis on polyacrylamide gel. As a rule both complementary chains obtained after denaturation of each subfragment and separation were analyzed. Their sequencing was performed by a modified Maxam-Gilbert procedure.

Investigation of the primary structure of the β subunit began with its limited tryptic hydrolysis. The analytic data showed that in this case the β subunit splits into five large fragments (M 62000, 52000, 37000, 24000 and I0000). However in the preparative experiment their yields were quite low and the mixture contained about 90 smaller peptides[8]. From the hydrolysate 55 low molecular weight peptides were isolated in all contained 453 amino acid residues. Isolation of the high molecular peptides proved difficult

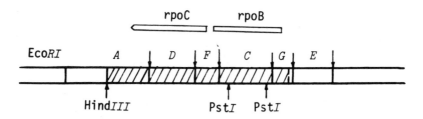

Fig. 1. EcoRI *restriction cleavage map of the* E.coli *DNA region including the structural genes (*rpoB *and* rpoC*) of the β and β' RNA polymerase subunits.*

because of both the little hydrolytic specificity and the low yield of most products.

As an alternative method, splitting the β subunit poly-peptide chain by digestion with *Staphylococcus aureus* pro-tease was chosen. Initial fractionation of the hydrolysate was performed by gel filtration on BioGel P-4. This yielded four fractions. Subsequent separation of the peptides was achieved by chromatography on the cation exchanger AG-50Wx4 and paper chromatography. Fraction I constituted a mixture of largest peptides. In order to facilitate separation and analysis of the peptides in this fraction the mixture was additionally digested with chymotrypsin. Altogether I2I pep-tides were isolated from fractions II-IV and 60 from frac-tion I.

In order to obtain the missing fragments, exhaustive tryptic digestion of the β subunit was carried out after modification of the lysine residues with citraconic anhyd-ride. The tryptic peptides were separated according to the same procedure as that used for staphylococcal peptides. After removal of the citraconic protection, the high-mole-cular-weight peptides were subjected to additional tryptic cleavage at the lysine residues.

For determination of the primary structure of the β' subunit its cyanogen bromide cleavage was carried out. The separation of the resulting peptides was achieved by gel filtration, paper and thin layer chromatographies, elec-trophoresis in acetate cellulose slabs and butanol extrac-tion of hydrofobic peptides. The tryptic digestion of the β' subunit was also carried out after modification of the lysine residues with citraconic anhydride.

The amino acid sequence of the peptide fragments deter-mined up to now constitutes about 85% of the total β subunit polypeptide chain and about 40% of the β' subunit.

358

Lipkin et al.

```
3160-3240   AAG AAA CTC GAA GCG AAA CGC CGC AAA ATC ACC CAG GGC GAC GAT CTG GCA CCG GGC GTG CTG AAG ATT GTT AAG GTA TAT
1027-1053   Lys-Lys-Leu-Glu-Ala-Lys-Arg-Arg-Lys-Ile-Thr-Gln-Gly-Asp-Asp-Leu-Ala-Pro-Gly-Val-Leu-Lys-Ile-Val-Lys-Val-Tyr-

3241-3321   CTG GCG GTT AAA CGC CGT ATC CAG CCT GGT GAC AAG ATG GCA GGT CGT CAC GGT AAC AAG GGT GTA ATT TCT AAG ATC AAC
1054-1080   Leu-Ala-Val-Lys-Arg-Arg-Ile-Gln-Pro-Gly-Asp-Lys-Met-Ala-Gly-Arg-His-Gly-Asn-Lys-Gly-Val-Ile-Ser-Lys-Ile-Asn-

3322-3402   CCG ATC GAA GAT ATG CCT TAC GAT GAA AAC GGT ACG CCG GTA GAC ATC GTA CTG AAC CCG CTG GGC GTA CCG TCT CGT ATG
1081-1107   Pro-Ile-Glu-Asp-Met-Pro-Tyr-Asp-Glu-Asn-Gly-Thr-Pro-Val-Asp-Ile-Val-Leu-Asn-Pro-Leu-Gly-Val-Pro-Ser-Arg-Met-

3403-3483   AAC ATC GGT CAG ATC CTC GAA ACC CAC CTG GGT ATG GCT GCG AAA GGT ATC GGC GAC AAG ATC AAC GCC ATG CTG AAA CAG
1108-1134   Asn-Ile-Gly-Gln-Ile-Leu-Glu-Thr-His-Leu-Gly-Met-Ala-Ala-Lys-Gly-Ile-Gly-Asp-Lys-Ile-Asn-Ala-Met-Leu-Lys-Gln-

3484-3564   CAG CAA GAA GTC GCG AAA CTG CGC GAA TTC ATC CAG CGT GCG TAC GAT CTG GGC GCT GAC GTT CGT CAG AAA GTT GAC CTG
1135-1161   Gln-Gln-Glu-Val-Ala-Lys-Leu-Arg-Glu-Phe-Ile-Gln-Arg-Ala-Tyr-Asp-Leu-Gly-Ala-Asp-Val-Arg-Gln-Lys-Val-Asp-Leu-

3565-3645   AGT ACC TTC AGC GAT GAA GAA GTT ATG CGT CTG GCT GAA AAC CTG CGC AAA GGT ATG CCA ATC GCA ACG CCG GTG TTC GAC
1162-1188   Ser-Thr-Phe-Ser-Asp-Glu-Glu-Val-Met-Arg-Leu-Ala-Glu-Asn-Leu-Arg-Lys-Gly-Met-Pro-Ile-Ala-Thr-Pro-Val-Phe-Asp-

3646-3726   GGT GCG AAA GAA GCA GAA ATT AAA GAG CTG CTG AAA CTT GGC GAC CTG CCG ACT TCC GGT CAG ATC CGC CTG TAC GAT GGT
1189-1215   Gly-Ala-Lys-Glu-Ala-Glu-Ile-Lys-Glu-Leu-Leu-Lys-Leu-Gly-Asp-Leu-Pro-Thr-Ser-Gly-Gln-Ile-Arg-Leu-Tyr-Asp-Gly-

3727-3807   CGC ACT GGT GAA CAG TTC GAG CGT CCG GTA ACC GTT GGT TAC ATG TAC ATG CTG AAA CTG AAC CAC CTG GTC GAC GAC AAG
1216-1242   Arg-Thr-Gly-Glu-Gln-Phe-Glu-Arg-Pro-Val-Thr-Val-Gly-Tyr-Met-Tyr-Met-Leu-Lys-Leu-Asn-His-Leu-Val-Asp-Asp-Lys-

3808-3888   ATG CAC GCG CGT TCC ACC GGT TCT TAC AGC CTG GTT ACT CAG CAG CCG CTG GGT GGT AAG GCA CAG TTC GGT GGT CAG CGT
1243-1269   Met-His-Ala-Arg-Ser-Thr-Gly-Ser-Tyr-Ser-Leu-Val-Thr-Gln-Gln-Pro-Leu-Gly-Gly-Lys-Ala-Gln-Phe-Gly-Gly-Gln-Arg-

3889-3969   TTC GGG GAG ATG GAA GTG TGG GCG CTG GAA GCA TAC GGC GCA GCA TAC ACC CTG CAG GAA ATG CTC ACC GTT AAG TCT GAT
1270-1296   Phe-Gly-Glu-Met-Glu-Val-Trp-Ala-Leu-Glu-Ala-Tyr-Gly-Ala-Ala-Tyr-Thr-Leu-Gln-Glu-Met-Leu-Thr-Val-Lys-Ser-Asp-

3970-4050   GAC GTG AAC GGT CGT ACC AAG ATG TAT AAA AAC ATC GTG GAC GGC AAC AT CAG GAG CCG GGC ATG CCA GAA TCC TTC
1297-1323   Asp-Val-Asn-Gly-Arg-Thr-Lys-Met-Tyr-Lys-Asn-Ile-Val-Asp-Gly-Asn-His-Gln-Met-Glu-Pro-Gly-Met-Pro-Glu-Ser-Phe-

4051-4129   AAC GTA TTG TTG AAA GAG ATT CGT TCG CTG GGT ATC AAC ATC GAA CTG GAA GAC GAG TAA TTC TCG CTC AAA CAG GTC A
1324-1342   Asn-Val-Leu-Leu-Lys-Glu-Ile-Arg-Ser-Leu-Gly-Ile-Asn-Ile-Glu-Leu-Glu-Asp-Glu  TER

4130-4210   CTG CTG TCG GGT TAA AAC CCG GCA GCG GAT TGT GCT AAC TCC GAC GGG AGC AAA TCC GTG AAA GAT TTA TTA AAG TTT CTG
1-8         Met-Lys-Asp-Leu-Leu-Lys-Phe-Leu-

4211-4291   AAA GCG CAG ACT AAA ACC GAA GAG TTT GAT GCG ATC AAA ATT GCT CTG GCT TCG CCA GAC ATG ATC CGT TCA TGG TCT TTC
9-35        Lys-Ala-Gln-Thr-Lys-Thr-Glu-Glu-Phe-Asp-Ala-Ile-Lys-Ile-Ala-Leu-Ala-Ser-Pro-Asp-Met-Ile-Arg-Ser-Trp-Ser-Phe-

4292-4372   GGT GAA GTT AAA AAG CCG GAA ACC ATC AAC TAC CGT ACG TTC AAA CCA GAA CGT GAC GGC CTT TTC TGC GCC CGT ATC TTT
36-62       Gly-Glu-Val-Lys-Lys-Pro-Glu-Thr-Ile-Asn-Tyr-Arg-Thr-Phe-Lys-Pro-Glu-Arg-Asp-Gly-Leu-Phe-Cys-Ala-Arg-Ile-Phe-

4373-4453   GGG CCG GTA AAA GAT TAC GAG TGC CTG TGC GGT AAG TAC AAG CTG AAA CAC CGT GGC GTC ATC TGT GAG AAG TGC GGC
63-89       Gly-Pro-Val-Lys-Asp-Tyr-Glu-Cys-Leu-Cys-Gly-Lys-Tyr-Lys-Arg-Leu-Lys-His-Arg-Gly-Val-Ile-Cys-Glu-Lys-Cys-Gly-

4454-4534   GTT GAA GTG ACC CAG ACT AAA GTA CGC CGT GAG ATG GGC CAC ATC GAA CTG ACT TCC CCG ACT GCG CAC ATC TGG TTC
90-116      Val-Glu-Val-Thr-Gln-Thr-Lys-Val-Arg-Arg-Glu-Arg-Met-Gly-His-Ile-Glu-Leu-Ala-Ser-Pro-Thr-Ala-His-Ile-Trp-Phe-

4535-4615   CTG AAA TCG CTG CCG TCC CGT ATC GGT CTG CTG CTC GAT ATG CCG CGC GAT ATC GAA CGC GTA CTG TAC TTT GAA TCC
117-143     Leu-Lys-Ser-Leu-Pro-Ser-Arg-Ile-Gly-Leu-Leu-Leu-Asp-Met-Pro-Leu-Arg-Asp-Ile-Glu-Arg-Val-Leu-Tyr-Phe-Glu-Ser-

4616-4696   TAT GTG GTT ATC GAA GGC GGT ATG ACC AAC TGG CAA CGT CAG CAG ATC CTG ACT GAA GGC AGA TAT CTG GAC GCG CTG GAA
144-170     Tyr-Val-Val-Ile-Glu-Gly-Gly-Met-Thr-Asn-Trp-Gln-Arg-Gln-Gln-Ile-Leu-Thr-Glu-Gly-Arg-Tyr-Leu-Asp-Ala-Leu-Glu-

4697-4777   GAG TTC GGT GAC GAA TTC GAC GCG AAG ATG GGG GCG GAA GCA ATC CAG GCT CTG CTG AAG AGC ATG GAT CTG GAG CAA CAG
171-197     Glu-Phe-Gly-Asp-Glu-Phe-Asp-Ala-Lys-Met-Gly-Ala-Glu-Ala-Ile-Gln-Ala-Leu-Leu-Lys-Ser-Met-Asp-Leu-Glu-Gln-Gln-

4778-4858   TGC GAA CAG CTG CGT GAA GAG CTG AAC GAA ACC AAG CGT AAA AAC CTG ACC AAG CGT ATC AAA CTG CTG
198-224     Cys-Glu-Gln-Leu-Arg-Glu-Glu-Leu-Asn-Glu-Thr-Asn-Ser-Glu-Thr-Lys-Arg-Lys-Asn-Leu-Thr-Lys-Arg-Ile-Lys-Leu-Leu-

4859-4939   GAA GCG TTC GTT CAG TCT GGT AAC AAA CCA CAG TGG ATG ATC CTG ACC GTT CTG CCG GTA CTG CCG CCA GAT CTG CGT CCG
225-251     Glu-Ala-Phe-Val-Gln-Ser-Gly-Asn-Lys-Pro-Gln-Trp-Met-Ile-Leu-Thr-Val-Leu-Pro-Val-Leu-Pro-Pro-Asp-Leu-Arg-Pro-

4940-5020   CTG GTT CCG CTG GAT GGT CGT TTC CGC GAT TCT GAC CTG AAC GAT CTG TAT CGT CGC GTC ATT AAC CGT AAC AAC CGT
252-278     Leu-Val-Pro-Leu-Asp-Gly-Gly-Arg-Phe-Ala-Thr-Ser-Asp-Leu-Asn-Asp-Leu-Tyr-Arg-Arg-Val-Ile-Asn-Arg-Asn-Asn-Arg-

5021-5101   CTG AAA CGT CTG CTG GAT CTG GCT GCG CCG GAC ATC ATC GTA CGT AAC GAA AAA CGT ATG CTG CAG GAA GCG GTA GAC GCC
279-305     Leu-Lys-Arg-Leu-Leu-Asp-Leu-Ala-Ala-Pro-Asp-Ile-Ile-Val-Arg-Asn-Glu-Lys-Arg-Met-Leu-Gln-Glu-Ala-Val-Asp-Ala-

5102-5182   CTG CTG GAT AAC GGT CGT CGC GGT CGT GCG ATC AAC GGT TCT AAC AAG CGT CCT CTG AAA TCT TTG GCC GAC ATG ATC AAA
306-332     Leu-Leu-Asp-Asn-Gly-Arg-Arg-Gly-Arg-Ala-Ile-Thr-Gly-Ser-Asn-Lys-Arg-Pro-Leu-Lys-Ser-Leu-Ala-Asp-Met-Ile-Lys-

5183-5263   GGT AAA CAG GGT CGT TTC CGT CAG AAC CTG CTC GGT AAG CGT GTT GAC TAC TCC GGT CGT TCT GTA ATC ACC GTA GGT CCA
333-359     Gly-Lys-Gln-Gly-Arg-Phe-Arg-Gln-Asn-Leu-Leu-Gly-Lys-Arg-Val-Asp-Tyr-Ser-Gly-Arg-Ser-Val-Ile-Thr-Val-Gly-Pro-

5264-5344   TAC CTG CGT CTG CAT CAG TGC GGT CTG CCG AAG AAA ATG GCA CTG GAA CTG TTC AAA CCG TTC ATC TAC GGC AAG CTG GAA
360-386     Tyr-Leu-Arg-Leu-His-Gln-Cys-Gly-Leu-Pro-Lys-Lys-Met-Ala-Leu-Glu-Leu-Phe-Lys-Pro-Phe-Ile-Tyr-Gly-Lys-Leu-Glu-

5345-5425   CTG CGT GGT CTT GCT ACC ACC ATT AAA GCT GAC AAA ATG GTT GAC CGC GAA GCA GCT GTC GTT TGG GAT ATC CTG GAC
387-413     Leu-Arg-Gly-Leu-Ala-Thr-Thr-Ile-Lys-Ala-Asp-Lys-Met-Val-Asp-Arg-Glu-Ala-Ala-Val-Val-Trp-Asp-Ile-Leu-Asp-

5426-5506   GAA GTT ATC CGC GAA CAC CCG GTA CTG CTG GAT CTG GCT CCG ACT CTG GAC GTT CTG GCA TTT GAA CCG GTA
414-440     Glu-Val-Ile-Arg-Glu-His-Pro-Val-Leu-Leu-Asn-Arg-Ala-Pro-Thr-Leu-His-Arg-Leu-Gly-Ile-Gln-Ala-Phe-Glu-Pro-Val-

5507-5587   CTG ATC GAA GGT AAA GCT ATC CAG CTG CAC CCG CTG GTT TGT GCG GCA TAT AAC GCC GAC TTC GAT GGT GAC GTG ATG GCT
441-467     Leu-Ile-Glu-Gly-Lys-Ala-Ile-Gln-Leu-His-Pro-Leu-Val-Cys-Ala-Ala-Tyr-Asn-Ala-Asp-Phe-Asp-Gly-Asp-Gln-Met-Ala-

5588-5668   GTT CAC GTA CCG ACG CTG GAA GCC CAG CTG GAA GCG CGT GCG CTG ATG ATG TCT ACC AAC AAC ATC CTG TCC CCG GCG
468-494     Val-His-Val-Pro-Leu-Thr-Glu-Ala-Gln-Leu-Glu-Ala-Arg-Ala-Leu-Met-Met-Ser-Thr-Asn-Asn-Ile-Leu-Ser-Pro-Ala-

5669-5749   AAC GGC GAA CCA ATC ATC GTT CCG TCT CAG GAC GTT GTA CTG GGT CTG TAC TAC ATG ACC CGT GAC TGT GTT AAC GCC AAA
495-521     Asn-Gly-Glu-Pro-Ile-Ile-Val-Pro-Ser-Gln-Asp-Val-Val-Leu-Gly-Leu-Tyr-Tyr-Met-Thr-Arg-Asp-Cys-Val-Asn-Ala-Lys-

5750-5830   GGC GAA GGC ATG GTG CTG ACT GGC CCG AAA GAA GCA GAA CGT CTG TAT CGC TCT GGT CTG GCT TCT CTG CAT GCG CGC GTT
522-548     Gly-Glu-Gly-Met-Val-Leu-Thr-Gly-Pro-Lys-Glu-Ala-Glu-Arg-Leu-Tyr-Arg-Ser-Gly-Leu-Ala-Ser-Leu-His-Ala-Arg-Val-

5831-5911   AAA GTG CGT ATC AAC GAG TAT GAA AAA GAT GCT AAC GGT GAA TTA GTA GCG AAA ACC GGT AAC AAC AAG CTG AAG ACG ACT GTT GGC
549-575     Lys-Val-Arg-Ile-Asn-Glu-Tyr-Glu-Lys-Asp-Ala-Asn-Gly-Glu-Leu-Val-Ala-Lys-Thr-Ser-Leu-Lys-Asp-Thr-Thr-Val-Gly-

5912-5992   CGT GCC ATT CTG TGG ATG ATT GTA CCG AAA GGT CTG CCT TAC TCC ATC GTC AAC GCG CTG GGT AAA AAA GTA TCC
576-602     Arg-Ala-Ile-Leu-Trp-Met-Ile-Val-Pro-Lys-Gly-Leu-Pro-Tyr-Ser-Ile-Val-Asn-Gln-Ala-Leu-Gly-Lys-Lys-Ala-Ile-Ser-

5993-6073   AAA ATG CTG AAC ACC TGC TAC CGC ATT CTC GGT CTG AAA CCG ACC GTT ATT TTT GAC GAC CAG ATC ATG TAC ACC GGC TTC
603-629     Lys-Met-Leu-Asn-Thr-Cys-Tyr-Arg-Ile-Leu-Gly-Leu-Lys-Pro-Thr-Val-Ile-Phe-Asp-Asp-Gln-Ile-Met-Tyr-Thr-Gly-Phe-

6074-6154   GCC TAT GCA GCG CGT TCT GGT GCA TCT GTT GGT ATC GAT GAC ATG GTC ATC CCG GAG AAG AAA CAC GAA ATC ATC TCC GAG
630-656     Ala-Tyr-Ala-Ala-Arg-Ser-Gly-Ala-Ser-Val-Gly-Ile-Asp-Asp-Met-Val-Ile-Pro-Glu-Lys-Lys-His-Glu-Ile-Ile-Ser-Glu-

6155-6235   GCA GAA GCA GAA GTT GCT GAA ATT CAG GAG CAG TTC CAG TCT GGT CTG GTA ACT GCG GGT CAG CGC TAC AAC AAA GTT ATC
657-683     Ala-Glu-Ala-Glu-Val-Ala-Glu-Ile-Gln-Glu-Gln-Phe-Gln-Ser-Gly-Leu-Val-Thr-Ala-Gly-Gln-Arg-Tyr-Asn-Lys-Val-Ile-

6236-6316   GAT ATC TGG GCT GCG GCG AAC GAT CGT GTA TCC AAA GCG ATG ATG GAT AAC CTG CAA ACT GAA ACC GTG ATT AAC CGT GAC
684-710     Asp-Ile-Trp-Ala-Ala-Ala-Asn-Asp-Arg-Val-Ser-Lys-Ala-Met-Met-Asp-Asn-Leu-Gln-Thr-Glu-Thr-Val-Ile-Asn-Arg-Asp-
```

```
6317-6397   GGT CAG GAA GAG AAG CAG GTT TCC TTC AAC AGC ATC TAC ATG ATG GCC GAC TCC GGT GCG CGT GGT TCT GCG GCA CAG ATT
711-737     Gly-Gln-Glu-Glu-Lys-Gln-Val-Ser-Phe-Asn-Ser-Ile-Tyr-Met-Met-Ala-Asp-Ser-Gly-Ala-Arg-Gly-Ser-Ala-Ala-Gln-Ile-

6398-6478   CGT CAG CTT GCT GGT ATG CGT GGT CTG ATG GCG AAG CCG GAT GGC TCC ATC ATC GAA ACG CCA ATC ACC GCG AAC TTC CGT
738-764     Arg-Gln-Leu-Ala-Gly-Met-Arg-Gly-Leu-Met-Ala-Lys-Pro-Asp-Gly-Ser-Ile-Ile-Glu-Thr-Pro-Ile-Thr-Ala-Asn-Phe-Arg-

6479-6559   GAA GGT CTG AAC GTA CTC CAG TAC TTC ATC TCC ACC CAC GGT GCT CGT AAA GGT CTG GCG GAT ACC GCA CTG AAA ACT GCG
765-791     Glu-Gly-Leu-Asn-Val-Leu-Gln-Tyr-Phe-Ile-Ser-Thr-His-Gly-Ala-Arg-Lys-Gly-Leu-Ala-Asp-Thr-Ala-Leu-Lys-Thr-Ala-

6560-6640   AAC TCC GGT TAC CTG ACT CGT CGT CTG GTT GAC GTG GCG CAG GAC CTG GTG GTT ACC GAA GAC GAT TGT GGT ACC CAT GAA
792-818     Asn-Ser-Gly-Tyr-Leu-Thr-Arg-Arg-Leu-Val-Asp-Val-Ala-Gln-Asp-Leu-Val-Val-Thr-Glu-Asp-Asp-Cys-Gly-Thr-His-Glu-

6641-6721   GGT ATC ATG ATG ACT CCG GTT ATC GAG GGT GGT GAC GTT AAA GAG CCG CTG CGC GAT CGC GTA CTG GGT CGT GTA ACT GCT
819-845     Gly-Ile-Met-Met-Thr-Pro-Val-Ile-Glu-Gly-Gly-Asp-Val-Lys-Glu-Pro-Leu-Arg-Asp-Arg-Val-Leu-Gly-Arg-Val-Thr-Ala-

6722-6802   GAA GAC GTT CTG AAG CCG GGT ACT GCT GAT ATC CTC GTT CCG CGC AAC ACG CTG CTG CAG GAA CAG TGG TGT GAC CTG CTG
846-872     Glu-Asp-Val-Leu-Lys-Pro-Gly-Thr-Ala-Asp-Ile-Leu-Val-Pro-Arg-Asn-Thr-Leu-Leu-Gln-Glu-Gln-Trp-Cys-Asp-Leu-Leu-

6803-6883   GAA GAG AAC TCT GTC GAC GCG GTT AAA GTA CGT TCT GTT GTA TCT TGT GAC ACC GAC TTT GGT GTA TGT GCG CAC TGC TAC
873-899     Glu-Glu-Asn-Ser-Val-Asp-Ala-Val-Lys-Val-Arg-Ser-Val-Val-Ser-Cys-Asp-Thr-Asp-Phe-Gly-Val-Cys-Ala-His-Cys-Tyr-

6884-6964   GGT CGT GAC CTG GCG CGT GGC CAC ATC ATC AAC AAG GGT GAA GCA ATC GGT ATT ATC GCG GCA CAG TCC ATC GGT GAA CCG
900-926     Gly-Arg-Asp-Leu-Ala-Arg-Gly-His-Ile-Ile-Asn-Lys-Gly-Glu-Ala-Ile-Gly-Ile-Ile-Ala-Ala-Gln-Ser-Ile-Gly-Glu-Pro-

6965-7045   GGT ACA CAG CTG ACC ATG CGT ACG TTC CAC ATC GGT GGT GCG GCA TCT CGT GCG GCT GAA TCC AGC ATC CAA GTG AAA
927-953     Gly-Thr-Gln-Leu-Thr-Met-Arg-Thr-Phe-His-Ile-Gly-Gly-Ala-Ala-Ser-Arg-Ala-Ala-Ala-Glu-Ser-Ser-Ile-Gln-Val-Lys-

7046-7126   AAC AAA GGT AGC ATC AAG CTC AGC AAC GTG AAG TCG GTT GTG AAC TCC AGC GGT AAA CTG GTT ATC ACT TCC CGT AAT ACT
954-980     Asn-Lys-Gly-Ser-Ile-Lys-Leu-Ser-Asn-Val-Lys-Ser-Val-Val-Asn-Ser-Ser-Gly-Lys-Leu-Val-Ile-Thr-Ser-Arg-Asn-Thr-

7127-7207   GAA CTG AAA CTG ATC GAC GAA TTC GGT CGT ACT AAA GAA GTA CCT TAC GGT GCG GTA CTG GCG AAA GGC GAT
981-1007    Glu-Leu-Lys-Leu-Ile-Asp-Glu-Phe-Gly-Arg-Thr-Lys-Glu-Val-Pro-Tyr-Gly-Ala-Val-Leu-Ala-Lys-Gly-Asp-

7208-7288   GGC GAA CAG GTT GCT GGC GGC GAA ACC GTT GCA AAC TGG GAC CCG CAC ACC ATG CCG GTT ATC ACC GAA GTA AGC GGT TTT
1008-1034   Gly-Glu-Gln-Val-Ala-Gly-Gly-Glu-Thr-Val-Ala-Asn-Trp-Asp-Pro-His-Thr-Met-Pro-Val-Ile-Thr-Glu-Val-Ser-Gly-Phe-

7289-7369   GTA CGC TTT ACT GAC ATG ATC GAC GGC CAG ACC ATT ACG CGT CAG ACA CTG ACC GGT CTG TCT TCG CTG GTG GTT
1035-1061   Val-Arg-Phe-Thr-Asp-Met-Ile-Asp-Gly-Gln-Thr-Ile-Thr-Arg-Gln-Thr-Leu-Thr-Gly-Leu-Ser-Ser-Leu-Val-Val-

7370-7450   CTG GAT TCC GCA GAA CGT ACC GCA GGT GGT AAA GAT CTG CGT CCG GCA CTG AAA ATC GTT GAT GCT CAG GGT AAC GAC GTT
1062-1088   Leu-Asp-Ser-Ala-Glu-Arg-Thr-Ala-Gly-Gly-Lys-Asp-Leu-Arg-Pro-Ala-Leu-Lys-Ile-Val-Asp-Ala-Gln-Gly-Asn-Asp-Val-

7451-7531   CTG ATC CCA GGT ACC GAT ATG CCA GCG CAG TAC TTC CTG CCG GGT AAA GCG ATT GTT CAG CTG GAA GAT GGC GTA CAG ATC
1089-1115   Leu-Ile-Pro-Gly-Thr-Asp-Met-Pro-Ala-Gln-Tyr-Phe-Leu-Pro-Gly-Lys-Ala-Ile-Val-Gln-Leu-Glu-Asp-Gly-Val-Gln-Ile-

7532-7612   AGC TCT GGT GAC ACC CTG GCG CGT ATT CCG CAG GAA TCC GGC GGT ACC AAG GAC ATC ACC GGT GGT CTG CCG CGC GTT GCG
1116-1142   Ser-Ser-Gly-Asp-Thr-Leu-Ala-Arg-Ile-Pro-Gln-Glu-Ser-Gly-Gly-Thr-Lys-Asp-Ile-Thr-Gly-Gly-Leu-Pro-Arg-Val-Ala-

7613-7693   GAC CTG TTC GAA GCA CGT CGT CCG AAA GAG CCG GCA ATC CTG GCT GAA ATC AGC GGT ATC GTT TCC TTC GGT AAA GAA ACC
1143-1169   Asp-Leu-Phe-Glu-Ala-Arg-Arg-Pro-Lys-Glu-Pro-Ala-Ile-Leu-Ala-Glu-Ile-Ser-Gly-Ile-Val-Ser-Phe-Gly-Lys-Glu-Thr-

7694-7774   AAA GGT AAA CGT CGT CTG GTT ATC ACC CCG GTA GAC GGT AGC GAT CCG TAC GAA GAG ATG ATT CCG AAA TGG CGT CAG CTC
1170-1196   Lys-Gly-Lys-Arg-Arg-Leu-Val-Ile-Thr-Pro-Val-Asp-Gly-Ser-Asp-Pro-Tyr-Glu-Glu-Met-Ile-Pro-Lys-Trp-Arg-Gln-Leu-

7775-7855   AAC GTG TTC GAA GGT GAA CGT GTA GAA GGT GGT GAC GTA ATT TCC GAC GGT CCG GAA GCG CCG CAC GAC ATT CTG CGT CTG
1197-1223   Asn-Val-Phe-Glu-Gly-Glu-Arg-Val-Glu-Gly-Gly-Asp-Val-Ile-Ser-Asp-Gly-Pro-Glu-Ala-Pro-His-Asp-Ile-Leu-Arg-Leu-

7856-7936   CGT GGT GTT CAT GCT GTT ACT CGT TAC ATC GTT AAC GAA GTA CAG GAC GTA TAC CGT CTG CAG GGC GTT AAG ATT AAC GAT
1224-1250   Arg-Gly-Val-His-Ala-Val-Thr-Arg-Tyr-Ile-Val-Asn-Glu-Val-Gln-Asp-Val-Tyr-Arg-Leu-Gln-Gly-Val-Lys-Ile-Asn-Asp-

7937-8017   AAA CAC ATC GAA GTT ATC GTT CGT CAG ATG CTG CGT AAA GCT ACC ATC GTT AAC GCG GGT AGC TCC GAC TTC CTG GAA GGC
1251-1277   Lys-His-Ile-Glu-Val-Ile-Val-Arg-Gln-Met-Leu-Arg-Lys-Ala-Thr-Ile-Val-Asn-Ala-Gly-Ser-Ser-Asp-Phe-Leu-Glu-Gly-

8018-8098   GAA CAG GTT GAA TAC TCT CGC GTC AAG ATC GCA AAC CGC GAA CTG GAA GCG AAC GGC AAA GTG GGT GCA ACT TAC TCC CGC
1278-1304   Glu-Gln-Val-Glu-Tyr-Ser-Arg-Val-Lys-Ile-Ala-Asn-Arg-Glu-Leu-Glu-Ala-Asn-Gly-Lys-Val-Gly-Ala-Thr-Tyr-Ser-Arg-

8099-8179   GAT CTG CTG GGT ATC ACC AAA GCG TCT CTG GCA ACC GAG TCC TTC ATC TCC GCG GCA TCG TTC CAG GAG ACC ACT CGC GTG
1305-1331   Asp-Leu-Leu-Gly-Ile-Thr-Lys-Ala-Ser-Leu-Ala-Thr-Glu-Ser-Phe-Ile-Ser-Ala-Ala-Ser-Phe-Gln-Glu-Thr-Thr-Arg-Val-

8180-8260   CTG ACC GAA GCA GCC GTT GCG GGC AAA CGC GAC GAA CTG CGC GGC CTG AAA GAG AAC CTT ATC GTG GGT CGT CTG ATC CCG
1332-1358   Leu-Thr-Glu-Ala-Ala-Val-Ala-Gly-Lys-Arg-Asp-Glu-Leu-Arg-Gly-Leu-Lys-Glu-Asn-Leu-Ile-Val-Gly-Arg-Leu-Ile-Pro-

8261-8341   GCA GGT ACC GGT TAC GCG TAC CAC CAG GAT CGT ATG CGT CGC CGT GCT GCG GAA GCT CCG GCT GCA CCG CAG GTG ACT
1359-1385   Ala-Gly-Thr-Gly-Tyr-Ala-Tyr-His-Gln-Asp-Arg-Met-Arg-Arg-Arg-Ala-Ala-Glu-Ala-Pro-Ala-Ala-Pro-Gln-Val-Thr-

8342-8426   GCA GAA GAC GCA TCT GCC AGC CTG CAG CTA CTG AAC GCA GGT CTG GGC GGT TCT GAT AAC GAG TAA TCGTTAATCCGCAAA
1386-1407   Ala-Glu-Asp-Ala-Ser-Ala-Ser-Leu-Gln-Leu-Leu-Asn-Ala-Gly-Leu-Gly-Gly-Ser-Asp-Asn-Glu-Ter

8427-8532   TAACGTAAAAACCCGCTTCGGCGGGTTTTTTTTATGGGGGGAGTTTAGGGAAAGAGCATTTGTCAGAATATTTAAGGAATTTCTGAATACTCATAATCAATGTAGAGA

8533-8639   TGACTAATATCCTGAAACTGACTGAACTAATTGAGTCAAACTCGGCAAGGATTCGATACTATTCCTGTGTAACTTTCTTAAGGAACGAGAATGAAACAGGAAGTGGA

8640-8746   AAAGTGGCGACCTTTTGCACATCCGGATGGTGATATTCGTGATTTATCATTTCTTGATGCTCATCAGGCTGTCTACGTTCAGCATCATGAGGGCAAAGAGCCTTTAG

8747-8853   AGTATCGCTTTTGGGTTACCTACTCTCTTCACTGCTTCACAAAAGATTATGAACATCAGACGAACGAAGAAAAACAATCGTTAATGTACCACGCGCCTAAAGAATCT

8854-8960   CGTCCCTTCTGCCAGCCACCGTTATAACTTAGCGCGCACACACTTAAAAAGAACTATTTTGGCGCTGCCAGAAAGCAACGTTATTCATGCCGGGTATGGTAGCTATGC

8961-9067   CGTGATTGAGGTGGACTTAGACGGAGGAGATAAGGCATTTTACTTTGTTGCGTTCAGGGAAAAGAAAAAACTCCGTTTGCATGTAACTAGCGCTTATC

9068-9174   CCATTTCTGAAAAACAGAAAGGTAAATCAGTGAAATTTTTCACCATTGCCTACAACTTATTGAGAAATAAGCAGCTTCCTCAGCCCTCAAAATAACAAAACCCACCT

9175-9281   TAAGGTGGGTTTCGCCAGAGAATTATCTCTGGTATTCAGAACGCCATTACCGGACTTTGCCTTGACCTTGCGATAATCGCAGGTTGCGGGATGTCTGAATTTCTTCA

9282-9388   GTCTGCTGCATCCTGGAAGATGAGAACATGTGTTCTTATTTTCGTCTCTATCATAGTTGAGTATTTACTCTCTTACAATCAGATCTCTTTCATTGCTCAACAGGCGA

9389-9450   TGGCTTCAGACTTTGCATTACGGAAT?TTTAAGAAAGGCAGGGCGAAACGAGGAAGAAGCTT
```

Fig. 2. *The nucleotide sequence of the* rpoBC *segment and the total amino acid séquences of the* β *and* β' *subunits of E.coli* RNA polymerase. The nucleotide sequence of the complementary DNA chain, equal to the sequence of mRNA, is given. The underlined amino acid sequences are those the structure of which has been determined from analysis of corresponding peptides. C* refer to 5-methylcytidine residues.

During this investigation the comparison of the nucleotide and amino acid sequences was carried out continually.
The search for correspondence between the amino acid sequences of the peptides and the nucleotide sequences of the
DNA fragments was carried out by means of a computer. A
conjunction of the methods of protein and nucleotide chemistry for the combined structural investigation of a protein
and DNA sharply accelerated and considerably simplified the
solution of both problems and enhanced the reliability of
the structural analysis.

As a result the continuous nucleotide sequence (9450
base pairs) embracing the entire *rpoB* and *rpoC* genes together with the total amino acid sequences of the β and β'
subunits comprising I342[2-3] and I407[5] residues were determined (Fig. 2).

At first we determined the primary structure of the
*rpoB*gene cloned from the *rif*[a] *rpoB*255 mutant DNA. To investigate the difference in the base sequence between this mutant gene and the wild-type one, the plasmid *pIB*I830 carrying the wild-type *rpoB* gene ha**s** been constructed. We determined the sequence of the central part of the *rpoB* gene
between the *Pst*I sites of the *Eco*RI-C fragment, where the
*rpoB*255 mutation had been localized by genetic methods. In
contrast with the wild-type gene, the *rpoB*255 mutant gene
was found to contain an A.T → T.A transvertion at position
I628 (Fig. 2), entailing the substitution of a valine residue in the mutant RNA polymerase β subunit for the aspartic
acid residue (5I6) of the wild-type β subunit[9].

REFERENCES
I. Ovchinnikov Yu.A., Lipkin V.M., Modyanov N.N., Chertov
   O.Yu., Smirnov Yu.V. (I977) FEBS Letters, 76, I08-III.
2. Ovchinnikov Yu.A., Monastyrskaya G.S., Gubanov V.V., Guryev S.O., Chertov O.Yu., Modyanov N.N., Grinkevich V.A.,
   Makarova I.A., Marchenko T.V., Polovnikova I.N., Lipkin
   V.M., Sverdlov E.D. (I980) Dokl. Akad. Nauk SSSR, 253,
   994-999.
3. Ovchinnikov Yu.A., Monastyrskaya G.S., Gubanov V.V., Guryev S.O., Chertov O.Yu., Modyanov N.N., Grinkevich V.A.,
   Makarova I.A., Marchenko T.V., Polovnikova I.N., Lipkin
   V.M., Sverdlov E.D. (I98I) Eur. J. Biochem., II6,
   62I-629.
4. Ovchinnikov Yu.A., Monastyrskaya G.S., Gubanov V.V., Salomatina I.S., Shuvaeva T.M., Lipkin V.M., Sverdlov E.D.
   (I98I) Bioorg. Khim., 7, II07-III2.

5. Monastyrskaya G.S., Gubanov V.V., Guryev S.O., Kalinina
   N.F., Salomatina I.S., Shuvaeva T.M., Lipkin V.M., Sverdlov E.D., Ovchinnikov Yu.A. Bioorg. Khim., in press.
6. Mindlin S.S., Ilyina T.S., Gorlenko Ch.M., Hachikyan
   N.A., Kovalev Yu.N. (I976) Genetika, I2, II6-I30.
7. Kirschbaum J.B., Konrad B.E. (I973) J. Bacteriology,
   II6, 5I7-526.
8. Marchenko T.V., Modyanov N.N., Lipkin V.M., Ovchinnikov
   Yu.A. (I980) Bioorg. Khim., 6, 325-33I.
9. Ovchinnikov Yu.A., Monastyrskaya G.S., Gubanov V.V.,
   Lipkin V.M., Sverdlov E.D., Kiver I.F., Bass I.A., Mindlin S.Z., Danilevskaya O.N., Khesin R.B. Mol. Gen.
   Genet., in press.

CORRELATION BETWEEN DNA AND PROTEIN SEQUENCES IN β-GALACTOSIDASE RELATED PROTEINS

AUDREE V. FOWLER AND IRVING ZABIN

Department of Biological Chemistry, UCLA
School of Medicine and Molecular Biology In-
stitute, University of California,
Los Angeles, CA 90024

THE LACTOSE OPERON

The lactose operon of Escherichia coli has been a valuable tool for studying protein-nucleic acid interactions and
correlations. Work in this laboratory is concerned not only
with gene protein correlations, but also with structure-
function relationships and evolution of proteins. The
emphasis has been on lac proteins, in particular β-galactosidase. In order to understand and to define this system,
it was important to determine the protein and nucleic acid
sequences of the complete operon. The four proteins coded
for by lacI, lacZ, lacY and lacA are the repressor (39,000
dalton subunit), β-galactosidase (116K), lac permease (46K)
and thiogalactosidase transacetylase (25K), respectively
(Fig. 1).

The amino acid sequence of all of these proteins are
known except for part of the transacetylase. The first
sequence completed was that of the repressor, by Beyreuther
et al. in 1973 (1). We published the sequence of β-galactosidase in 1977 (2). Both primary structures were determined by conventional protein methods. At about that time,
the nucleic acid sequence of the lacI gene was reported (3)
and agreed with the amino acid sequence although a small
correction of the latter was indicated. The complete nucleotide sequence of the lacY gene coding for the lac permease was reported in 1980 (4) including the intergenic

363

LAC OPERON

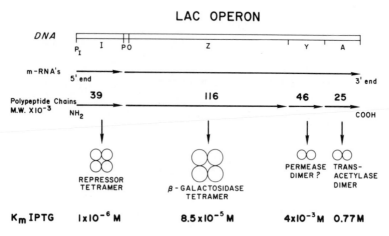

Fig. 1.   The Lactose Operon

regions between lacZ and lacY and between lacY and lacA, as well as the region coding for the amino terminus of the lacA gene product (acetylase). The sequence of the amino terminal region of the in vivo and in vitro products of the lacY gene (the permease) agreed with that predicted from the DNA sequence (5). The sequence of the amino terminal region of acetylase predicted from the lacA DNA sequence agreed with our published amino acid sequence (6). Interestingly, the amino terminal analysis of acetylase never yielded stoichiometric levels of any amino acid. The results indicated about one-half mole of Asx and one-half mole of Met per 2 moles of subunit (7). When the protein was sequenced directly, both Met and Asn were found at step 1, and Asn and Met at step 2, indicating that the methionine was only partially removed. these results are explained by the presence of three methionine residues at the amino terminus of acetylase (Met-Asn-Met-Pro-Met-Thr-). This region must be exposed so that some of the methionines are oxidized, and therefore ambiguous results were obtained.

EVOLUTION AND DOMAINS IN β-GALACTOSIDASE

Two other regions of the lac operon DNA, both in lacZ, have been sequenced. The first of these was the beginning of lacZ and was useful in the final sequence determination

of β-galactosidase (2). This sequence analysis also helped
to define a second low affinity repressor binding site within lacZ (8,9). The second region came from a study by Calos
and Miller (10) on deletion formation mediated by transposon
Tn9. This region is a segment of DNA about one-third of the
way into lacZ, and is of special interest in regard to the
evolution of β-galactosidase. When the sequence analysis of
β-galactosidase was completed, we wondered if we could determine something about the origin of the very large lacZ
gene and of the lac operon. Thus far no evolutionary relationship can be discerned among the different lac proteins
by comparison of amino acid sequences. However, when the
sequence of β-galactosidase was compared to itself, a number
of striking similarities were seen. Specifically, eight
long pairs of segments, each separated by about 390 residues,
were found to have a large number of amino acid identities
The overall number of identities was found to be about 25%
(11). This is considerably more than expected by chance,
which is about 10 or 11%. The results indicated that the
first three-eighths was homologous to the second three-
eighths of the polypeptide and suggested that the first
three-fourths of lacZ was derived by duplication and fusion
of a smaller gene. For control comparisons, a number of
other proteins were also examined. Curiously, another protein from E. coli, the small protein, dihydrofolate reductase containing 156 residues, was found to be homologous to
the carboxyl-terminal fourth of β-galactosidase (11). The
results clearly imply that fusion to an earlier dihydrofolate reductase gene occurred at some time in lacZ's evolutionary history. When the DNA sequence determined by Calos
and Miller (10) was inspected, the lacZ region coding for
the second homologous part of the protein was found to have
a nucleic acid sequence similar to that of a normal translation start sequence, with a Shine-Delgarno ribosome binding sequence ten residues upstream from the AUG specifying
amino acid residue 377 of β-galactosidase (Fig. 2). The
initiating codon here as well as for lacZ and lacY is the
normal AUG but lacI uses GUG and lacA uses UUG as initiating
codons. This same kind of observation for gene fusion and
internal start sequence was seen in the thrA gene of E. coli
by Kalinka et al. (12). The thrA protein, like β-galactosidase, is a tetramer. However, it is a bifunctional enzyme
carrying discrete domains for aspartokinase and homoserine
dehydrogenase activities. The beginning of the second
domain, containing the dehydrogenase activity, could be cor-

related to an internal start in the thrA gene. The DNA
sequence in lacZ at the region homologous to the dehydro-
folate reductase is not yet known.

    Other genetic, biochemical and immunological data indi-
cate that, in fact, the β-galactosidase polypeptide may be
made up of three domains. This was first indicated by in-
tracistronic complementation studies of Ullmann, Jacob and
Monod (13). They demonstrated that amino terminal polypeptide fragments from nonsense mutants in lacZ could complement with extracts of strains containing deletion mutations
in the amino terminal region to restore enzyme activity.
We were able to show that some of these nonsense fragments
of β-galactosidase were cross-reactive with anti-β-galacto-
sidase (14). The shortest cross-reactive species was about
40,000 daltons. This size corresponds to the first three-
eighths of β-galactosidase. Another indication of a second
smaller ancestral gene as well as for the third domain was
evidence from several laboratories of two restarts within

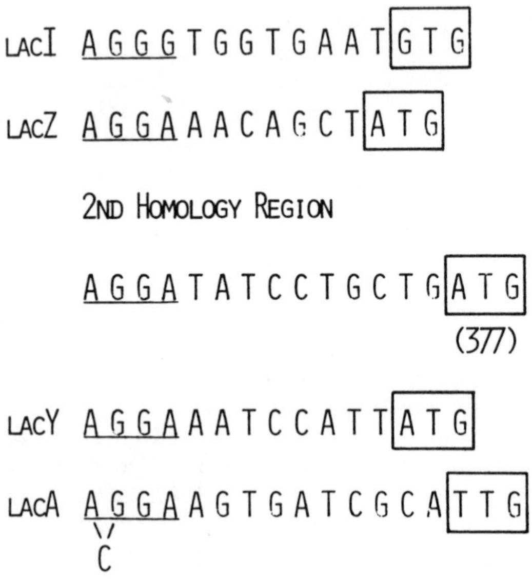

LacI A G G G T G G T G A A T G T G

LacZ A G G A A A C A G C T A T G

          2ND HOMOLOGY REGION

        A G G A T A T C C T G C T G A T G
                                   (377)

LacY A G G A A A T C C A T T A T G

LacA A G G A A G T G A T C G C A T T G
          \/
          C

Fig. 2. Ribosome binding regions in the lactose operon.

lacZ (15-18). These results indicated that polypeptide
fragments can be initiated at two points within the polypeptide chain, at the "β" barrier and at the "ω" barrier.
Genetic mapping indicated that these barriers correspond to
regions of the protein around amino acid residues 350-400
and about amino acid 800, respectively, of β-galactosidase.
Also, indications that the ω domain exists comes from
studies of deletion mutants at the carboxyl terminus of β-
galactosidase. Evidence for ω globules was first obtained
by complementation studies (19). Goldberg suggested that
the ω globules are present in both ω-complemented and native
enzyme (20). Finally, the ω segment was isolated from a
mutant strain and found to cross-react with anti-β-galac-
tosidase (21).

                        HYBRID PROTEINS

     Another kind of evolution involving β-galactosidase is
artificial evolution or "tinkering with DNA". This technique involves genetic fusions of "foreign" proteins to a
site near the amino terminus of β-galactosidase, producing
hybrid proteins retaining β-galactosidase enzyme activity.
Since β-galactosidase is so easy to assay, this procedure
facilitates analysis of other proteins that are ordinarily
difficult to study. This has been an especially useful
technique to define and study membrane and periplasmic proteins. Some of the proteins involved with maltose transport into the cell were first characterized by such gene
fusions. With the information that became available,
several new maltose operon proteins were identified. The
role of leader or signal sequences in the transport of maltose proteins to the membranes or the periplasmic space was
also defined by mal-lac gene fusions.

     The maltose operon is proving to be a very important
system for defining the mechanism of transport of sugars.
In contrast to lactose transport, which apparently requires
only one gene product, maltose (and maltodextrin) transport
has several protein components. Since these gene products
are in different parts of the cell (outer membrane, periplasmic space and cytoplasmic membrane), this system also
serves as an ideal model to study transport of proteins
to various sites in the cell. The gene products are
specified by the genes malE, lamB, malK, malF, and malG,

all part of the <u>malB</u> region. Therefore, to define the system, the nucleotide sequence determination of the complete
<u>malB</u> region is underway (22) and all of the proteins are
being characterized.

One of the proteins of the maltose transport system
that was analyzed by the genetic fusion technique is the
λ receptor. This protein is coded by the <u>lamB</u> gene and is
an outer membrane protein. When the <u>lamB</u> locus was fused
to <u>lacZ</u> in order to determine the sequence of the leader,
a quite interesting result was obtained. The hybrid protein
contained the entire signal sequence plus 16 residues of
the mature λ receptor protein fused to residue 20 of β-
galactosidase (Fig. 3), yet the protein remained cytoplasmic (23). This result was one of the earliest to indicate
that the signal sequence is not sufficient for transport of
proteins to membranes, and that other factors, in addition
to the signal sequence, must therefore influence the translocation process.

The first maltose transport protein studied by the
genetic fusion technique was the maltose binding protein
(specified by the <u>malE</u> gene), a key component of the maltose
transport system. The sequence of a hybrid protein from a
<u>malE</u>-<u>lacZ</u> fusion strain was determined (24). The hybrid

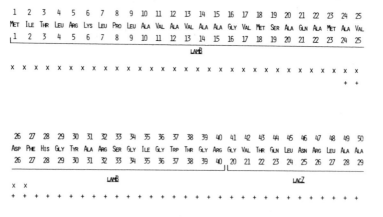

Fig. 3. Amino-terminal sequence of the 52-4 <u>lamB</u>-<u>lacZ</u>
hybrid protein. X, amino acid residues identified in the
intact protein; +, amino acid residues identified in CNBr
2* derived from the protein.

protein was cytoplasmic and contained 14 residues of the
leader. Another fusion strain was analyzed and the protein
was found to have the complete leader plus 23 amino acid
residues of the mature protein before bonding to β-galactosidase (deleting residues 1-26 of β-galactosidase). This
strain contained a mutation in the leader region (Met→Arg
at 19 of the leader). It also had a second change (Ser→Asp
at residue (23) of the leader). Using a combination of DNA
andprotein sequencing techniques, the complete signal sequence and part of the amino terminal sequence of the mature
periplasmic binding protein were determined (24). The wild-
type signal sequence was found to resemble those of other
proteins in that it had many hydrophobic residues, and
threonine and serine, but no acidic amino acids (Fig. 4).
Certain mutant proteins containing amino acid substitutions
are not transported to the periplasmic space and accumulate
in the precursor form in the cytoplasm. These studies indicated that the change of a single hydrophobic or uncharged
amino acid to a charged amino acid within the signal
sequence is sufficient to block the secretion process.
These studies are now being extended to study the complete
sequence of the maltose binding protein. So far, five
cyanogen bromide peptides ranging in size from 10 to 154
amino acid residues have been isolated by Sephadex G-50
and G-100 gel filtration. These five peptides account for
the whole protein and have been partially sequenced. Correlation to the DNA sequence of malE will facilitate the
rapid sequence determination of the protein.

This work was supported by U.S. Public Health Service Grant
AI-04181 and National Science Foundation Grant PCM-7819974.

ATG AAA ATA AAA ACA GGT GCA CGC ATC CTC GCA TTA TCC GCA TTA ACG ACG ATG ATG TTT TCC GCC TCG GCT CTC
MET LYS ILE LYS THR GLY ALA ARG ILE LEU ALA LEU SER ALA LEU THR THR MET MET PHE SER ALA SER ALA LEU
            5   10   15   20   25

GCC AAA ATC GAA GAA GGT AAA CTG GTA ATC TGG ATT AAC GCC GAT AAA GGC TAT AAC GGT CTC GCT GAA -
ALA LYS ILE GLU GLU GLY LYS LEU VAL ILE TRP ILE ASN GLY ASP LYS GLY TYR ASN GLY LEU ALA GLU -
        30   35   40   45

Fig. 4. DNA and amino acid sequences at the beginning
of MalE and the precursor maltose binding protein. The
arrow shows the site of cleavage between leader and
mature protein.

REFERENCES

1.) Beyreuther, K., Weber, K., Geisler, N. and Klemm, A., Proc. Nat. Acad. Sci. (1973) 70, 3570-3580.
2.) Fowler, A.V. and Zabin, I., Proc. Nat. Acad. Sci. (1977) 74, 1507-1510.
3.) Farabaugh, P.J., Nature (1978) 274, 765-769.
4.) Buchel, D.E., Gronenborn, B. and Müller-Hill, B., Nature (1980) 283, 541-545.
5.) Ehring, R., Beyrether, K., Wright, J.K. and Overath, P., Nature (1980) 283, 537-540.
6.) Zabin, I. and Fowler, A.V., in The Operon, J.A. Miller and W.S. Reznikoff (Eds.), Cold Spring Harbor Laboratory, New York (1978) pp. 89-121.
7.) Brown, J.L., Koorajian, S. and Zabin, I., J. Biol. Chem. (1967) 242, 4259-4262.
8.) Reznikoff, W.S., Winter, R.B. and Hurley, C.K., Proc. Nat. Acad. Sci. (1974) 71, 2314-2318.
9.) Gilbert, W., Gralla, J., Majors, J. and Maxam, A., in Protein-Ligand Interactions, (Ed. H. Sund and G. Blauer), Walter de Gruyter and Co., Berlin (1975) pp. 193-210.
10.) Calos, M.P. and Miller, J.H., Nature (1980) 285, 38-41.
11.) Hood, J.M., Fowler, A.V. and Zabin, I., Proc. Nat. Acad. Sci. (1978) 75, 113-116.
12.) Katinka, M. et al., Proc. Nat. Acad. Sci. (1980) 77, 5730-5733.
13.) Ullmann, A., Jacob, F. and Monod, J., J. Mol. Biol. (1967) 24, 339-343.
14.) Fowler, A.V. and Zabin, I., J. Mol. Biol. (1968) 33, 35-47.
15.) Grodzicker, T. and Zipser, D., J. Mol. Biol. (1968) 38, 305-314.
16.) Newton, A., J. Mol. Biol. (1978) 125, 449-466.
17.) Michels, C. and Zipser, D., J. Mol. Biol. (1969) 41, 341-347.
18.) Manley, J.L., J. Mol. Biol. (1978) 125, 449-466.
19.) Ullmann, A., Perrin, D., Jacob, F. and Monod, J., J. Mol. Biol. (1965) 12, 918-923.
20.) Goldberg, M.E. (1970) The lactose operon (ed. J.R. Beckwith and D. Zipser), pp. 273-278, Cold Spring Harbor, New York.
21.) Celada, F., Ullmann, A. and Monod, J., Biochemistry (1974) 13, 5543-5547.

22.) Raibaud, O., Clements, J.M., and Hofnung, M., Molec.
     Gen. Genet. (1979) 174, 261-267.
23.) Bedouelle, H., Bassford, P.J., Jr., Fowler, A.V.,
     Zabin, I., Beckwith, J. and Hofnung, M., Nature
     (1980) 285, 78-81.
24.) Moreno, F., Fowler, A.V., Hall, M., Silhavy, T.J.,
     Zabin, I. and Schwartz, M., Nature (1980) 286, 356-359.

THE MOLECULAR CONSEQUENCES OF FORMALDEHYDE AND ETHYL

METHANESULFONATE MUTAGENESIS IN <u>DROSOPHILA</u>: ANALYSIS OF

MUTANTS IN THE ALCOHOL DEHYDROGENASE GENE

Allen R. Place,* Cheeptip Benyajati,† and

William Sofer#

*Department of Biology, University of

Pennsylvania, Philadelphia, PA. 19104;

†Frederick Cancer Research Center, P.O. Box

B, Frederick, MD. 21701; and #Waksman

Institute, Rutgers University, Piscataway,

N.J. 08854

INTRODUCTION

Chemical mutagenesis is a process whose analysis is complicated by the large number of possible reactions of mutagens and their byproducts with the genetic material, and the variety of pathways by which a mutational lesion can be repaired (1,2). One simplying factor, however, is that in some organisms the final result of a mutagenic pathway can be deduced, or directly assessed, by examination of the gene or gene product affected by the mutation. For example, by isolating and sequencing a protein that is a product of a mutant gene, one can deduce from the amino acid sequence, the nature of the nucleotide lesion. Similarly, by comparing the nucleotide sequence of a mutant gene with its wild-type counterpart, one can directly determine the nature of the nucleotide defect.

In higher eukaryotes, this kind of approach has
generally not been feasible. One reason is that in many
genetic schemes used to detect mutations in eukaryotes, the
gene product affected by the mutagen is unknown.
Conversely, where gene products have been characterized, it
is often difficult to screen for mutants that affect them.
However, in the alcohol dehydrogenase (ADH:alcohol:NAD$^+$
oxidoreductase, EC 1.1.1.1) system of Drosophila, detection
of mutants is facilitated by a chemical screening method
that selects Adh-null activity mutants (3,4).   Moreover,
both the amino acid (5) and nucleotide sequences (6) have
been determined for the Drosophila ADH. Hence by examining
the amino acid and nucleotide sequences of a number of
chemically induced Adh-negative mutants, we hope to gain
insight into the mode of action of several environmental
mutagens in higher organisms.   In the present report we
concentrate on two mutagens: formaldehyde and ethyl
methanesulfonate (EMS).

FORMALDEHYDE MUTAGENESIS

Slizynska (7) found that formaldehyde mutagenesis
resulted in the production of a relatively large proportion
of deletions among first-instar larvae of Drosophila.   Of
114 chromosomal abnormalities examined, 25% were deletions.
The remainder were largely dulpications, with only a small
proportion of inversions and translocations.

Our own data on the Adh mutants generated with
formaldehyde substantiates these findings.   Of 18
formaldehyde generated lesions 14 were found to be deletions
recognizable by the absence of salivary chromosome bands in
the region around the cytological position of the Adh-gene,
and by the inability of these mutants to complement a number
of markers known to be close to the Adh-gene (8). The four
remaining formaldehyde generated mutants, designated
Adh^{fn4}, Adh^{fn6}, Adh^{fn23}, and Adh^{fn24} are mutants
that appear to exhibit no detectable loss of genetic
material when their salivary chromosomes are examined, nor
do they show a lack of complementation with adjoining genes.
Only by examination of the gene or gene product directly can
the type of lesion in these mutants be determined.

Of these four mutants, only Adh^{fn23} synthesizes detectable ADH-like protein. When cross-reacting-material (CRM) was immunoprecipitated with goat anti-ADH antibody and analyzed by SDS-PAGE in high percentage acrylamide slab gels, the CRM for Adh^{fn23} migrated faster than the wild type protein from $\overline{Adh}^D$. We estimate that the apparent molecular mass of Adh^{fn23} is 26,700, while the wild type subunit has an apparent M_R of 28,000.

To determine the basis of the observed difference in molecular mass between wild type and mutant proteins, the ADH-CRM from Adh^D and Adh^{fn23} were subjected to peptide fingerprinting (9). Adult males of each strain were labeled in vivo with ^{3}H-tryptophan (10), isolated by indirect immunoprecipitation (11), digested with trypsin and the peptides resolved on thin layer silica plates (9). In wild type ADH, there are five tryptophan residues, each located in a different tryptic peptide (6). Three of the tryptophan containing peptides are located at the C-terminus and are represented by a cluster of three acidic spots on the fingerprint (see Figure 1). The lesion in Adh^{fn23} appears to involve these three C-terminal peptides. Fingerprints of tryptic peptides from Adh^{fn23} are missing the last two C-terminal peptides. Fingerprints obtained with CRM from ADH^{fn23} labeled with a mixture of amino acids designed to label vitually all the tryptic peptides were found to be missing only these two peptides.

To determine if the lesion extends beyond the C-terminus, DNA from Adh^{fn23} was cloned into lambda Charon 4 and phage containing the Adh gene isolated. Based on restriction enzyme digestion of these clones the gene for Adh^{fn23} has a deletion of 60 to 70 bases in the C-terminal protein coding region of the Adh gene. The 3' limit is between the Hinf (nucleotide 1443) and Dde I (nucleotide 1455) restriction sites. The 5' limit is around nucleotide 1380 to 1390. This extent of missing nucleotides would be expected to eliminate the last two tryptic peptides, as observed in the peptide maps. It is unclear at this time whether the original termination codon is used or whether a new termination codon has been formed. DNA sequencing of the clone should answer this and other questions concerning the nature of this deletion.

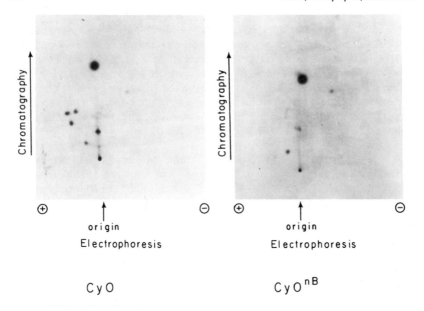

FIGURE 1. Tryptic peptide maps of alcohol dehydrogenase
    isolated from Adh^{Cyo} and Adh$^{Cyo\ nB}$.
Fluorograph for protein isolated from 20 individuals labeled
in vivo with ^{3}H-tryptophan. Note the three missing spots
in Adh$^{Cyo\ nB}$.

ETHYL METHANESULFONATE MUTAGENESIS

Ethyl methanesulfonate (EMS) is a potent mutagen in
Drosophila. It is easily applied by feeding adult males
(12) and per locus mutation frequencies as high as 2% have
been reported (13). The vast majority of EMS induced nulls
in Drosophila do not appear to involve gross chromosomal
rearrangements (1). In microorganisms, it appears to induce
predominantly single base alterations (14).

Of 22 EMS induced ADH-mutants, none could be
unequivocally classified as a deletion. In fact, all but
five were found to yield an ADH-like protein. When this CRM
was analyzed by SDS gel electrophoresis it was found, in all
but two cases, to migrate with the same molecular mass as
wild type ADH. Two of these mutants have been analyzed at

the amino acid and nucleotide level. In the electrophoretic positive mutant, Adh^D, a glycine (Gly 232) is changed to glutamic acid (a GGA to GAA change), while in the negative mutant, Adh^{n11}, a glycine (Gly 11) to aspartic change has occurred (a GGC to GAC change). In both cases a G to A transition has been induced.

Adh$^{Cyo\ nB}$, one of the two EMS nulls which behaves as a shorter protein on SDS gels, was found to be a C-terminal deletion. Figure 1 is a photograph of the tryptic peptide fingerprint for ^{3}H-tryptophan labeled protein. Note that the three acidic C-terminal peptides are missing in Adh$^{Cyo\ nB}$. Preliminary evidence indicates the change of a tryptophan codon (Trp 235) to a termination codon (TGG to either TGA or TAG) in Adh$^{Cyo\ nB}$. Thus in all three instances, EMS has induced G to A transitions.

These findings are in accord with data derived from other organisms. In T4 (15) and E. coli. (16) the major consequences of EMS mutagenesis are transitions from GC to AT base pairs. Moreover, although EMS reacts primarily with the N-7 of guanine (2), most evidence suggests that the alkylation of the O-6 of guanine is the main couse of mutatagenesis (2). However, some studies of EMS mutagenesis in fungi (17) and more recently in Drosophila (18), have been inconsistent with these results and its has been questioned whether chromatin from eukaryotes is affected in the same way by EMS as is the genetic apparatus of prokaryotes. Our studies of the Adh-mutants while limited in scope, suggest that for the Drosophila Adh-locus, EMS may cause mutations by the mechanisms postulated for prokaryotes. More data obtained by sequencing the protein and gene for additional mutants will help to verify this conclusion.

In summary, formaldehyde appears to induce largely deletions at the Adh locus. Even Adh^{fn23}, which was not detected as a deletion by genetic or cytological methods, prove to be a deletion when analyzed by restriction enzyme digestion. EMS, on the other hand, appears to induce predominantly point mutants, with a strong propensity for causing G to A transitions. It is apparent from these sets of studies that the ADH system of Drosophila will prove extremely valuable in characterizing the mode of action of mutagens at the molecular level.

ACKNOWLEDGEMENTS

This work was supported by Grants GM-28791 from the National Institutes of Health, ES-02920-01 from the National Institute of Environmental Health Sciences, and Contract #DE-AC02-81EV10566 from the Department of Energy to W. Sofer. This work was also aided by Grant #IN-135 from the American Cancer Society to A. Place.

REFERENCES

1. Auerbach, C. (1976) "Mutation Research", John Wiley & Sons, Inc., New York.

2. Singer, B. (1977) J. Toxicology and Environmental Health 2, 1279-1295.

3. Sofer, W. and Hatkoff, M.A. (1972) Genetics 72, 545-549.

4. O'Donnell, J., Gerace, L., Leister, F., and Sofer, W. (1975) Genetics 79, 73-83.

5. Thatcher, D.R. (1980) Biochem. J. 187, 875-886.

6. Benyajati, C., Place, A.R., Powers, D.A., and Sofer, W. (1981) Proc. Natl. Acad. Sci. USA 78, 2717-2721.

7. Slizynski, B.M. (1945) Proc. Roy. Soc. Edin. 62B, 114-119.

8. O'Donnell, J., Mandel, H.C., Krauss, M., and Sofer, W. (1977) Genetics 86, 553-566.

9. Fishbein, J.C., Place, A.R., Ropson, I.J., Powers, D.A., and Sofer, W. (1980) Anal. Biochem. 108, 193-201.

10. Reddy, A.R., Pelliccia, J.G., and Sofer, W. (1980) Biochem. Genet. 18, 338-351.

11. Gough, N.M. and Adams, J.M. (1978) Biochemistry 17, 5560-5566.

12. Lewis, E. and Bacher, F. (1968) Dros. Inf. Serv. 43, 193.

13. Jenkins, J.B. (1967) Mutation Res. 4, 90-92.

14. Drake, J.W. (1970) "The Molecular Basis of Mutation", Holden-Day, San Francisco.

15. Bautz, E. and Freese, E. (1960) Proc. Natl. Acad. Sci. USA 46, 1585-1594.

16. Osborn, M., Person, S., Phillips, S., and Funk, F. (1967) J. Mol. Biol. 26, 437-448.

17. Malling, H. V. and de Serres, F.J. (1970) Mutation Res. 6, 181-193.

18. Shukla, P.T. and Aurbach, C. (1981) Mutation Res. 83, 81-89.

Use of High Performance Liquid Chromatrography (HPLC) for the Separation of Proteins, Peptides, and Amino Acids

Protein Microsequencing by HPLC Peptide Purification and

Solid Phase Edman Degradation

James J. L'Italien & Richard A. Laursen

Yale University   Boston University
Dept. Internal Medicine   Dept. of Chemistry
New Haven, CT   Boston, MA

Introduction

The rate determining step in the total sequence analysis of proteins, by protein chemical means, is generally found to be the isolation of purified fragments in sufficient quantity for sequence analysis. The amount of material necessary for automated sequence analysis is dependent upon both the level at which the sequenator can function and the level at which the individual amino acid residues resulting from the sequencing process can be identified. In addition, if the sequencing method used is solid-phase Edman degradation then the level at which fragments can be immobilized to a solid-support must also be considered. The application of High Performance Liquid Chromatography (HPLC) to the problem of peptide purification (1-3) has greatly reduced the time required for the isolation of purified material while extending the lower working limits into the picomole range using totally non-destructive means (UV absorbance) for detection. Identification of PTH-amino acids by HPLC using UV detection at 254 nm has lowered the level of detection into the 1-100 pmole range (4,5) while allowing each cycle to be easily quantified. We have recently examined peptide immobilization at levels of material which are commensurate with those obtained by HPLC peptide purification during the elucidation of the sequence of the polypeptide elongation factor Tu (EF-Tu) from E. coli (5,6). Here we report

383

further refinements in the integration of peptide purification via HPLC and solid-phase sequencing to form a
system capable of the determination of complete sequence
information on a microscale without the use of radiolabels
or modified PITC. The salient points of this system are
given in Table 1.

Table 1   Microsequencing Strategy

1. Screen methods and conditions of cleavage (chemical
    and proteolytic) via SDS-PAGE and RP-HPLC.
2. Optimization of fragment separation via RP-HPLC.
3. Preparative purification of peptides via RP-HPLC using
    optimized profile.
4. Collection of peaks as discrete fractions following
    direct UV detection at both 206-214 nm and 254 nm.
5. Assessment of peak purity by dansylation and/or amino
    acid analysis.
6. Direct immobilization of purified peptides to solid-
    supports without removal of HPLC buffer salts used in
    purification.
7. Solid-phase sequencing of immobilized peptides using
    65 minute sequenator program and automated conversion
    of PTH amino acids.
8. Direct analysis (i.e. no extraction) of PTH amino
    acids (>10pmole) by RP-HPLC.

Materials and Methods

Protein Isolation and Fragmentation

    The polypeptide elongation factor Tu (EF-Tu) was
isolated from E. coli B cells as described by Miller and
Weissbach (7). One mg of the protein was reduced and
alkylated by the method of Konigsberg (8). Following dial-
ysis and lyophilization, a cyanogen bromide digest was
performed on 600μg of protein in 250μl of 70% formic acid
by adding 1.5 mg of CNBr (Aldrich) and allowing the
reaction to proceed for 24 hours in the dark. The sample
was then dried under N_2, dissolved in water and lyophilized. The remaining 200μg of carboxyamidomethyl EF-Tu was
digested with trypsin-TPCK (Worthington) as previously
described (5).

HPLC Separation of Peptides

    Peptide separations were performed on either an Altex

Model 324 Chromatography System (2 Model 100A pumps,
Model 421 Microprocessor System Controller, a dynamically
stirred gradient mixing chamber, a Rheodyne Model 7125
sample injector, a Kipp and Zonen dual pen chart recorder,
a Model 153 Altex detector, and a Gilson Holochrome
detector) or a Waters Chromatography System (2 Model 6000A
pumps, a U6K injector, a model 720 System controller, a
Model 730 Data Module, a Model 440 detector and a Gilson
Holochrome detector).

CNBr fragments were dissolved in 6M̲ GnCl (Pierce) and
applied directly to a μBondapack C-18 column (Waters
Assoc.) equilibrated with 0.05% trifluoroacetic acid
(Sequemat, Inc.) in water. The column was then developed
using gradients of acetonitrile (Baker, HPLC grade) which
was also 0.05% in trifluoroacetic acid. The purity of the
fractions was assessed by 10-30% gradient SDS-PAGE (9).
The large CNBr fragments were tryptically digested as previously described (5). Tryptic peptide mixtures (from
EF-Tu and the large CNBr fragments of EF-Tu) were dissolved
in 50% acetic acid (Baker reagent) and applied to a
μBondapack C-18 column (preceded by a precolumn described
in 5) equilibrated in 20mM̲ potassium phosphate (pH 2.5).
Peptides were eluted with acetonitrile gradients which
were generally optimized for each mixture. The column
eluates were monitored nondestructively by direct UV
detection at both 206-214nm and 254nm which allowed assessment of the aromatic content of each peptide while providing a second fingerprint for comparative purposes. Each
peak was collected as a discrete fraction, neutralized
with triethylamine (redistilled) and dried prior to further
characterization.

Immobilization of Peptides

All peptides (<30 residues) were immobilized to aminopolystyrene via activation of their C-terminal carboxyl
group by a water soluble carbodiimide. The procedure used
was modified from that of Wittmann-Liebold et al. (10)
for coupling in the 1-5nmole range. Five hundred μl of
1.0M̲ pyridine chloride (pH5.0) (pyridine redistilled from
phthalic anhydride) was added to 10 mg of aminopolystyrene
(Sequemat) and the resin was stirred at 30°C for 10 minutes
or until a color transition of the resin was observed. The
resin was then washed 2 times with 3 mls of water and 2
times with 500 μl of dimethylformamide (redistilled from

P_2O_5). The lyophilized peptide was dissolved in 100 µl of
1.0 $\underline{M}$ pyridine chloride (pH 5.0) and transferred to the
swollen resin. The original sample tube was rinsed with
100 µl of 1.0 $\underline{M}$ pyridine chloride followed by 100 µl of dimethylformamide (DMF) and both rinses were added to the
peptide resin mixture. A fresh solution of 1-ethyl-3,3'-di-
methylaminopropylcarbodiimide·HCl (Sigma) was prepared such
that 3 mg of carbodiimide in 75 µl of water–DMF (1:4) was
added to each tube. The reaction was allowed to proceed at
30°C for 1 hour with stirring, after which the sample was
centrifuged and the supernatant removed. The pH of the
resin was raised to ∿9 by adding 100 µl 1.0 $\underline{M}$ N–methyl-
morpholine (Pierce) and 200 µl of DMF, and allowing the
resin to stir at 30°C for 10 minutes. The sample was then
centrifuged, the supernatant removed and the excess amino
groups were blocked by adding 150 µl DMF, 50 µl of N–methyl-
morpholine, and 50 µl of phenylisothiocyanate (Pierce). The
blocking reaction was allowed to proceed at 30°C for 1 hour.
The excess reagents were removed by washing the support
with 4 mls of methanol (Baker, reagent) 3 or 4 times. The
last traces of methanol were removed by drying under vacuum
(water aspirator) and the coupled peptide was stored under
refrigeration until sequenced. Other methods of immobilization used were as previously described (5).

Solid–Phase Edman Degradation

    All peptides were sequenced by Edman degradation on a
Sequemat Mini-15 solid-phase sequenator equipped with a
P-6 Autoconvertor. A 65 minute sequenator program (Table
2) was developed using the standard reagents. The solvents
were methanol and dichloroethane (both Baker reagent), the
coupling buffer was 1.0 $\underline{M}$ N-methylmorpholine-trifluoroace-
tic acid (pH 8.1): pyridine (2:3) (Sequemat), the PITC
(Pierce) concentration was 10% in acetonitrile (Baker, HPLC
grade) and trifluoroacetic acid (Sequemat). The autoconvertor reagents were 2$\underline{M}$ methanolic·HCl which was prepared from
acetylchloride (Sequemat) and the wash solvent was dichloro-
ethane-methanol (7:3) (both Baker reagent). PTH's were collected in centrifuge tubes and dried under N_2. The samples
were then dissolved in 300 µl of methanol (Baker, HPLC
grade) and dried under N_2 again to remove the last traces of
a volatile pyridinium salt which was present in the samples.
The PTH's were then dissolved in 50 µl of methanol (Baker
HPLC grade) prior to application to HPLC for identification.

Table 2. Single Column Program (65 minutes)

Step	Functions	Channel	Time
0	Start	8	0-1
1	MeOH	1,8	1-5
2	Buffer	3,8	5-6
3	Buffer/PITC	3,4,8	6-19
4	Buffer	3,8	19-23
5	MeOH	1,8	23-31
6	DCE	2,8	31-35
7	MeOH	1,8	35-39
8	DCE	2,8	39-43
9	MeOH/F.C.	1,9,8	43-45
10	TFA/Collect/Reag.Part.	5	45-54
11	TFA/Collect/Channel 15/Reag.Part.	5,15	54-56
12	TFA/Collect/Reag.Part.	5	56-62
13	MeOH/Collect	1	62-63
14	Rest	8	63-64
15	MeOH/Collect	1	64-65
16	End	E.P.	65-0

This single column sequencing program has been used on an otherwise unmodified Sequemat Mini-15 Solid Phase Sequenator. Channel numbers are those referred to in the Sequemat Mini-15 Manual. The reagents used are described in the text. Channel 15 is the signal for the P-6 Autoconvertor. The drying time of the unconverted sample in the Autoconvertor was shortened by 4 minutes because of the decreased volume of solvent to be dried.

Results and Discussion

The separation of CNBr fragments of EF-Tu (Fig.1) demonstrates that complex peptide mixtures over a broad range of fragment sizes can be separated on a single chromatographic run, detected non-destructively and collected for further characterization in a reasonable time period. The speed and sensitivity of the technique allows the elution profile to be optimized on minute amounts of material (5-20μg of peptide mixtures). The use of trifluoroacetic acid in both solvents appears to maintain the solubility of even large fragments at the high acetonitrile concentrations sometimes needed to elute them from reverse phase columns (concentrations of organic solvent which might otherwise cause precipitation). Furthermore, the addition of trifluoroacetic acid to both solvents greatly reduces

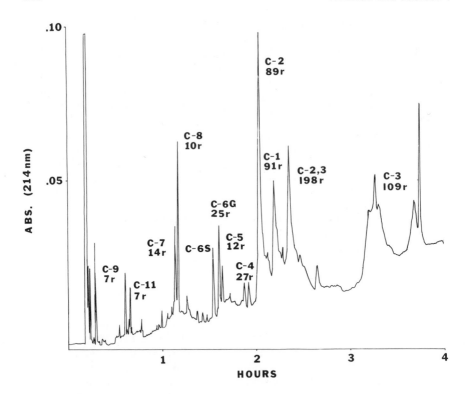

Fig. 1.   HPLC separation of 40μg of EF-Tu CNBr fragments.
The lyophilized fragments were solubolized in 6M guanidine
hydrochloride and applied to a μBondapack C-18 column
(Waters Assoc.) equilibrated in 0.05% trifluoroacetic acid
in water (v/v). Peptides were eluted with the following
successive linear gradients of acetonitrile (which was
also 0.05% in trifluoroacetic acid): 0-50 min (0-25%
acetonitrile), 50-150 min (25-50% acetonitrile), 150-170
min (50-70% acetonitrile). The initial column backpressure
was 900 psi at the 1.0 ml per min flow rate used for this
chromatogram. The eluent was monitored at both 214nm and
254nm (trace not shown). Each peak is designated by a C-
number which refers to CNBr fragment numbers previously
reported (ref.6). The number preceding the r refers to
the number of residues in that particular fragment.

the problem of baseline drift which can be troublesome when
monitoring by direct UV (206-214nm) at high sensitivity.
In addition, this buffer system is completely volatile
which enables subsequent chemical characterization and/or
proteolytic digestion without desalting. All 11 of the
major CNBr fragments of EF-Tu, with a range of fragment
size from 2-109 residues, were detected in this chromatogram as was a partial cleavage product of 198 residues.
The nomenclature of the fragments is consistent with that
previously reported (6) and the actual number of residues
in each peptide is designated on the chromatogram. The one
site of microheterogeneity in EF-Tu is readily discernable
on this profile as two distinct peaks. This Gly-Ser substitution in the 25 residue peptide CB-6 was designated as
CB-6S for the serine containing peptide and CB-6G for the
glycine containing fragment. The serine form of the peptide eluted earlier from the column than did the glycine
form as might be expected by methods of prediction of peptide elution (11,12). The order of elution of the peptides
from the column was not based strictly on size, although as
a group all peptides less than 30 residues did elute earlier
than peptides containing greater than 80 residues.

     The resolution and reproducibility of HPLC for the
separation of complex peptide mixtures is illustrated in
Figure 2. In this figure we have compared the elution of
tryptic peptides from both intact EF-Tu and its large CNBr
fragments using the same column and gradient conditions.
This comparison allowed the identification of the methionine and homoserine containing peptides, which were necessary for establishing overlaps, as well as aiding in the
establishment of linear domains. As seen in this figure
the proteolytic digestions of the CNBr fragments gave
simpler profiles which made it easier to purify each of the
peptides but this advantage must be tempered against the
material cost of obtaining the purified CNBr fragments. In
recent studies on variant specific glycoproteins (VSG's)
from T.congolense, it has been observed that this material
tradeoff may not be favorable as the efficiency of tryptic
digestion and separation via HPLC (even if a peak may have
to be rechromatographed) is much greater than isolation of
CNBr fragments for tryptic digestion to simplify the profile (13). This finding is perhaps due to both the efficiency of proteolytic digestion vis-a-vis chemical cleavage
and the efficiency of recovery of fragments less than 50
residues from reverse phase columns. Finally, we prefer to

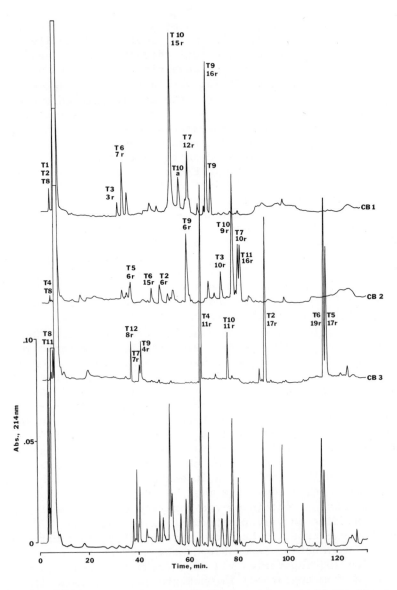

Fig.2.   HPLC separation of EF–Tu tryptic peptides. These
traces represent the tryptic peptide profiles of 1 nmole of
peptide mixture from intact EF–Tu (lowest trace) and the
three large CNBr fragments (from Fig.1), CB 1 (91 residues)
CB 2 (89 residues) and CB 3 (109 residues). Each of the

tryptic peptide mixtures were dissolved in 50% acetic acid and applied directly to a µBondapack C-18 column equilibrated in 20 mM potassium phosphate (pH 2.5). All of the profiles were developed with the same linear gradient of acetonitrile (0–40%) over 120 minutes at a flow rate of 1 ml per min. The eluent was monitored at both 214 nm and 254 nm (trace not shown).

develop all proteolytic peptide mixtures at low pH to give maximum peptide retention and resolution. We have, however, altered the pH of the aqueous buffer from pH 2.2, as previously reported (5), to pH 2.5 because it appears to extend column life without adversely affecting resolution.

Perhaps the greatest concern in the use of solid-phase sequencing at any sequencing level has been immobilization of peptides. The earliest reports by Laursen of C-terminal activation by carbonyldiimidazole (14) had caused many protein chemists to shy away from the methods of C-terminal attachment because of problems with sidechain coupling leading to termination of the sequencable peptide. These problems were circumvented through the development of the DITC (15) and lactone (16,17) procedures which rely upon the specific chemistry of the sidechain at the C-terminal residue of the peptide for immobilization. Nevertheless, there remained a need for a more universal method of attachment of peptides to solid supports. A number of carbodiimide coupling procedures have been proposed in recent years (10,18,19) and demonstrated using model peptides, the most useful of which in our hands was that described by Wittmann-Liebold et al. (10).   Numerous papers have reported that coupling yields by this procedure are low but to our knowledge there has been no extensive quantative data published on coupling yields of peptides from proteolytic digestion using this method. Of the last 60 peptides coupled and sequenced from the three large CNBr fragments of EF-Tu: 48 were coupled by carbodiimide to aminopolystyrene, 6 were coupled DITC to aminopolystyrene, 1 was attached to DITC-Glass, 4 were coupled via homoserine to triethylenetetra-amine (prior to HPLC of PTH's) and 1 was attached via homoserine to β-aminopropyl glass (after HPLC of PTH's). These peptides were coupled and sequenced from the 50 nmole level (prior to HPLC of PTH's) down to the 5-10 nmole level (following implementation of HPLC for peptide separation and PTH analysis) with slight modification (5) of the Wittmann Liebold et al. immobilization procedure (10).   Further-

more it has been used successfully in over 40 cases in the
structural studies of VSG's from T.congolense, while DITC
to aminopolystyrene has been used 3 times (to selectively
couple lysine peptides from mixtures of lysine and arginine peptides eluting in a single peak from HPLC), and
DITC-Glass 5 times for intact protein and large CNBr fragments. In every situation in which the carboddimide method was used it appeared to be highly selective for the C-
terminal carboxyl of the peptide as both aspartic and glutamic acid residues were readily discernable and there was
no detectable loss of peptide in the cycle following
either of these residues. In Table 3 we list the coupling
yields for 27 peptides (from 5 to 22 residues) and give
average coupling yields for 13 more peptides. The average
coupling yields for these 40 peptides was 57.3% with the
average being somewhat lower for peptides 16-22 residues
long.

Table 3.   Carbodiimide Immobilization of Peptides Isolated
           via HPLC Using Phosphate Eluting Buffers

Peptide	Coupling Yield	Nmole*
AIVGGWGNPTTPDESGLPTTFK	40%	1.7
AGDNGSGTLDNNGDNTGKPAK	77%	2.8
SAKAKFTKAIVGGWGNPTTPDE	43%	4.5
XLVFDIACLCTTSDSASSSTK	25%	1.7
LSFTEPSAVVTTLDGTR	66%	2.2
AGEGLKEEDWLPCAGK	39%	3.7
XTINTXHVEYDTPTR	51%	5.0
XYAHVDCPGHADYVK	36%	5.0
XTPELISTXLVIELIXXR	25%	0.7
YELQNSASTR	54%	4.0
TAASIEDMFVK	31%	1.5
ELVAALQAR	37%	2.5
DWLPCAGKAACE	65%	6.5
FESEVYILSK	48%	5.0
GYRPQFVFR	59%	6.0
AFDQIDNAPEEKAR	48%	7.5
MAEGDLR	53%	3.4
YAGGNGK	87%	1.7
AQIAWE	63%	7.6
KARGITINTSHVE	53%	7.5
SYIPEPE	69%	9.0
GLLTAK	29%	2.8
YTCGPK	51%	4.1

Table 3, continued

Peptide	Coupling Yield	Nmole*
HTPFFK	66%	6.5
AQQLK	74%	3.0
ANFTK	47%	3.0
GCIDYK	46%	3.7

The average coupling yield for 10 peptides between 16 and 22 residues was 48.6% with a range of 25-77%.

The average coupling yield for 30 peptides between 5 and 15 residues was 60.2% with a range of 29-87%.

*The column titled nmole refers to the number of nmoles placed on the sequenator (which was the total amount of peptide coupled minus that percentage used to determine the coupling yield).

All of these peptides were sequenced in their entirety using the 65 minute sequenator program previously described.

Table 4. Microsequencing of Tryptic Peptide Mixture

Peptide	Rec. in[a] pmole	CY[b] (pmole)	Seq.[c] (pmole)	Rep.[d] Yield
ALEGDAEWEAK	900	55%(500)	60%(300)	91±1%
VGEEVEIVGIK	900	50%(450)	60%(270)	91±1%
ELSVYDFPGDDTPIVR	900	61%(550)	n.d.	n.d.
ILELAGFLDSYIPEPER	500	60%(350)	57%(200)	91±1%
AIDKPFLLPIEDVFSISGR	250	60%[e]	n.d.	n.d.

a   Recovery refers to the amount of each peptide isolated following application of 1 nmole of tryptic peptide digest to the HPLC. This value is the average of two samples which were measured in pmole following amino acid analysis.

b   The percent coupling yield refers to the amount of peptide immobilized to the support divided by the amount of initial peptide. This value is the average of two samples as measured by amino acid analysis. The number in parenthesis is the average number of pmoles coupled.

c   The percent sequencable peptide is that percentage of the immobilized peptide which is capable of being sequenced as judged by PTH analysis on HPLC. The number in parenthesis refers to the amount of peptide in pmoles which was actually sequenced.

d   The repetitive yield was calculated as per Smithies (20)
    from PTH analysis on HPLC.
e   This value was determined from a previous experiment
    starting with 1 nmole of peptide. The amount of this
    peptide coupled in this experiment fell below the limit
    of detection of the amino acid analyser.

    To demonstrate the feasability of the total sequencing
method which we have proposed here, we took 1 nmole of the
109 residue fragment from EF-Tu (CB-3) which had been isolated via HPLC (as in Fig.2) and digested it with trypsin
(6 samples were handled in parallel). Following digestion
each mixture was dissolved in 50% acetic acid and applied
to a µBondapack C-18 column. The profile of this chromatogram is shown in Figure 3. The individual peaks were collected as discrete fractions following direct UV detection,
neutralized, and dried. Two samples of each peptide were
hydrolysed prior to attachment, and the remaining four samples of each peptide were immobilized to aminopolystyrene
via carbodiimide activation. Two of the four coupled peptide samples were hydrolysed to give the coupling yields
and the remaining samples were sequenced. The results are
listed in Table 4.
    The recovery of the first three peptides listed in
Table 4 was quite good. Recovery of the last two peptides,
which eluted as a doublet (Fig.3) was reduced as a result
of the necessity of shaving peaks in order to minimize cross
contamination. The coupling yields for all of the peptides
were between 50-60% while the sequencable yields were found
to be 60% of the immobilized peptide. Thus, on an average,
we were sequencing 30-36% of the initial peptide amount.
These levels proved sufficient to sequence through 16 cycles
starting with 500 pmole (before coupling) of the peptide
ILELAGFLDSYIPEPER (Table 4). The data for peptides less
than thirty residues compares quite favorably with the recently described Gas-Liquid Phase Sequenator (21), which
required 1.4 nmole to sequence 14 cycles of Somatostatin
and 500 pmole to sequence 8 cycles of Angiotensin. Furthermore, the data listed in Tables 3 and 4 show that significant sequence information can be obtained on virtually any
peptide less than 30 residues starting with 500 pmole to 10
nmole of peptide using carbodiimide immobilization.
    The immobilized peptides were sequenced using the 65
minute program previously described on a Sequemat Mini-15
Sequenator. An internal standard (500 pmole PTH Norleucine)

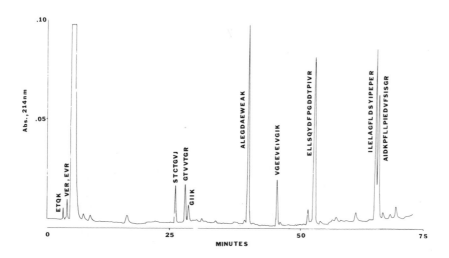

Fig.3. HPLC separation of 1 nmole of EF-Tu CB-3 Tryptic
Digest. The peptide mixture was dissolved in 50% acetic
acid and applied to a μBondapack C-18 column equilibrated
in 10 mM potassium phosphate (pH 2.5). The peptides were
eluted from the column using a series of linear gradients
of acetonitrile (solvent B): 0-10 min (0.5%B), 10-55 min
(5-35%B), 55-70 min (35-50%B). The initial column back
pressure was 2000 psi at the 1 ml per min flowrate used to
develop the chromatogram. The column eluate was monitored
at 214 nm. The sequence of each peptide is given by the 1
letter code adjacent to its corresponding peak. All 109
residues of this fragment are accounted for in this chromatogram.

was added to each cycle prior to drying and direct application of the sample to HPLC for identification. The
actual tracing of one of these samples (ALEGDAEWEAK from
Table 4) is shown in Fig.4. The major PTH amino acid peak
is clearly visible in each cycle. There are three major
peaks of UV absorbing contaminants from the sequencing
process which elute: 1) between PTH Arg and PTH Ala, 2)
between PTH Pro and PTH Trp and 3) between PTH Trp and PTH
Phe. These UV absorbing peaks do not interfere with the
identification of any of the PTH-amino acids. In Fig.4
cycle 1 there is an unsymmetrical peak between 5 and 8
minutes on the trace. This peak is the pyridine peak

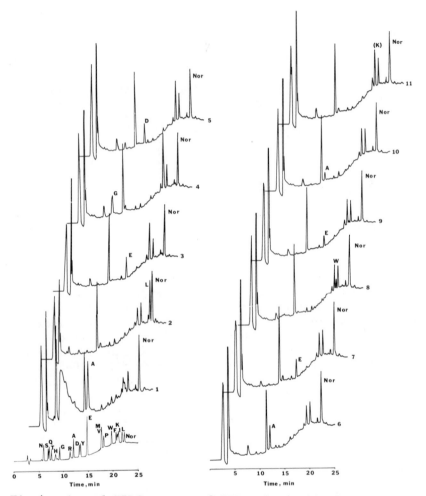

Fig.4. Actual HPLC traces of PTH amino acids from the sequencing of 500 pmole of the peptide ALEGDAEWEAK coupled to aminopolystyrene via carbodiimide as described in the text and Table 4. Each cycle from the Sequenator (after conversion) was dried twice under N_2 following the addition of 500 pmole of PTH Norleucine. Each sample was then dissolved in 50 µl of methanol and applied to an Altex Ultrasphere RP-18 column (5 micron particle size). PTH's were eluted at room temperature with a series of linear gradients of acetonitrile (solvent B) into 100 mM sodium acetate (pH 4.1) (Solvent A): 0-10min (27-45%B), 10-15 min (45-50% B), 15-20 min (isocratic at 50% B), 20-25 min (50-60%B),

25-27 min (isocratic at 60%B), 27-33 min (60-27%B). The
PTH standard shown is 50 pmole except for PTH Glu which is
100 pmole. Both PTH Asp and PTH Glu appear as methyl
esters resulting from conversion to PTH's in 2$\underline{M}$ Methanolic-
HCl. The cycle number of each step is given immediately
following each trace. Detection is by direct UV at 254nm.

referred to in the Materials and Methods section which can
be removed (as seen in the remaining cycle traces) by
simply redrying the sample under N$_2$ after adding a small
volume of methanol.

Summary

The principle advantages of the sequencing strategy
outlined here are:

1. It allows for the development of a logical fragmentation strategy by permitting the screening of methods
and conditions on minute amounts of material. Thus, those
methods and conditions which will provide a maximum of information on limited samples can be determined and used
while those methods determined to be less useful can be
eliminated without wasting much material.

2. It eliminates handling steps (and thus loss of
material) by direct application of peptide samples to HPLC
and collection of peaks as discrete fractions following
nondestructive UV detection at both 206-214nm and 254nm.
In addition, once an optimized elution profile is established scale up to preparative (i.e. collection) runs is
generally found to be linear.

3. Immobilization of peptides without removal of
buffer salts (potassium phosphate) both eliminates losses
from desalting and streamlines the number of steps involved
from peptide mixture to sequenator. Furthermore, attachment of peptides (<30 residues) by carbodiimide activation
permits the identification of every residue in the peptide.

4. The use of the 65 minute program on an otherwise
unmodified commercially available solid-phase sequenator
allows the determination of 22 cycles per day with low
purchase and reagent cost.

5. Direct analysis of PTH amino acids via HPLC without

manipulation or extraction allows for the quantative assessment of all PTH's in every cycle which is important
for sequencing at low levels (≤1nmole) where extraction
may preferentially remove some PTH's in addition to PTHArg
and PTH His.

Acknowledgements

We gratefully acknowledge the contribution of Dr.
James E. Strickler for helpful discussions during the preparation of this manuscript. We thank Charlene Sullivan
for typing this manuscript. This work was supported in
part by USPHS grants AI09614-14, AI15530-03 and a grant
from the American Heart Association 81-661 to Frank F.
Richards and NSF grant PCM 79-04910 to Richard A. Laursen.

References

1. Hancock, W.S., Bishop, C.A. and Hearn, M.T.W. (1976)
    FEBS Lett. 72, 139.
2. Fullmer, C.S. and Wasserman, R.H. (1979) J. Biol. Chem.
    254, 7208-7212.
3. Pearson, J., Mahoney, W.C., Hermondson, M.A. and
    Regnier, F.E. (1981) J. Chromatogr. 207, 325-332.
4. Johnson, N.D., Hunkapiller, M.W. and Hood, L.E. (1979)
    Anal. Biochem. 100, 335-338.
5. L'Italien, J.J. and Laursen, R.A. (1981) J. Biol. Chem.
    256, 8092-8101.
6. Laursen, R.A., L'Italien, J.J., Nagarkatti, S. and
    Miller, D.L. (1981) J. Biol. Chem. 256, 8102-8109.

7. Miller, D.L. and Weissbach, H. (1970) Arch. Biochem.
    Biophys. 141, 26-37.
8. Konigsberg, W.H. (1972) Methods Enzymol. 25, 185-188.
9. Strickler, J.E. and L'Italien, J.J. (manuscript in
    preparation)
10. Wittmann-Liebold, B., Braver, D., and Dognin, J.M.
    (1977) in Solid Phase Methods in Protein Sequence
    Analysis (Previero, A. and Coletti-Previero, M.-A.,
    eds.) p. 219-232, North Holland, Amsterdam.
11. Molnar, I. and Horvath, C. (1977) J. Chromatogr. 142,
    623-640.
12. Meek, J.L. (1980) Proc. Nat. Acad. Sci.(U.S.A.) 77,
    1632-1636.
13. L'Italien, J.J., Strickler, J.E. and Richards, F.F.
    (manuscript in preparation).
14. Laursen, R.A. (1971) Eur. J. Biochem. 20, 89-102.

15. Laursen, R.A., Horn, M.J. and Bonner, A.G. (1972) FEBS Lett. 21, 67-70.
16. Horn, M.J. and Laursen, R.A. (1973) FEBS Lett. 36, 285-289.
17. Wachter, E. and Worhahn, R. (1979) Anal. Biochem. 97, 56-64.
18. Previero, A., Derancourt, J., Coletti-Previero, M.-A. and Laursen, R.A. (1973) FEBS Lett. 33, 135-138.
19. Beyreuther, K. (1977) in Solid Phase Methods in Protein Sequence Analysis (Previero, A. and Coletti-Previero, M.-A., eds.) p. 107-120, North Holland, Amsterdam.
20. Smithies, O., Gibson, D., Fanning, E.M., Goodfliesh, R.M., Gilman, J.G. and Ballantyne, D.L. (1971) Biochem. 10, 4912-4921.
21. Hewick, R.M., Hunkapiller, M.W., Hood, L.E. and Dreyer, W.J. (1981) J. Biol. Chem. 256, 7990-7997.

REVERSE PHASE HIGH-PERFORMANCE LIQUID CHROMATOGRAPHY FOR

PROTEIN PURIFICATION: INSULIN-LIKE GROWTH FACTORS, Ca^{2+}-

BINDING PROTEINS AND METALLOTHIONEINS

KENNETH J. WILSON, MARTIN W. BERCHTOLD[+], PETER
ZUMSTEIN, STEPHAN KLAUSER AND GRAHAM J. HUGHES[+]

Biochemisches Institut der Universität Zürich,
CH-8028 Zürich and [+]Laboratorium für Bioche-
mie der Eidgenössischen Technischen
Hochschule, CH-8092 Zürich, Switzerland

INTRODUCTION

Utilization of high-performance liquid chromatography (HPLC) in peptide/protein chemistry has become widespread. The technique is being used for amino acid analysis (1,2,3), for peptide isolation from natural sources (4,5) or from enzymatic/chemical cleavages (6,7), for the identification of products arising from Edman degradations (8,9) and for determining the specificity of chemical cleavage methods (10). Its application in the chromatography of larger peptide fragments or proteins has been limited, in part, by the lack of suitable column packings. Until only recently most commercially available reverse phase packings exhibited nonspecific adsorption and/or reduced resolution. However, some proteins have been chromatographed under a variety of conditions on quite different column supports.

Cytochrome c, albumin, collagen and tyrosinase, representing a molecular weight range from 11'700 to 128'000,

[+]Permanent address: Institut für Pharmakologie und Biochemie, Tierspital, CH-8057 Zürich

have been chromatographed using pyridine formate buffers
and 1-propanol as eluant (11). Using 100 Å and 500 Å pore
sizes, 10 μm particle size, RP-8 packings yields exceeding
80% were observed. Hemoglobin chains have been separated
and recovered in yields >70% using RP-18 packings and pyridine formate-propanol buffers (7,12). Diol and RP-8 resins
(13) were used in the chromatography of ribonuclease (50 -
100% yields) and CN packings (14) for collagen samples (>80%
yields). Numerous column types were eluted with 2-propanol/
1-butanol mixtures in volatile buffers at low pH during the
purification of interferon (15). The phenyl-derivative has
proven useful in the chromatography of Ca^{2+}-binding proteins
in phosphate-acetonitrile buffers at pH 6.1 (16). Of the examples given only the work of Friesen et al. (15) aimed at
using HPLC as an isolation method for interferon. The remaining investigations reported the chromatographic characteristics of the indicated proteins on the various supports.

In a recent study (17) we have also investigated the
chromatographic behaviour of a number of proteins on both
100 Å and 300 Å pore size RP-18 packings. Using phosphate-
perchlorate buffers at low pH and either acetonitrile or
2-propanol as the organic modifier (18) recoveries for proteins with molecular weights exceeding ca 15'000 were higher
on the larger pore size packing. Conversely, the yields were
usually higher for the smaller proteins (<15'000) on the
100 Å support. On the basis of these observations it was
then possible to determine the usefulness (or lack thereof)
of reverse phase HPLC in the purification of a number of
smaller proteins, some of which are found in quite limited
quantities. This report deals with our preliminary results
on the isolation of insulin-like growth factors (IGF),
Ca^{2+}-binding proteins and metallothioneins using HPLC as an
integral step in the purification procedure.

MATERIALS AND METHODS

Chemicals. Acetonitrile and 2-propanol (HPLC) quality
were purchased from J.T. Baker, water was quartz bidistilled.
All other chemicals were of reagent grade quality from either
Fluka or Merck and used without further purification.
Apparatus. The HPLC instrument consisted of two Altex

model 100 pumps, a Rheodyne model 7125 injector (100 µl or 600 µl sample loops), a Kontron model 200 microprocessor and an Uvikon 725 spectrophotometer with an 8 µl flow-through cell was used for UV detection. Samples were chromatographed at room temperature on 4.6x250 mm columns with packings of 10 µm particle diameter. The supports used were LiChrosorb RP-18 (100 Å pore size) and Aquapore RP 300 (300 Å pore size) from Brownlee Labs. The column effluent was collected manually and the protein recovered by either vortex-evaporation or lyophilization. The buffers used for chromatography are indicated in the legends to each figure.

Methods. Prior to HPLC the proteins were partially purified by various means. To this end IGFs were enriched by chromatographing human serum under acidic conditions over Sephadex G-75 and pooling the fractions containing IGF activity, measured by radioimmunoassays (19). Following concentration by rotary evaporation the material was lyophilized and the resulting proteins dissolved in buffer A (see Fig. 1) and used for HPLC. For the Ca^{2+}-binding proteins an extraction of rat brain was performed at neutral pH, the cellular debris removed by centrifugation and the supernatant heated for 30 min at $80^{\circ}C$. Following centrifugation, the proteins contained within the supernatant were concentrated by TCA precipitation, dissolved and dialyzed at neutral pH. Subsequently, a chromatographic step on Sephadex G-50 superfine was carried out. The low molecular weight (5'000-30'000) fraction was utilized. Partially purified rabbit liver metallothioneins were prepared essentially as outlined by Kimura et al. (20). In short, liver material from cadmium-exposed rabbits was extracted in Tris-ethanol-chloroform at low temperature. Following centrifugation, the proteins in the supernatant were precipitated with ethanol and collected by centrifugation. Gel chromatography on Sephadex G-50 at neutral pH was then employed for further purification. The low molecular weight, Cd^{2+}-containing fractions (determined by atomic absorption spectroscopy) were pooled and lyophilized. This material was dissolved in buffer A (see Fig.4) for HPLC.

For two-dimensional isoelectric focusing-SDS gel electrophoresis samples were labeled by reductive methylation (21) using $NaCNBH_3$ and [^{14}C]formaldehyde and detected by fluorography according to Laskey and Mills (22). Protein re-

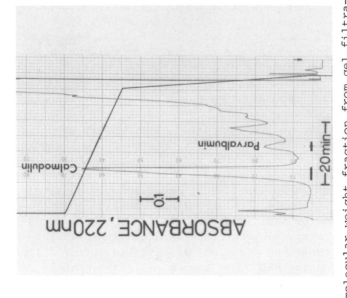

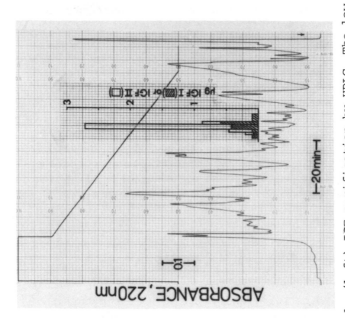

Fig.1. (left) IGF purification by HPLC. The low molecular weight fraction from gel filtration of serum was chromatographed on a LiChrosorb RP-18 (100 Å) column using a gradient formed between buffers A, 0.1% H₃PO₄-10mM NaClO₄, and B, 60% acetonitrile in A buffer; arrows indicate point of injection. Fig.2. (right) HPLC purification of Ca²⁺-binding proteins. Pre-purified material from Sephadex chromatography of rat brain extract (see Methods) was chromatographed under the conditions given in Fig.1; buffer B contained,however,60% (v/v)2-propanol.

coveries from HPLC were estimated by comparing peak areas,
by the recovery of counts following the inclusion of a radio-
actively-labeled sample of the same protein or by amino acid
analysis on a Durrum D-500 analyzer following hydrolysis in
6M HCl under vacuum for 24 h.

RESULTS AND DISCUSSION

The reverse phase chromatography of a serum sample containing IGF activity is illustrated in Fig.1. Based on radio-
immune assays recoveries for each factor were in excess of
85% of the amounts injected. Although a clear separation of
the factors, IGF I and II, is not apparent in the illustrated chromatogram, it can be accomplished (23). HPLC is presently being used in an attempt to replace some of the previously used methods employed in IGF purification which were
extremely laborious, time-consuming and suffered from low
yields (24).

Various Ca^{2+}-binding proteins have also been isolated
by HPLC. Their extreme stability towards both heat and exposure to acidic conditions, allows a preferential purification to be carried out prior to chromatography. Thus, the
number of components as well as the actual sample concentration can be reduced prior to injection. Fig.2 shows the elution pattern of a rat brain extract and two of the peaks
have been identified as calmodulin and parvalbumin. Rechromatography in a dilute TFA buffer system using acetonitrile
as organic eluant indicated homogeneity of calmodulin. This
was further attested by twodimensional electrophoresis (Fig.3).
The parvalbumin fraction was homogeneous, as confirmed by
electrophoresis (results not shown), only after rechromatography in the TFA buffer system.

The third example where HPLC was instrumental in the
purification of smaller proteins is illustrated in Fig.4.
Although the fraction from gel chromatography (see Materials)
contained predominantly metallothionein this step allowed not
only the separation of the isoproteins previously designated A and B by Kimura et al. (20), but also separation of
others. To date the number of such liver isoproteins induced
by Cd^{2+}-exposure, which can be separated by HPLC to give

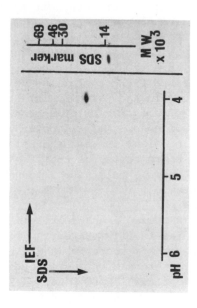

ABSORBANCE, 220nm

Fig.3. (above)Two-dimensional gel electrophoresis of
HPLC-purified calmodulin. Indicated are both the pH
and molecular weight (MW) ranges on the gel (15% acrylamide); SDS marker spot is rat muscle parvalbumin
(MW=11'700 (25)).

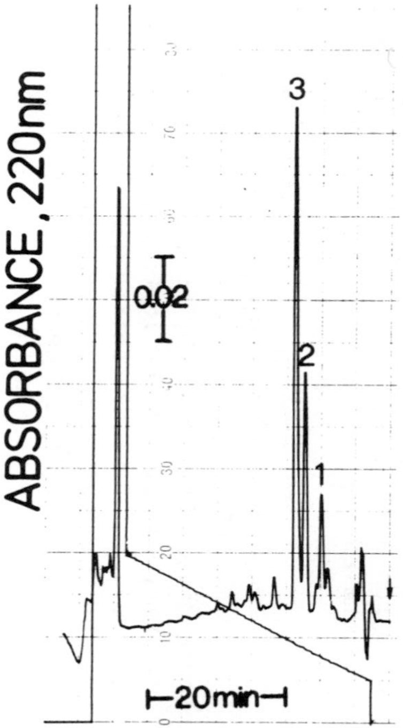

Fig.4. (right) Metallothionein isolation. A high Cd^{2+}-containing fraction from gel chromatography was injected on an Aquapore RP-300 (300 Å) column and eluted using as buffer A,
50mM Tris, pH 7.5,and B, 60% acetonitrile in A buffer. The peaks designated 1,2, and 3 were
found by amino acid analysis to be metallothioneins.

significantly different amino acid compositions, are mini-
mally four (26). Thus, reverse phase chromatography substi-
tutes for the final ion-exchange step, where separation is
based solely on charge differences, used in the purification
of most metallothioneins. The utilization of a buffer system
at neutral pH for metallothionein HPLC allows chromatography
to be carried out on the native protein, i.e. conditions
under which the thiolate complexes are stable.

In summary, the usefulness of reverse phase HPLC as an
integral step in the purification of relatively low molecular weight proteins has been demonstrated. The possibility
of carrying out the chromatography under conditions where
the proteins can be recovered in an active form (the IGFs)
or so their physiological properties remain unaltered, such
as metal content (Ca^{2+}-binding proteins, metallothioneins),
is of upmost importance. A preliminary work up of the sample
by such methods as heat treatment, precipitation and/or gel
filtration often achieves a pre-purification of the sample
to a degree that only a single separation, or at the most
two, were required to reach homogeneity.

REFERENCES

1. Schuster, R. (1980) Anal.Chem. 52, 617-620.
2. Radjai, M.K. and Hatch, R.T. (1980) J.Chromatogr. 196,
      319-322.
3. Hughes, G.J., Winterhalter, K.H., and Wilson, K.J.,
      manuscript submitted.
4. Chang, R.C.C., Huang, W.-Y., Redding, T.W., Arimura, A.,
      Coy, D.H. and Schally, A.V. (1980) Biochim.Biophys.
      Acta 625, 266-273.
5. Kimura, S., Lewis, R.V., Stern, A.S., Rossier, J.,
      Stein, S. and Udenfriend, S. (1980), Proc.Natl.
      Acad. Sci. USA 77, 1681-1685.
6. Hughes, G.J., Winterhalter, K.H. and Wilson, K.J. (1979)
      FEBS Lett. 108, 81-86.
7. Hughes, G.J., De Jong, C., Fischer, R.W., Winterhalter,
      K.H. and Wilson, K.J. (1981) Biochem.J. 199, in
      press.
8. Johnson, N.D., Hunkapiller, M.W. and Hood, L.E. (1979)
      Anal.Biochem. 100, 335-338.

9. Wilson, K.J.,Rodger, K. and Hughes, G.J., (1979) FEBS
   Lett. 108, 87-91.
10. Honegger, A., Hughes, G.J. and Wilson, K.J. (1981)
    Biochem.J. 199, in press.
11. Lewis, R.V., Fallon, A., Stein, S., Gibson, K.D. and
    Udenfriend, S. (1980) Anal.Biochem. 104, 153-159.
12. Petrides, P.E., Jones, R.T. and Böhlen, P. (1980)
    Anal.Biochem. 105, 383-388.
13. Rubinstein, M. (1979) Anal.Biochem. 98, 1-7.
14. Fallon, A., Lewis, R.V. and Gibson, K.D. (1981) Anal.
    Biochem. 110, 318-322.
15. Friesen, H.-J., Stein, S., Evinger, M., Familletti, P.C.,
    Moschera, J., Meienhofer, J., Shively, J. and
    Pestka, S. (1981) Arch.Biochem.Biophys. 206,
    432-450.
16. Klee, C.B., Oldewurtel, M.D., Williams, J.F. and Lee,
    J.W. (1981) Biochem.Intl. 2, 485-493.
17. Wilson, K.J., van Wieringen, E., Klauser, S., Berchtold,
    M.W. and Hughes, G.J., manuscript submitted.
18. Wilson, K.J., Honegger, A. and Hughes, G.J. (1981)
    Biochem.J. 199, in press.
19. Zapf, J., Walter, H. and Froesch, E.R. (1981) J. Clin.
    Invest., in press.
20. Kimura, M., Otaki, N. and Imano, M. (1979) in Metallo-
    thionein (Kägi, J.H.R. and Nordberg, M., eds.) pp.
    163-168, Birkhäuser, Basel, Boston, Stuttgart.
21. Jentoft, N. and Dearborn, D.G. (1979) J.Biol.Chem. 254,
    4359-4365.
22. Laskey, R.A. and Mills, A.D. (1975) Eur.J.Biochem. 56,
    335-341.
23. Wilson, K.J. and Hughes, G.J. (1981) Chimia 9, in press.
24. Rinderknecht, E. and Humbel, R. (1976) Proc.Natl. Acad.
    Sci. USA 73, 2365-2369.
25. Berchtold, M.W., Heizmann, C.W. and Wilson, K.J., un-
    published data.
26. Klauser, S. and Wilson, K.J., unpublished data.

REVERSED PHASE (RP) HPLC OF PROTEINS AND PEPTIDES AND ITS IMPACT ON PROTEIN MICROSEQUENCING

Louis Henderson, Ray Sowder, and Stephen Oroszlan

Frederick Cancer Research Center

P.O. Box B, Frederick, Maryland 21701

In the past, amino acid sequences of proteins have proven fundamental to a wide variety of biochemical and genetic studies. Recent developments in nucleic acid sequencing have revealed gene structures at the molecular level and shown gene splicing to be an important phenomenon. The rapidly growing body of known DNA structures require protein structures for full understanding of gene organization and expression. Today, more than ever before, there is a pressing demand for this fundamental information.

The increased demand for sequence information has greatly increased the need for more rapid and sensitive techniques. Edman degradation is the primary method for determining amino acid sequences in most laboratories. Recent advances have greatly extended its application and reduced the time and labor involved.

Purification of adequate amounts of protein and peptides for sequence studies has always been one of the most difficult and time consuming problems. This report will deal with the specific application of rp-hplc to protein and peptide purifications with special reference to sequence studies.

Commercially available rp-supports are made by covalently bonding organic functional groups, such as an octadecyl (C_{18}) or propylphenyl (phenyl) to porous silica. Variables in the manufacturing processes used by different manufacturers can produce rp-supports with the same func-

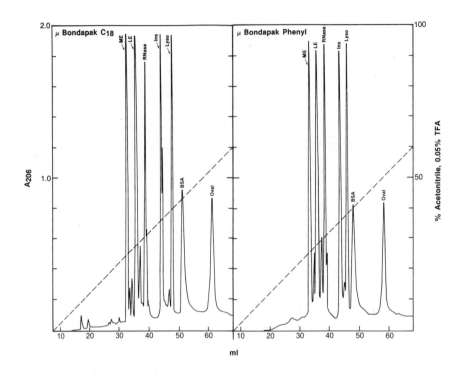

Fig. 1. Separation of 50 ug each of Met-enkephlin
(ME), Leu-enkephlin (LE), ribonuclease (RNase), insulin
(Ins), Lysozyme (Lyso), bovine serum albumin (BSA), and
ovalbumin (Oval) by rp-hplc on u-Bondapak C_{18} and
phenyl with linear gradient from 0.05% TRA in water to 60%
(v/v) acetonitrile containing 0.05% TFA (1,2,3) over 60
min at 1.0 ml/min.

tionality but widely different properties with respect to
protein and peptide chromatography. We have tested a variety of rp-supports for their ability to separate proteins
and large peptides. The test mixture contained 50 ug each
of Met-enkephlin (ME), Leu-enkephlin (LE), ribonuclease
(RNase), insulin (Ins), lysozyme (Lyso), bovine serum albumin (BSA), and ovalbumin (Oval). Many rp-supports failed
to give a sharp well-defined elution peak for one or more
of the proteins in the test mixture; however, some commercially available rp-supports gave excellent separation.

Example test separations obtained on u-Bondapak C_{18} and
u-Bondapak phenyl (Waters Associates, Milford, MA) are
shown in Fig. 1, where all components of the test mixture
were recovered with a yield of 90 $\pm$ 10%. Similar excellent
results were obtained on Syncropak RP-P supplied by
Syncropak (Linden, IND).

Test separations on each of the previously cited rp-
supports were performed replacing acetonitrile with other
organic solvents. These included methanol, ethanol, 1-pro-
panol, and 2-propanol. Gradient elution with methanol (0
to 100%) gave sharp well-defined peaks for all components
similar to the separation shown in Fig. 1. Gradient elution
with ethanol, 1-propanol, or 2-propanol (0 to 60%) gave
broad and poorly defined peaks for Lyso, BSA, and Oval. We
prefer to use methanol or acetonitrile for the organic solvent wherever possible; however, some cases may require the
use of less polar solvents, such as the propanols. (A case
will be cited later).

We have applied rp-hplc technology for the purification
of viral proteins used for structural studies. Rauscher (R)
murine leukemia virus (MuLV) is a typical mammalian type C
retrovirus. It is composed of a lipid outer envelope surrounding a RNA protein inner core structure. Three viral
coded proteins are associated with the lipid envelope:
gp70 (70,000 MW glycoprotein); p12(E) (19,000 MW transmembrane protein); and p2(E) (16 residue peptide). The RNA
protein inner core contains p15 (15,000 MW), p12 (12,000 MW
phosphoprotein), p30 (30,000 MW), p10 (6,500 MW nucleic
acid binding protein). In addition to the seven proteins
cited above, most preparations of the virus contain minor
amounts of posttranslationally modified forms of the above
proteins and genetic variants of some of these proteins.
The seven major viral proteins and variant forms of some of
these proteins have been purified from whole disrupted
virus with a single rp-hplc step as shown in Fig. 2. The
conditions were chosen to optimize the separation of the
viral proteins on u-Bondapak C_{18}. The elution positions of the viral proteins are indicated in Fig. 2. The
hydrophobic transmembrane protein, p12(E), and viral associated lipids were eluted by replacing acetonitrile with
1-propanol as indicated in the figure. Two variants of p10
were isolated as indicated in Fig. 2. The earlier eluting
p10 is composed of 56 amino acid residues and its complete
sequence has been reported (4). The later eluting p10 is
composed of 60 residues of which the first 56 residues are
identical in sequence to the structure mentioned above.

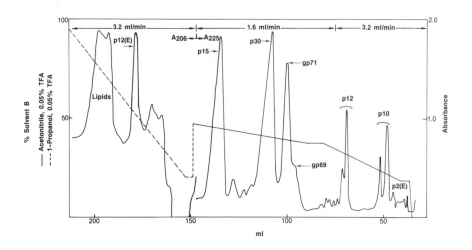

Fig. 2.   Separation of R-MuLV protein by rp-hplc.
Whole virus (17 mg) was disrupted and solubilized in
10 ml of 6M Gu-HCl at pH 2.0 (TFA) and injected onto a
0.8 x 30 cm column of u-Bondapak C_{18}.   Gradients and
flow rates were run as indicated.   Solvents as in Fig. 1
were used from 0 to 150 ml, followed by a gradient using
1-propanol in place of acetonitrile.   Detection was
accomplished at 225 nm from 0 to 150 ml and at 206 nm after
150 ml.   See text for description of the viral proteins.

Two p12s were isolated as indicated in Fig. 2.   The
complete amino acid sequence of the earlier eluting p12 has
been determined (5) and the later eluting p12 appears to
differ from the first by a single amino acid substitution
(Leu to Ile).   Two glycoproteins, gp69 and gp71, were
isolated as indicated in Fig.   2.   The gp69 was only
partially purified on the C_{18} rp-support and was
rechromatographed on a u-Bondapak phenyl column which
gave better resolution between gp69 and gp71.   Amino acid
sequence studies on these two proteins are currently in
progress.   The complete amino acid sequence of p30 and
partial sequences of p15 and p12 (E) have been determined.
Other minor viral associated proteins were also isolated
with this procedure by rechromatography of some of the

smaller peaks seen in Fig. 2. In all, 13 viral associated
proteins have been purified from R-MuLV by rp-hplc.
Similar separations have been achieved for a variety of
mammalian type C and B viruses.

Isolation of sufficient amounts of pure protein for
sequence analysis is merely the first step in a structure
determination. Specific cleavage of the protein and subsequent isolation of peptides is often the most difficult
and time consuming aspect of complete amino acid sequence
analysis. We have used a variety of classical enzymic and
chemical cleavage methods to fragment viral proteins. Many
of the denatured viral proteins were only sparingly soluble
in aqueous buffer at pH 7 (condition used for many enzymic
cleavages). The denatured proteins were usually much more
soluble in 0.1 M Tris-HCl, pH 7.0, containing 30 to 40%
acetonitrile. Trypsin and endoprotease Lys-C (obtained
from Bolinger) were found to be active in this solvent
system. Viral p30 (2.3 mg) was dissolved in 5.0 ml of 0.1
M Tris-HCl, pH 7.0, containing 35% (v/v) acetonitrile. The
protein solution was heated to 100° for 5 min, cooled to
room temperature before adding 0.02 mg of endoprotease
Lys-C. Small aliquots (0.1 to 1 nmole) of the digested
protein were taken at varying time intervals and separated
by rp-hplc at 1.0 ml/min using a linear gradient from 0 to
60% acetonitrile (0.05% TFA) over 20 min. Detection was
accomplished at 206 nm on the 0.2 scale. Digestion was
continued until a stable rp-hplc peptide map was obtained.
The resulting peptide mixture was separated by rp-hplc as
shown in Fig. 3.

Homogenous peptides isolated from the rp-hplc
separation are listed in Table 1, together with the
acetonitrile concentration at the midpoint of elution. All
peptides were recovered with greater than 50% yield. However, in some cases, the same peptide was recovered in more
than one peak as indicated in Table 1. Some of the peaks
in Fig. 3 represent unresolved mixtures. In most cases,
these mixtures could be resolved by rechromatography with a
more shallow gradient (data not shown). Endoprotease Lys-C
cleaves specificity at Lys residues. The data presented in
Fig. 3 and Table 1 indicate that under the digesting
conditions used here the specificity of the enzyme is
maintained.

Trypsin is also active in 30 to 40% acetonitrile. Its
activity can be restricted to Arg residues by specific
modification of lysine ε-amino groups of the substrate
protein. Ethyl acetimidate is a convenient reagent for

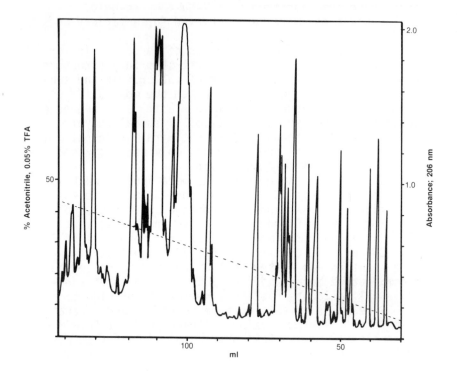

Fig. 3. Separation of peptides from an endoprotease
Lys-C digest of viral p30 (2.3 mg) by rp-hplc on a 0.29 x
30 cm column of u-Bondapak C_{18} was accomplished at
1.0 ml/min using a 0 to 60% gradient over 180 min. The
sample was solubilized and injected in 2.0 ml of 6 M
Gu-HCl, pH 2.0 (TFA). Solvents were as in Fig. 1.

this modification since ε(N)-acetimido-lysine is stable
and retains its positive charge at acid pHs used in the
subsequent rp-hplc. The separation of tryptic peptides
from viral p30 (modified with ethyl acetimidate) was similar
to that shown in Fig. 3 (data not shown). The peptides
obtained from these two digests gave sufficient information
to deduce most of the sequence of viral p30.

Table I.   Peptides Isolated from Separation in Fig. 3

Percent Acetonitrile	Peptide
6	E R D R R R H R R E M S K (Met as sulfoxide)
8	E E R R R A E D E Q R E K
10	E R D R R R H R R E M S K
12	I F N K
13	E E R R R A E D E Q R E K E R D R R R H R E M S K
15	Q R V L L Q A R K
19	R E T P E E R E D R I R R E T E E K
20	E D R I R R E T E E K
22	T L G D L V R E A E K
26	G I T Q G P N Q E P S A F L E R L K
30	A V R G E D G R P T Q L P N D I N D A F P L E R P D W D Y N T Q R G R N H L V H Y R Q L L L A G L Q N A G R S P T N L A K
32	P L R L G G N G Q L Q Y W P F S S S D L Y N W K
34	E A Y R R Y T P Y D P E D P G Q E T N V S M S F I W Q S A P D I G R K
39	N N N P S F S E D P G K L I A L I E S V L L I H Q P T W D D C Q Q L L G T L L T G E E K

Several important points regarding the practical use of this chromatographic techniques should be mentioned. The solvents are completely volatile and UV transparent at wave lengths as low as 206 nm where 1 ug of eluting peptide or protein can be detected. Most salts, free amino acids, and nucleic acids are not retained by the rp-support and are easily separated from the sample protein. Protein samples solubilized in 6 M guanidine hydrochloride (Gu-HCl) can be applied and separated with good resolution and recovery. In general, proteins and peptides are recovered quantitatively, in some cases aggregated or insoluble protein may be retained by the column but can often be removed with Gu-HCl-containing solvents as previously described (2). The amount of total protein that can be

applied to the column depends upon the nature of the
proteins involved, but protein loads up to 10 mg can be
separated on a 0.29x30 cm column. There is a small but
significant difference in selectivity between the C_{18} and
phenyl supports (see Fig. 1). The previously mentioned
organic solvents give slightly different selectivities when
used with the same support (data not shown). Selectivity
advantages may also be gained by varying gradient shape,
flow rate, and column temperature. These properties make
rp-hplc an extremely flexible and powerful technique for
the purification and preparation of samples (peptide or
protein) for subsequent automated Edman degradation.

The examples given in this report dramatically
demonstrate the power of rp-hplc applied to protein and
peptide separations. In the case of the viral proteins it
was possible to isolate each protein and peptides derived
from the proteins using rp-hplc as the only purification
technique. With these techniques proteins and peptides
were purified much faster than they could be sequenced by
an automated Edman-type sequenator. Thus in many cases the
rate limiting steps in complete sequence analysis may be
the cycle time for Edman degradation and the number of
cycles required.

REFERENCES

1. Bennett, H.P.J., Browne, C.A., Goltzman, D., and
    Solomon, S. (1979) in Proceedings of the 6th American
    Peptide Symposium (Gross, E., and Meienhofer, J., eds.)
    pp. 121-124, Pierce, Rockford, IL.
2. Henderson, L.E., Sowder, R., and Oroszlan, S. (1981)
    in Chemical Synthesis and Sequencing of Peptides and
    Proteins (Liu, T.-Y., Schechter, A.N., Heinrikson, R.,
    and Condliffe, P.G., eds) pp. 251-260, Elsevier,
    Amsterdam.
3. Mahoney, W.C., and Hermodson, M.A. (1980). J. Biol.
    Chem. 255, 11199-11203.
4. Henderson, L.E., Copeland, T.D., Sowder, R.C.,
    Smythers, G.W., and Oroszlan, S. (1981). J. Biol.
    Chem. 256, 8400-8406.
5. Versteegen, R.J., Copeland, T.D., and Oroszlan, S.
    (1982). J. Biol. Chem., in press.

PREPARATIVE ISOLATION OF Ia ANTIGEN MEMBRANE PROTEIN COMPONENT POLYPEPTIDES ON C_{18} REVERSE PHASE HPLC

D. J. McKean and M. Bell

Department of Immunology, Mayo Clinic

Rochester, MN 55905

Considerable interest in the past several years has focused on the biochemical characterization of integral cell membrane proteins that are involved in regulating cell-cell interactions. Since these membrane proteins are readily available in only microgram quantities, both analytical and preparative separations of these polypeptides have often been done with SDS PAGE. Polypeptides preparatively isolated from SDS PAGE, however, are denatured and frequently recovered in low yields. The development of alternative polypeptide separation procedures is constrained by the inherent physicochemical properties of membrane proteins. These polypeptides are often heterogeneous with respect to both size and charge, due principally to the presence of variable amounts of carbohydrate. The polypeptides are also amphiphilic molecules which, after being solubilized from the membrane lipid bilayer, are soluble only in detergent-containing solutions or in strongly denaturing solvents.

Recent improvements in high pressure liquid chromatography columns have provided a potential alternative to SDS PAGE for separating these polypeptides. The successful application of reverse phase HPLC columns for the separation of small peptides has indicated that similar procedures might be useful for separating large polypeptides. It has been reported (1) that although the

417

retention orders of small peptides (< 15 residues) can be predicted on the basis of their hydrophobic properties, larger polypeptides do not conform to the prediction rules. The separation of relatively large proteins ($\geq$ 20,000 daltons) by reverse phase chromatography has generally not been successful primarily because of low recoveries of the proteins.

In this report we will describe chromatogrpahic procedures which can be utilized for the rapid, high yield separation of integral membrane protein component polypeptides. This procedure uses a C_{18} reverse phase HPLC column with a mobile phase gradient that consists of the nonionic detergent Triton X-100 in a triethylamine-trifluoroacetic acid solution and acetonitrile. This procedure will resolve the component polypeptides of murine I-E subregion Ia antigens which are approximately 28,000 (β) and 33,000 (α) daltons.

MATERIALS AND METHODS

Radiolabeled Ia antigens were isolated from A/J or BALB/c splenocytes which had been intrinsically radiolabeled with ^{3}H leucine in in vitro cell culture (2). The ^{3}H I-A^k and I-E^d subregion Ia antigens were solubilized from the cell membranes with 0.5% Lubrol WX 0.15M NaCl 0.1M Tris, 1% Trasylol 2mM iodoacetamide, pH 7.4, and the lysate was centrifuged to remove debris. Ia antigens were isolated from the detergent lysate using monoclonal hybridoma immunoadsorbents (2). The immunoadsorbent was extensively washed with 0.1M Tris 0.15M NaCl 0.2% Triton X-100, pH 7.2, and 2mM HEPES 0.2% Triton X-100, pH 7.2. The Ia antigens were eluted from the immunoadsorbents with 3M KSCN 0.2% Triton X-100 and the eluates precipitated with acetone (10 v/v) in the presence of 2 μl normal mouse serum at 0°C for 16 hours. The protein sample was redissolved in 10 M urea 0.5% TFA 0.2% Triton and injected directly in the HPLC column. The Ia samples used in these HPLC studies were analyzed for purity on 20 cm SDS PAGE (2).

A Waters model 204 liquid chromatograph with one model 6000A pump, one model M45 pump and a model 740 systems controller were used for all polypeptide separations. Samples were run on Waters C_{18} μ-Bondapak (4 mm x 30 cm) columns with a gradient system consisting of 0.1% triethylamine 0.2% Triton X-100, pH adjusted to 3.0 with trifluoroacetic acid (A buffer) and acetonitrile (B buffer). The radiolabeled polypeptides were chromatographed at 1 ml/min and fractions were collected from the C_{18} column directly into scintillation vials by connecting the column outlet tubing to an Isco model 328 fraction collector. Fractions from analytical runs were counted in a liquid scintillation counter. For preparative runs, aliquots (10%) of each fraction were counted and the contents of each peak were pooled, lyophilized and electrophoresed on 20 cm SDS polyacrylamide gels (2). At the end of each run the HPLC column was flushed with 5M urea 0.2% Triton X100, pH 3.0.

RESULTS

Murine Ia antigens are integral membrane glycoproteins expressed on populations of B lymphocytes, T lymphocytes, macrophages and epidermal cells. At least two different molecular species of these antigens are encoded within the I-A and I-E subregions of the H-2 major histocompatibility complex. Primary structural differences exist between the Ia antigens encoded in both I subregions as well as polymorphic differences between mouse strains that are genetically disparate at the I-A and I-E subregions. In a separate report (2) we described procedures utilizing monoclonal immunoadsorbents for the high yield, preparative isolation of Ia antigens which are free of contaminating radiolabeled proteins. I-A and I-E subregion Ia antigens isolated from splenocytes are each comprised of two different populations of molecules (3). One population is located primarily on the plasma membrane and consists of a polymorphic 33,000 dalton α polypeptide noncovalently associated with a polymorphic 28,000 dalton β polypeptide. The intracellular, less glycosylated, Ia α/β precursors are associated with a 31,500 dalton, nonpolymorphic invariant (I) polypeptide (Fig. 1). Lubrol WX solubilizes only the plasma membrane fraction of the α/β Ia complex (3). Since this is presumably the immunologically functional Ia

molecule, all Ia preparations used in this study originated
from Lubrol-solubilized splenocytes.

In order to analytically or preparatively separate the
intrinsically radiolabeled Ia component polypeptides we
have previously had to rely on SDS PAGE. This procedure
is, however, time consuming and generally results in low
preparative recoveries of the component polypeptides. We
were, therefore, interested in developing conditions
whereby HPLC technology could be utilized to accomplish
this separation. We initially tried to separate the
polypeptides on molecular sieve HPLC columns (Waters I-125,
Altex TSK 2000 and 3000 columns) in denaturing solvents but
were unable to completely resolve the two polypeptides
(data not shown).

Intrinsically radiolabeled Ia antigens were
subsequently chromatographed on a μ-Bondapak C_{18} column
using a 60 minute, linear gradient from 100% of 0.2% Triton
X-100 1% triethylamine (TEA) 1% trifluoracetic acid (TFA),
pH 2.2 or 3.0 to 100% acetonitrile (Figure 1). The Triton
X-100 was present to maintain the solubility of the Ia
component polypeptides. The pH was maintained at 2.2 with
TFA and triethylamine in order to provide acidic anions to
ion-pair with the polypeptides and to have the mobile phase
below the pI of the free silanol groups (approximately
pH 4) in the column matrix and thus minimize potential
absorptive interactions between the polypeptides and the
free silanol groups. The pH of the A buffer was later
increased during the course of these experiments to pH 3.0
in order to minimize potential hydrolytic damage to the
column matrix. This pH change produced no alteration in
elution profile of the Ia antigen polypeptides. The linear
gradient eluted one major and several minor peaks of
radioactivity. Approximately 70% of the radioactivity
injected onto the column was recovered in the column
eluant, indicating that the Ia polypeptides were remaining
soluble in these solvents and were not being irreversibly
bound to the column matrix.

Based on the retention times of the Ia polypeptides
eluted with the linear gradient, a step gradient was
developed that would resolve three major peaks of
radioactivity (Fig. 2B). Each of the three peaks was
preparatively isolated from the C_{18} column.

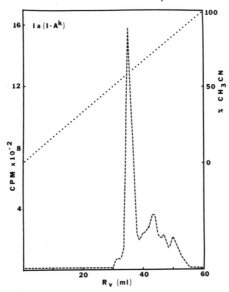

Figure 1: Chromatography of ^{3}H leu I-A^k Ia antigens on a C$_{18}$ HPLC column using buffers described in the text.

An SDS PAGE analysis of the contents of the peaks (Figure 2C) demonstrated that peak A was comprised of low molecular weight contaminants, peak B was comprised of the β polypeptide and peaks C and D were comprised of the α polypeptde. Of the amount of radioactivity injected on the C$_{18}$ column 70-95% is usually recovered in the toal column eluate. We estimate that this recovery yield approaches the theoretical since some eluant is lost when the fraction collector changes vials.

DISCUSSION

Ia antigens are integral membrane proteins that are directly involved in macrophage antigen presentation to helper T cells, and act as antigenic determinants in the mixed lymphocyte reaction. A rapid, high yield procedure for the separation of these polypeptides was needed to facilitate the biochemical characterization of these glycoproteins. Chromatographic procedures are presented in

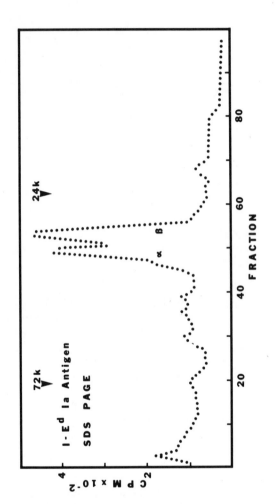

Figure 2A: SDS PAGE analysis of ^3H leu I-E^d Ia antigens.
^{14}C molecular weight markers were coelectrophoresed with the
^3H Ia sample.

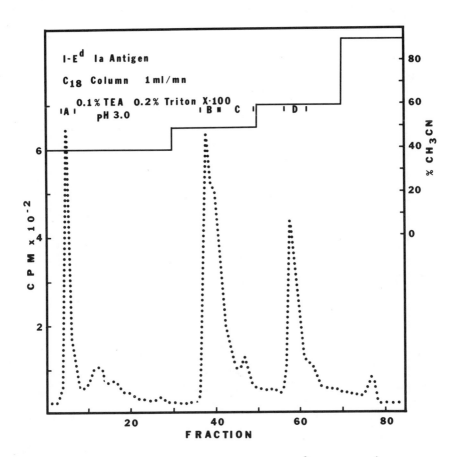

Figure 2B: Preparative separation of the ^3H leu I-E^d Ia antigen component polypeptide on a C$_{18}$ HPLC column.

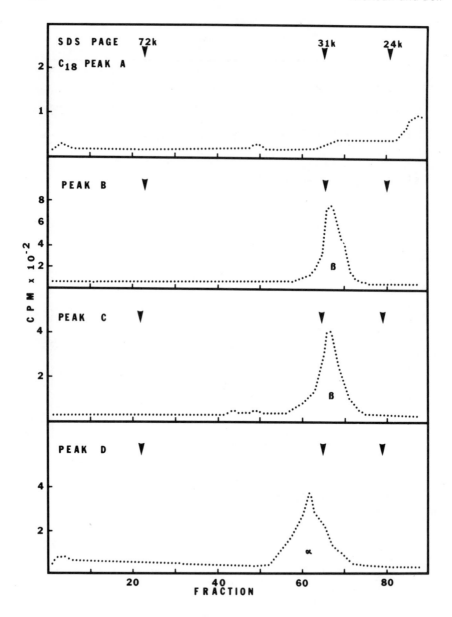

Figure 2C: SDS PAGE analysis of the HPLC fractions preparatively isolated from the (Fig. 2B) C_{18} column run.

this paper for the rapid isolation of pure Ia α and β polypeptides on C_{18} reverse phase high pressure liquid chromatography columns with acidic aqueous solvents that contain Triton X-100.

Nonionic detergents such as Triton X-100 are soluble amphiphiles consisting of polyoxyethylene hydrophilic heads and actylphenol hydrophobic tails. The hydrophobic tails of Triton X-100 are thought to interact predominantly with the hydrophobic regions of integral membrane proteins which span the lipid bilayer. The hydrophilic parts of the amphiphilic protein probably do not bind the detergent molecules (5) and thus the hydrophilic domain should remain accessible for ion-pairing with the anions in the mobile phase of the chromatographic system. It is not clear if the detergent interacts with only the polypeptides or if it also interacts with the stationary phase. The direct effects of Triton X-100 in the mobile phase are not easily demonstrated with these Ia polypeptides because Ia antigen polypeptides are not soluble in the absence of detergents.

When the anionic detergent sodium dodecyl sulfate (0.5% or 0.05%) was substituted for the nonionic Triton X-100 in the mobile phase and the Ia antigens chromatographed with a linear gradient (0-100% acetonitrile), essentially no radioactivity was recovered in the column eluant. The use of SDS in mobile phase containing acetonitrile is restricted since SDS precipitates in high concentrations of acetonitrile. We have also observed that if the Ia antigen samples are dissolved in SDS containing buffers and acetone precipitated prior to column injection, they will not chromatograph with the Triton/TFA/acetonitrile mobile phase in a reproducible pattern. Presumably, variable amounts of SDS remain bound to the protein molecules.

This separation of the Ia polypeptides is very pH dependent. When the pH of the mobile phase was changed from 3.0 to 6.0 by increasing the amount of triethylamine, essentially none of the three labeled polypeptides were recovered in the column effluent. At least some of this loss of recovery could be attributed to the polypeptides being tightly bound to the charged silanol groups on the stationary phase.

Our previous experiences with chromatographing proteins in 6M urea 0.5M acetic acid on molecular sieve HPLC columns (Water's I-125) have indicated that denatured proteins greater than about 16,000 daltons are totally excluded from the 125 A column matrix pores (6). Thus, the Ia polypeptide-detergent micelle complex should be totally excluded from the pores of the C_{18} column particulate matrix with the chromatographic conditions described here. Limiting the interactions of the Ia polypeptides to only the surface of the particulate matrix would be expected to significantly decrease the column's efficiency. A thorough analysis of the amount of protein which can be chromatographed on these μ-Bondapak columns has not been completed.

The data presented in this paper demonstrate that radiolabeled membrane protein component polypeptides can be successfully separated in high yield on μ-Bondapak C_{18} columns with mobile phases containing Triton X-100 nonionic detergent and low concentrations of TFA. The chromographic system is, however, limited to radiolabeled polypeptides since the mobile phase cannot be optically monitored at light wavelengths which are absorbed by proteins.

REFERENCES

1. O'Hare, M. J. and Nice, E. C. J. of Chrom. 171, 209-226, 1979.

2. Moosic, J. P., Nilson, A., Hammerling, G. J. and McKean, D. J. J. Immunol. 125, 1463-1469, 1980.

3. Moosic, J. P., McKean, D. J., Sung, E. and Jones, P. (submitted).

4. McKean, D. J., Melvold, R., and David, C. Immunogenetics, in press.

5. Helinius, A. and Simons, K. Biochim. Biophys. Acta 415, 29-79, 1975.

6. McKean, D. J. Biological/Biomedical Applications of Liquid Chromatography, Vol. 11, Dekker, in press.

CHROMATOGRAPHY AND RECHROMATOGRAPHY IN HPLC SEPARATION OF PEPTIDES

KRATZIN, H., YANG, C.Y. & HILSCHMANN, N.

Max-Planck-Institut für Experimentelle Medizin

Hermann-Rein-Str. 3, D-3400 Göttingen

In former times the separation of the enzymatically or chemically produced splitting products of a protein has been
the most cumbersome time- and material consuming part in
the elucidation of the primary structure of a protein. The
introduction of reverse-phase HPLC-chromatography has created a revolution in this respect. The amount of material
necessary for a separation could be reduced to milligrams
and the time needed to hours instead of days. Besides, the
resolving power of this method is much higher than with the
formerly used procedure. This has made reverse-phase HPLC
to a preferentially used method in protein chemistry (1-7).
We described a method which enabled us to determine the
structure of two immunoglobulin light chains (M.W. 23.000),
one with only 3 mg of starting material (8). From protein
WES all tryptic peptides could be separated in practically
one run in less than one hour. From protein DEN peptides,
which could not be isolated in pure form in system I, could
be easily separated by rechromatography under different
conditions. Such rechromatography systems (system II and
III) have been developed by us recently (9). The general
usefulness of these systems, their advantages and their
limitations shall be discussed in this paper.

MATERIAL AND METHODS

Isolation and cleavage with trypsin.

The Bence Jones proteins were isolated from the urine of plasmocytoma patients as described earlier (10). After·reduction with dithiothreitol and carboxymethylation they were split with TPCK-treated trypsin (Merck, Darmstadt) within 4.5 h at a pH of 8.5 and an enzyme/substrate ratio of 1:35.

HPLC-Separation

The tryptic hydrolysates were separated with the use of a Dupont High-Pressure Liquid Chromatograph (model 850) with reverse-phase columns. For protein WES a prepacked Zorbax C-8 (Dupont), for protein DEN a selfpacked Hypersil-ODS (Shandon) column was used. The oven-temperature in all chromatographies was 60° C.

The following three buffer systems were used with linear gradients in which the concentration of buffer B was increased during 40 or 60 min from 0 to 70% at a flowrate of 1.9 ml/min. Peptides were detected by absorbance at 220nm. The effluent under each peak was collected manually and if necessary dried by a stream of nitrogen.

System 1:   A 25   mM ammoniumacetate, pH 6.0
             B 40% 50 mM ammoniumacetate, pH 6.0,
                60% acetonitrile

System 2:   A   5 mM KH$_2$PO$_4$ /K$_2$H PO$_4$, pH 6.0
             B 40% A, 60% acetonitrile

System 3:   A Water, adjusted with TFA to pH 2.15
             B 40% A, 60% acetonitrile

The hydrolysed (5.7 M HCL, containing 0.23% phenol,24h, 110° C) samples were analysed with a Durrum-aminoacid analyser (Model D 500, Durrum Instrument Corporation, Palo Alto). Aliquots of the dried samples were reacted with 4-Dimethyl-amino-azobenzol-4'-isothiocyanate according to Chang (11-14) and Yang (15). For the identification of the substituted thiohydantoins thin layer sheets (3x3 cm) were used. The identification of 20 pmol of the derivative was still possible.

RESULTS

The general usefulness and reliability of a method for the separation of peptides is proved if it allows the separation of <u>all</u> splitting products generated by enzymatical or chemical digestion.

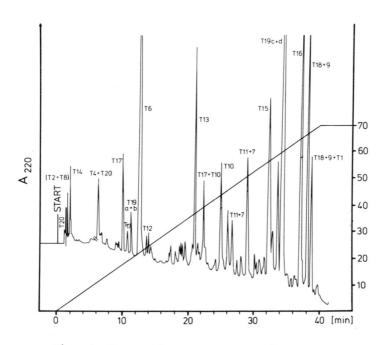

Fig. 1: Separation of the tryptic hydrolysate of Bence Jones protein WES by reverse-phase chromatography on Zorbax C-8, column dimensions 250 x 4,6 mm; buffer A: 50 mM ammoniumacetate, buffer B: 40% of buffer A + 60% acetonitile, gradient from 0-70% buffer B in 40 min.

Fig. 1 shows the separation of the tryptic hydrolysate of Bence Jones protein Westermark (WES), a κ-type immunoglobulin L-chain. The buffer system used allows photometric detection of peptide material by measuring the absorbance at 220 nm. So the effluent under each peak could be collected during the run. After less than one hour the whole mixture was separated into its components.

The high resolving power of reverse-phase HPLC is clearly
demonstrated by amino acid analyses as 18 fractions yielded
analyses with integer numbers for each amino acid. Only in
one peak a mixture of two fragments was identified (T2+T8).
These two as well as the two missing peptides T5 and T3 were
isolated under different conditions (8). 3 mg of starting
material were sufficient to perform amino acid analyses and
a modified form of the Edman degradation (14,15) by which the
amino acid sequence of the peptides and thus of the L-chain
could be elucidated. Just one fragment of 46 residues (T18+
9) had to be digested by α-chymotrypsin to yield smaller
pieces which were again isolated by reverse-phase HPLC.

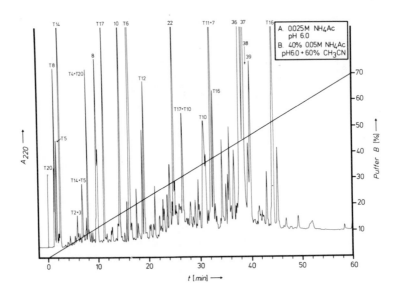

Fig. 2: Separation of the tryptic hydrolysate of Bence Jones
protein DEN by reverse-phase chromatography on Hypersil-ODS
with buffer system I.

The separation of the tryptic digest of another κ-type
L-chain, Bence Jones protein DEN, is shown if fig. 2.
Instead of C-8 here C-18 material was used and what others
call "optical titration" was done by using 25 mM ammonium-
acetat as buffer A. Consequently no baseline drop is observed.

Although in this chromatography the excellent resolving power of reverse-phase chromatography is demonstrated
again we were not able to isolate all the tryptic peptides
of protein DEN from this run. While the peaks signed with
Tn (where n means the number of the tryptic peptide according
to the nomenclature used in connection with the isolation of
peptides with the aid of ion-exchangers) consisted of really
pure peptides those marked only by numbers contained a mixture of tryptic fragments.

Other attempts to isolate all the peptides of this protein in one run did non succeed. Changing the buffer salt,
the pH of the buffer, the organic solvent and the bonded
organic phase resulted in different separations but always
a few splitting products coeluted. Therefore we decided to
chromatograph and rechromatograph these complex peptide mixtures.

Seven fractions had to be rechromatographed (8,10,22,
36,37,38,39). Two examples are shown in fig. 3 and 4.

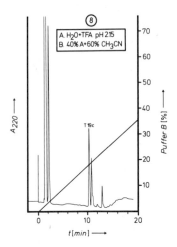 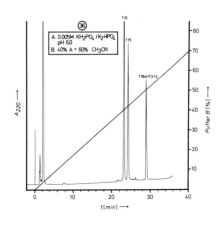

Fig. 3: Rechromatography   Fig. 4: Rechromatography of
of fraction 8 from fig.2   fraction 36 from fig.2 with
with buffer system III.   buffer system II.

The tryptic peptide T18 was isolated in pure form as demonstrated in fig.4 and was digested into smaller pieces
afterwards. With the chymotryptic fragments isolated by
reverse-phase chromatography also the complete primary
structure of protein DEN was established. Like protein
WES it consists of 214 residues. Details of this work have
been published (9).

DISCUSSION

The complete primary structures of two proteins could be
established by isolating all the tryptic fragments. This
clearly demonstrates the general usefulness of reverse-phase
HPLC in peptide purification. To our knowledge proteins WES
and DEN are the first in which all enzymatic fragments were
isolated by HPLC.

The buffer systems used are well suited if Edman degradation and other procedures have to be carried out after
having isolated the peptides. Systems I and III are completely volatile and with system II, because of its low salt concentration, no interference with subsequent reactions (amino
acid analysis at the 1 n Mol level, modified Edman degradation, enzymatic digestions) occurred.

Many separation problems can be solved with these buffers. The first chromatography of a peptide mixture might be
performed under acidic (system III) or nearly neutral
(system I) conditions. Together with system II two powerful
rechromatography systems are available in each case. If an
enzymatic digest is soluble at pH 6.0 system I should be
preferred for the initial separation. It has a good buffer
capacity which guarantees an excellent reproducibility. The
deviations of the k' values are negligible from run to run.
Therefore in repetitive chromatographies more material of
desired fractions can easily be produced. With all three
systems the UV-detection at 220 nm is another adavantage.
No ninhydrin or any other identification reaction is
necessary. Material and time is saved in all cases.

Among the bonded phases tested so far best results
were obtained with Hypersil ODS (5 μ). Because reverse-phase
chromatography is a kind of hydrophobic interaction chromatography besides the bonded phase the buffer ions contribute considerably to the separation. That is why we succeeded
mostly in peptide purifications when using buffer system
II for rechromatographies. The phosphate ion has a marked
influence on the separation as demonstrated in fig.4. As
discussed by others (16) an ion-pair mechanism has to be
considered.

Whether this or any other mechanism is the decisive one
was not investigated in this work. Rather complex interactions between the solutes, the organic, the bonded phase
and the buffer ions must be responsible for these extremely
effective separations.

When system I is the first chromatography system we
prefer to rechromatograph fractions which have been eluted
just after T_0 with system III and those with higher k'-
values with system II. For both a general trend was observed: while the acidic buffer shifts the peptides to
higher k'-values the reverse was observed for the phosphate
system.

Recently HPLC is often used for fingerprinting (17,18).
Probably it is at the time the most powerful tool in this
respect. In our hands minimal buffer changes influence the
separation strongly. Therefore the following assumptions
must be fullfilled, if fingerprints of proteins are to be
compared: the chromatogramms must be performed with the
same column and buffer just one after the other on the same
day. Only these conditions guarantee the absolute identity
of conditions.

With this sensitive and rapid method the main time and
material consuming step in protein sequence work - the isolation of numerous splitting products - could be minimized.
At the time only soluble hydrophilic fragments were investigated and gave yields between 30 and 80 percent. Whether
the buffer systems described are also qualified for hydrophobic molecules is under investigation.

LITERATURE

1   Molnár, I. & Horvath, C.
     J. Chromatogr., 142, 623-640 (1977)

2   Hancock, W.S., Bishop, C.A., Prestidge, R.C. &
     Harding, D.R.K.; Science 200,1168-1170 (1977)

3   Nice, E.C. & O'Hare, M.J.
     J. Chromatogr. 162, 401-407 (1979)

4   O'Hare, M.J. & Nice, E.C.
     J. Chromatogr. 171, 209-226 (1979)

5   Burgus, R. & Rivier, J.
     Proc.Eur. Peptide Symp.,
     in: A. Loffett, Peptides 85 (1976)

6   Mönch, W. & Dehnen, W.
     J. Chromatogr. 147, 415-418 (1978)

7   Lundanes, E. & Greibrock, T.
     J. Chromatogr. 149, 241-254 (1978)

8   Kratzin, H., Yang, C.Y., Krusche,J.U. & Hilschmann,N.
     Hoppe-Seyler's Z. Physiol. Chem. 361, 1591-1598 (1980)

9   Yang, C.Y., Pauly, E., Kratzin, H. & Hilschmann, N.
     Hoppe-Seyler's Z. Physiol. Chem., 362, 1131-1146 (1981)

10   Eulitz, M., Götze D. & Hilschmann, N.
     Hoppe-Seyler's Z. Physiol. Chem. 355, 819-841 (1974)

11   Chang, J.Y., Creaser, E.H. & Bentley, K.W.
     Biochem. J. 153, 607, 611 (1976)

12   Chang, J.Y.; Biochem. J. 163, 517-526

13   Chang, J.Y., Creaser, E.H. & Hughes, G.J.
     FEBS Lett. 78, 147-150 (1977)

14   Chang, J.Y., Brauer, D. & Wittmann-Liebold, B.
     FEBS Lett. 93, 205-214 (1978)

15   Yang, C.Y.
     Hoppe-Seyler's Z. Physiol. Chem. 360, 1673-1675 (1979)

16   Snyder, L.R. & Kirkland, J.J.
     Introduction to modern liquid chromatography

17   Klareskog, L., Rask, L., Fohlman, J. & Peterson,P.A.
     Nature 275, 763-764 (1978)

18   Rivier, J.E.; J. Liquid Chromatogr. 1, 343-366 (1978)

SEQUENCE ANALYSIS OF MEMBRANE PROTEIN FRAGMENTS ISOLATED

BY HIGH PRESSURE LIQUID CHROMATOGRAPHY

Francis S. Heinemann and Juris Ozols

Department of Biochemistry, University of
Connecticut Medical School
Farmington, Connecticut   06032

INTRODUCTION

Sequence analysis of membrane proteins has been
severely limited by the lack of effective chromatographic
techniques for the separation of hydrophobic polypeptides.
The crux of this problem is the insolubility of these
peptides except under conditions which preclude the use of
conventional chromatographic methods. Some progress has
been made with gel permeation chromatography using organic
solvents or concentrated organic acids, but these methods
require large amounts of protein - something not usually
available for membrane proteins. The recent advent of
reverse phase high pressure liquid chromatography (RPHPLC)
has revolutionized the field of peptide separation. Polar
peptides in particular can be separated in good yield, with
high resolution, and sensitivity in the picomole range by a
variety of solvent systems. The application of RPHPLC to
the isolation of hydrophobic peptides, however, is less
well developed. We have previously shown that a 44 residue
hydrophobic segment of cytochrome b_5 can be isolated by
RPHPLC using dilute phosphoric acid and acetonitrile (1,2).
This method has the disadvantage that peptides must be
desalted to remove phosphate prior to sequence analysis.
In addition this method fails to elute some of the hydrophobic peptides we have examined (unpublished observation).
Isolation of the CNBr fragments of bacteriorhodopsin by
RPHPLC using a mobile phase of dilute formic acid and
ethanol has been reported, however one very hydrophobic

segment could not be isolated by this method (3). The
latter method also fails to resolve hydrophobic peptides
generated from the membranous segment of cytochrome b_5
(4). This method does have the advantage of a completely
volatile solvent system. Neither of these two methods
has high resolution capability, and thus both methods have
serious limitations.

Therefore, the purpose of the present work has been
further development of RPHPLC methods for the isolation of
membrane protein fragments. To do this we have used
cytochrome b_5 as a model membrane protein. Cytochrome b_5
is a suitable protein for this purpose since we have
sequenced it from five mammalian species, and it yields a
variety of structurally related peptides of known sequence
upon fragmentation (5,6,7,8). We presented the initial
results of this investigation at the third meeting of this
conference where we discussed the effect of pH, ionic
strength, and triethylammonium on the HPLC separation of
polar peptides using phosphate buffer and acetonitrile
(1). In this paper we report several methods for the
isolation of hydrophobic peptides by RPHPLC and the
application of these methods to the sequence analysis of
cytochrome P-450 isozymes.

EXPERIMENTAL

Materials – Cytochrome b_5 was isolated from microsomal
membranes of bovine, and rat liver as previously described
(9,10). The cytochrome P-450 isozymes were provided by
Dr. Eric Johnson, Scripps Clinic, La Jolla, California.
Trypsin (TPCK treated) was obtained from Worthington.
Staphylococcus aureus V8 protease was obtained from Miles
Laboratories. Chlorosulfonic acid was obtained from Sigma.
Solvents for HPLC were obtained from Burdick and Jackson.
Heptafluorobutyric acid (HFBA, sequenal grade) was
purchased from Beckman. Ammonium Hydroxide (Ultrex grade)
was obtained from J.T. Baker. Ammonium Acetate (HPLC
grade) was from Fisher.

Methods – The methods for proteolytic and chemical
cleavage of cytochrome b_5 have been published (11,12).
Cytochrome P-450 was fragmented using the same procedure
except that P-450 preparations were lyophilized rather than
precipitated in acetone.

HPLC was performed on a system consisting of the following components from Waters: two M6000A solvent delivery systems, a 660 solvent programmer, a U6K injector, a 440 variable wavelength detector, a μBondapak C-18 column (0.4 x 30 cm), and a chart recorder. One separation (fig. 5) employed a Synchropak RP-8 column (4.1 x 100 mm) obtained from Synchrom Inc., Linden, Indiana. The solvent system and gradient used for each separation are described in the legend of the appropriate figure. Peptides isolated with phosphate buffers were desalted on a Biogel P-2 column (0.9 x 12 cm) in 50% acetic acid prior to sequence analysis. Peptides isolated in ammonium acetate or ammonium heptafluorobutyrate were lyophilized, redissolved in 88% formic acid, and applied to the sequencer cup directly.

The amino acid compositions of the peptides were determined on acid hydrolysates of the samples using an updated single column Beckman 121 automatic amino acid analyzer.

Automated sequence analysis of the peptides was performed on a Beckman 890C Sequencer using the 0.1 M quadrol program #030176. 3 mg of polybrene and 100 nmole glycylglycyl glycine, were applied to the cup and subjected to three precycles of Edman degradation prior to sample application for each sequencer run. Conversion of the thiazolinone derivatives to PTH-amino acids was carried out in-line with a Sequamat P-6 autoconverter using 1.5 N acetylchloride in methanol (13). PTH-amino acids were identified by HPLC as previously described (10). The identity of some residues was reconfirmed by amino acid analysis following back hydrolysis in hydroiodic acid vapor (14). Yields of peptides were based on the amount of cytochrome digested.

RESULTS AND DISCUSSION

The peptides used in this study were fragments of the COOH-terminal region of cytochrome b5 (Fig. 1). This seg-ment of the cytochrome b5 molecule has a high percentage of hydrophobic amino acids, and is required for membrane bind-ing of the cytochrome (15). Peptides were generated by cleavage of the intact protein, and in one case by cleavage of tryptic membranous segment isolated by HPLC.

AMINO ACID SEQUENCE OF COOH—TERMINAL SEGMENTS OF CYTOCHROME b_5

```
            90                          100                              110
Equine  - Ser—Lys—Ile—Ala—Lys—Pro—Val—Glu—Thr—Leu—Ile—Thr—Thr—Val—Asp—Ser—Asn—Ser—Ser—Trp—Trp—Thr—Asn—Trp —Val—Ile
Bovine          Thr              Ser       Ser—Ile              Ile              Pro                        Leu
Porcine         Ala              Ser                            Glu
Rabbit      Leu—Ser              Met
Rat                              Ser                            Glu
```

```
            120                         130
Equine  —Pro—Ala—Ile—Ser—Ala—Val—Val—Val—Ala—Leu—Met—Tyr—Arg—Ile—Tyr—Thr—Ala—Glu—Asp—COOH
Bovine                  Leu—Phe              Ile       His—Leu           Ser       Asn
Porcine                 Val          Ser              His—Phe            Ser
Rabbit                  Leu—Ile                        Leu      Met      Asp—Asn
Rat                     Leu                            Leu      Met
```

Fig. 1. Composite of membranous segment amino acid sequences of five species of cytochrome b_5.

The tryptic peptides of bovine cytochrome b_5 including the 44 residue hydrophobic peptide, can be completely separated by a single HPLC run using 5 mM ammonium acetate as the aqueous buffer and acetonitrile as the organic modifier (Fig. 2). The advantages of this isolation method are the simplicity of a single HPLC run and a completely volatile buffer system. The yield of the large hydrophobic peptide is 58%. However, when a tryptic digest of rat cytochrome b_5 was chromatographed by this method, the yield of the membranous tryptic peptide was less than 10%. The gradient conditions which maximize the yield and resolution of the large hydrophobic segment from this species of cytochrome b_5 are shown in Figure 3. Two large hydrophobic peptides identical in structure except for the loss of three residues at the amino terminus by cleavage of the Lys-Pro bond at residue 93 are separated by this method. The smaller peptide is more strongly retained by the reverse phase column presumably because one of the four charged amino acids has been removed. The difference in the chromatographic behavior of the bovine and rat tryptic membranous peptides appears to result from loss of the C-terminal acidic hexapeptide in the rat species because of an arginine substitution for histidine at position 127.

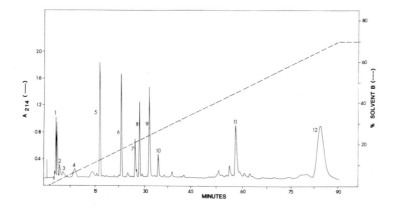

Fig. 2. RPHPLC of bovine cytochrome b₅ tryptic digest.
Buffer A was 5 mM NH₄Ac pH 6.0, Buffer B was acetonitrile,
flow rate was 1.0 ml/min. The column was a C-18
μBondapak. The 44 residue membranous segment eluted in
70% acetonitrile in peak 12.

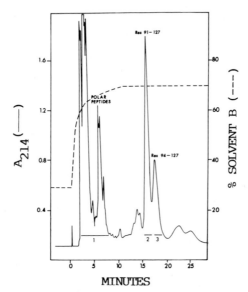

Fig. 3. RPHPLC of rat cytochrome b₅ tryptic digest using
a rapid logarithmic gradient. Buffer A was 20 mM potassium
phosphate pH 2.4, Buffer B was acetonitrile, flow rate 1.5
ml/min, C-18 μBondapak (0.39 x 30 cm).

In order to render the hydrophobic segment of cyto-
chrome $\underline{b}_5$ more polar, and therefore more readily sequenced,
sulfonation of residues 94-127 was attempted, since it has
been reported that chlorosulfonic acid reacts with the
aliphatic hydroxyl group of Ser, Thr, and Tyr yielding
O-sulfate esters (16). The peptide was not only sulfonated,
it was cleaved selectively and quantitatively at the
tryptophan residues. The sulfonated fragments were
separated by HPLC (fig. 4). The peak labeled T-11b-SO$_3$II
contained residues 113-127 of the cytochrome and as such
consisted of fourteen consecutive hydrophobic amino acids
with serine at residue 6 and a COOH terminal arginine.
The sulfonated form of this peptide sequenced fifteen
cycles with a repetitive yield of 94%. This is a
significant improvement over the results obtained with
similar peptides in the unsulfonated form.

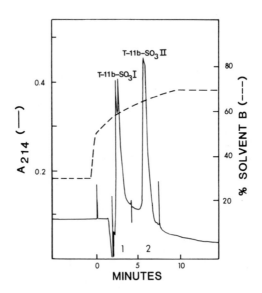

Fig. 4. RPHPLC of the sulfonated tryptophan cleavage
products of rat cytochrome $\underline{b}_5$ tryptic membranous segment
(res. 93-127). Conditions were identical to those in
Figure 3.

Cleavage of native holo-cytochrome b_5 by S. aureus
protease gave four peptides: a NH$_2$-terminal blocked
dipeptide, a large polar peptide spanning residues 3-96, a
linking peptide comprising residues 97 to 103, and a
hydrophobic peptide comprising residues 104-132. The
peptide representing residues 104-132 could not be
recovered from reverse phase columns when potassium
phosphate or ammonium acetate buffers were used as the
aqueous solvent. However, this peptide was recovered in
good yield using dilute heptafluorobutyric acid and
propanol as shown in Figure 5. The key ingredient in this
isolation is the perfluorinated carboxylic acid since
trifluoroacetic acid and acetonitrile will also elute the
peptide, albeit in somewhat lower yield. The separation
shown in figure 5 employs a large pore octyl-silane
column. When a µBondapak C-18 column was used, the
hydrophobic peptide eluted in a higher concentration of
propanol.

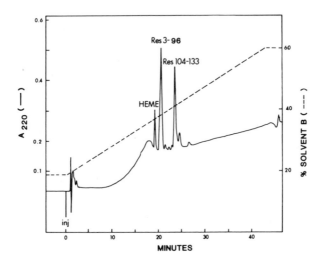

Fig. 5. RPHPLC of rat cytochrome b_5 Staph. protease
digest. Buffer A was 0.1% NH$_4$HFBA pH 2.5, Buffer B was 75%
1-propanol containing 0.1% NH$_4$HFBA pH 2.4, flow rate 1.0
ml/min, column was a Synchropak RP-8 (4.1 x 100 mm).

We have also used RPHPLC for comparative peptide
mapping of cytochrome P-450 isozymes (2,17). More
recently we have reported the preparative use of RPHPLC
for the isolation of an analogous tryptic peptide from two
P-450 isozymes (18). Figure 6 compares the 214 nm peak
profiles of preparative tryptic peptide separations from
two forms of cytochrome P-450. These patterns are
extremely reproducible for a single digest, and reasonably
reproducible between digests. This characteristic of
RPHPLC allows one to pool peaks from different runs without
the need for confirmation of peak identity by amino acid
analysis. Although there are many peaks with similar
retention times in these profiles, the patterns generated
are quite different. The dissimilarity in the maps is
more clearly seen by comparing the 280 profiles (Fig. 7).
Some of the peaks in Figure 6 are broad or unsymmetrical,
and obviously contain more than one peptide. However, peak
shape is not a reliable indicator of purity, and even very
sharp peaks may contain more than one peptide. Figure 8
shows the rechromatography of single sharp peaks from one
of the preparative runs shown in Figure 6. We have found
that rechromatography at pH 6.0 is the most reliable
method for separating a group of peptides which coelute at
pH 2.5. Ion pairing reagents have also been of some use in
this regard. Peptides rechromatographed in 20 mM NH4Ac pH
6.0 may be applied directly to the sequencer cup. One
difficulty with this method is that some peptides may not
elute under these conditions. We found this particularly
true for peptides eluting in greater than 50% acetonitrile
at pH 2.5. These peptides are best rechromatographed using
perfluorinated carboxylic acids.

The method of choice for preparative RPHPLC of simple
mixtures of polar peptides is ammonium acetate/acetoni-
trile. The small decrement in peak resolution compared to
phosphate systems is unimportant for simple peptide
mixtures. The solvent system is compatible with long
column life, and its complete volatility allows one to
apply peptides to the sequencer.

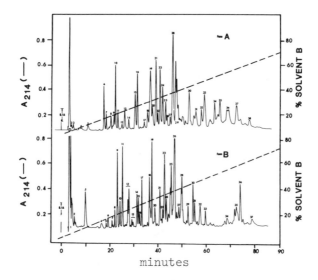

Fig. 6.   Preparative tryptic peptide maps of two forms of cytochrome P-450.   Both runs were done using 20 mM potassium phosphate pH 2.4 as Buffer A and acetonitrile as Buffer B at a flow rate of 1.0 ml/min on a C-18 µBondapak column.

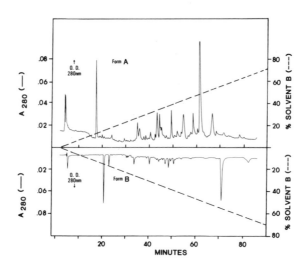

Fig. 7.   Comparative 280 nm peptide mapping of cytochrome P-450 tryptic peptides.   Conditions were identical to those in Figure 6.

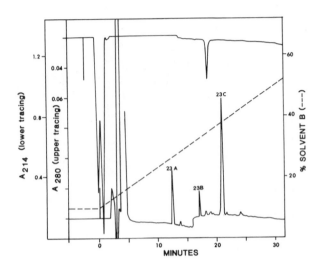

Fig. 8. Rechromatography of peak #23 from the preparative
tryptic peptide map of P-450 Form B shown in Fig. 6.
Buffer A was NH₄Ac pH 6.0, Buffer B was acetonitrile, flow
rate was 1.0 ml/min, and the column was a C-18 µBondapak.

Complex peptide mixtures cannot be completely resolved
by a single RPHPLC run.   It is necessary therefore to plan
a rechromatography step.   The initial separation should
have a high resolution capacity in order that the mixtures
for rechromatography be as simple as possible.   The second
step should be completely volitile to allow direct
sequencing of peaks.   The system we have used successfully
is potassium phosphate pH 2.4/acetonitrile as the initial
step with rechromatography in ammonium acetate pH 6.0/
acetonitrile.

The best solvent system for resolving very hydrophobic
peptides is still a matter trial and error in most cases.
Rapid logarithmic gradients to a high concentration of
organic solvent using potassium phosphate as aqueous buffer
have been useful for certain separations.   This method has
the advantages of high resolution, rapidity, and high
yield.   However, peptides must be desalted prior to
sequence analysis.   Our initial results suggest that dilute

TABLE I

Amino Acid Compositions

	Polypeptide Fragments of Cytochrome b_5 Membranous Segment					Cytochrome P-450 Form B Peptides	
Species Residues Fig./Peak	Bovine 91-133 2 / 12	Rat 91-127 3 / 2	Rat 94-127 3 / 3	Rat 104-132 5 /	Rat 113-127 4 / 2	Tryptic peptide 23A Figure 8	Tryptic peptide 23C Figure 8
Amino Acid							
Lysine	1.0(1)[a]	0.8(1)	0.1(0)	0 (0)	0 (0)	0 (0)	0 (0)
Histidine	0.9(1)	0 (0)	0 (0)	0 (0)	0 (0)	0 (0)	0.8(1)
Arginine	0 (0)	1.1(1)	1.9(2)	1.3(1)	1.0(1)	0.7(1)	0.9(1)
Aspartic acid	3.5(4)	2.1(2)	2.0(2)	2.1(2)	0 (0)	0.8(1)	0.9(1)
Threonine	4.3(5)	3.8(4)	4.2(4)	0.7(1)	0 (0)	0 (0)	0.9(1)
Serine	4.9(6)	4.4(5)	4.6(5)	2.8(4)	0.6(1)[c]	0 (0)	0 (0)
Glutamic acid	1.9(2)	1.9(2)	2.2(2)	0.8(1)	0 (0)	2.1(2)	1.8(2)
Proline	3.0(3)	2.2(2)	2.2(2)	1.2(1)	1.0(1)	0 (0)	1.3(1)
Glycine	0 (0)	0 (0)	0 (0)	0 (0)	0 (0)	0 (0)	0 (0)
Alanine	3.1(3)	3.7(4)	2.8(3)	3.6(4)	3.2(3)	2.1(2)	0.9(1)
Valine	1.2(1)	3.8(4)	3.9(4)	2.8(3)	2.8(3)	1.0(1)	0.5(1)
Methionine	0.3(1)	0.9(1)	0.9(1)	1.2(2)	0.8(1)	0 (0)	0.7(1)
Isoleucine	6.2(7)	3.8(4)	3.1(3)	2.0(2)	1.7(2)	0 (0)	1.2(2)
Leucine	3.9(4)	3.1(3)	3.3(3)	2.8(3)	1.9(2)	1.0(1)	0 (0)
Tyrosine	1.8(2)	0.8(1)	1.0(1)	1.2(1)	0.4(1)[c]	0 (0)	0.9(1)
Phenylalanine	1.0(1)	0 (0)	0 (0)	0 (0)	0 (0)	1.2(1)	0 (0)
Tryptophan	ND (3)[b]	ND (3)	ND (3)	ND (3)	ND (0)	ND (0)	ND (0)
Yield[d]	58%	30%	20%	72%	40%	65%	65%

a   The value in parentheses is the number obtained from the sequence.
b   Tryptophan was not determined (ND).
c   Serine and Tyrosine were sulfonated.
d   Yields were based on the amount of cytochrome fragmented.

HFBA/propanol is a powerful method for isolating hydrophobic peptides. The hydrophobic counter ion apparently maintains the solubility of very hydrophobic peptides in the concentration of organic solvent necessary to elute them from reverse phase columns. As seen in Table I, the hydrophobic peptide generated by S. aureus protease cleavage of cytochrome b_5, residues 104-132, was recovered in good yield using the HFBA system although it could not be recovered using phosphate buffers. This method has all the advantages of high yield, high resolution, and volatility, and may be used either for initial peptide separation or rechromatography.

ACKNOWLEDGEMENTS

This investigation was supported in part by grants from United States Public Health Service GM 26351, National Science Foundation PCM 78-0822, and American Cancer Society NP-134.

REFERENCES

1. Ozols, J., Heinemann, F.S. and Gerard, C. (1980) In Methods in Peptide and Protein Sequence Analysis, Proceedings of the 3rd International Conference. Birr, C. (ed.) pp. 419-433, Elsevier-Holland Biomedical Press, Amsterdam.
2. Ozols, J., Heinemann, F.S., Johnson, E.F. and Muller-Eberhard, U. (1980) Fed. Proc. 39, Vol. 6, 2053.

3. Gerber, G.E., Anderegg, R.J., Herlihy, W.C., Gray,
    C.P., Biemann, K. and Khorana, H.G. (1979) Proc.
    Natl. Acad. Sci. U.S.A. 76, 227-231.
4. Takagaki, Y., Gerber, G.E., Nihei, K. and Khorana,
    H.G. (1980) J. Biol. Chem. 255, 1536-1541.
5. Ozols, J. and Gerard, C. (1977) Proc. Natl. Acad.
    Sci. U.S.A. 74, 3725-3729.
6. Ozols, J. and Gerard, C. (1977) J. Biol. Chem. 252,
    8549-8553.
7. Fleming, P.J., Dailey, H.A., Corcoran, D. and
    Strittmatter, P. (1978) J. Biol. Chem. 253, 5369-
    5372.
8. Kondo, K., Tajima, S., Sato, R. and Narita, K. (1979)
    J. Biochem. (Tokyo) 86, 119-1128.
9. Kondo, K., Tajima, S., Sato, R. and Narita, K. (1979)
    J. Biochem. (Tokyo) 86, 1119-1128.
10. J. Ozols, F.S. Heinemann. (1981) Biochimica et
    Biophysica Acta (in press).
11. Nobrega, F.G. and Ozols, J. (1971) J. Biol. Chem.
    246, 1706-1717.
12. Ozols, J. and Gerard, C. (1977) J. Biol. Chem. 252,
    5896-5989.
13. Horn, M.S. and Bonner, A.G. (1977) Solid Phase
    Methods in Protein Sequence Analysis. Previero, A. and
    Colletti, M.A. (eds.) pp. 163-176, Elsevier North-
    Holland, Amsterdam.
14. Ozols, J., Gerard, C. and Nobrega, F.G. (1976) J.
    Biol. Chem. 251, 6767-6774.
15. Strittmatter, P., Rogers, M.J. and Spatz, L. (1972)
    J. Biol. Chem. 247, 7188-7199.
16. Previero, A., Coletti-Previero, M-A., and Roundouin, G.
    (1980) In Methods in Peptide and Protein Sequence
    Analysis, Proceedings of the 3rd International Confer-
    ence. Birr, C. (ed.) pp. 349-358, Elsevier North-
    Holland Biomedical Press, Amsterdam.
17. F.S. Heinemann and J. Ozols. (1980) Fed. Proc.
    39(3), 642.
18. J. Ozols, E.F. Johnson, F.S. Heinemann. (1981) J.
    Biol. Chem. 256 (22) In press.

PREPARATION OF PEPTIDES AND PROTEINS FOR SEQUENCE ANALYSIS AT THE LOW NANOMOLE TO SUBNANOMOLE LEVEL BY REVERSE-PHASE HIGH-PERFORMANCE LIQUID CHROMATOGRAPHY: RESULTS FOR CYTOCHROMES P450 AND FIBRONECTIN

John E. Shively, Hema Pande, Pau-Miau Yuan, and David Hawke

City of Hope Research Institute

Duarte, California 91010

A number of key developments have led to the routine microsequence analysis (low nanomole to subnanomole range) of peptides and proteins in a number of laboratories. In spinning cup microsequence analysis, the developments were improved instrument design and performance (1-4), better methods for solvent and reagent purification (2, 5), automated conversion of amino acid anilinothiozolinone to phenylthiohydantoin (PTH) derivatives (6), high sensitivity separation and quantitation of PTH derivatives by reverse-phase HPLC (2, 7-9), and retention of µg amounts of proteins and peptides in the spinning cup with polybrene (2, 10). The ability to perform routine microsequence analysis on µg amounts of peptides and proteins has led to structural studies on relative rare substances which possess important biological properties. In our laboratory we have used this methodology to obtain structural information on human leukocyte and fibroblast interferons (11-15) and on bovine adrenal opioid peptides (16-18). Over the past two years these studies have led us to consider another critical aspect of microsequence analysis, namely the compatibility of sample preparation with microsequence analysis. The major requirements for the purification of µg amounts of proteins or peptides which are present in complex mixtures are high resolution, high sensitivity of detection, and high recovery. For peptides, and increasingly so for proteins, reverse-phase HPLC has become the method of choice. It is important that the solvent system used in this application be

volatile and unreactive with amino acid functional groups.
In this report we compare two systems, one of which can be
used to directly detect peptides or proteins at 200-220 nm
(19,20). The second system is compatible with post-column
fluorescence detection via sample stream splitting (21,
22).

     A comparison of the two systems is shown in Figure 1,
using a tryptic digest of cytochrome C as a standard. The
first system consists of an alkylphenyl column, a
trifluoroacetic acid/acetonitrile solvent, and a 206 nm
detector. Complete resolution of the major tryptic
fragments is obtained in less than 80 min. The second
system consists of an alkylphenyl or RP8 column, a
pyridine-acetate/1-propanol solvent, and post-column
fluorescence detection. Poor resolution was observed on

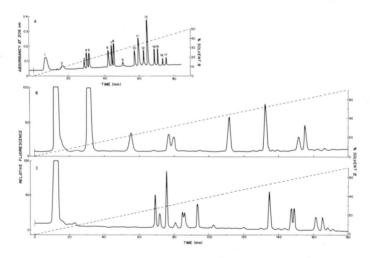

Figure 1. Tryptic maps of horse heart cytochrome C on
reverse phase HPLC. Seven nmoles of tryptic digest were
applied to the following systems. A: μ-Bondapak phenyl
column, a linear gradient from solvent I (.09% TFA) to
solvent II (10%, I 90% CH_3CN), flow rate of 48 mL/hr,
delection at 206 nm. B: μ-Bondapak phenyl column, a
linear gradient from solvent I (0.9 M acetic acid, 0.25 M
pyridine, pH 4.0) to solvent II (40% I, 60% 1-propanol),
flow rate 20 mL/hr, 15% of sample reacted post column with
fluorescamine. C: same as B, but used an RP8 column.

the alkylphenyl columns, and even on the RP8 column a
number of peptides were either unresolved or undetected.
The run time on this system was over 3 hrs. A further
problem with this system is that it was originally
described for a pyridine-formate buffer, a buffer which
occasionally modifies amino acid functional groups. We
have found that the formic acid can N-formylate amino
groups and react with the indole ring of tryptophan
residues. These anamolous side reactions can effectively
block NH_2-terminal residues or result in the misidenti-
fication of lysine or tryptophan residues (5). Based on
these considerations we have adopted the first HPLC system
as the method of choice for purification of peptides. The
TFA/acetonitrile system is compatible with most reverse-
phase supports, and when used with 300A pore size supports
can give good resolution of proteins. Careful attention
to the purification of the solvents is important. We
redistill TFA over chromium trioxide, and purify CH_3CN
by passing it over RP18 before use. These steps reduce
the UV absorbance of the solvents and the effects of
contaminants on subsequent microsequence analysis.

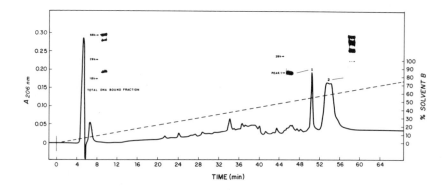

Figure 2. Reverse-phase HPLC separation of DNA-binding
tryptic fragments from human plasma fibronectin. The
HPLC system is similar to that described in Figure 1A. A
SDS-gel profile of the sample applied is shown on the left
and of the peaks obtained on the right.

An example of the use of reverse-phase HPLC with
TFA/CH3CN for the separation of large peptides is shown
in Figure 2. The water soluble protein, human plasma
fibronectin (subunit size 220K daltons), was trypsinized
and subjected to affinity chromatography on DNA-cellulose.
The DNA-binding fragments (>10K daltons) were purified on
an alkylphenyl column. This system gave a number of
peaks, one of which was a single polypeptide (M_r of 24K
daltons) as analyzed by SDS gel electrophoresis. The 24K
dalton peptide was sequenced through 54 cycles of Edman
degradation. Chromatograms for the first 7 cycles,
together with a standard separation of the PTH amino
acids, is shown in Figure 3. These chromatograms
demonstrate the compatibility of this HPLC purification
method with microsequence analysis. The high signal to
noise ratio permits confident identification of PTH amino
acids at the subnanomole level. An overall repetitive
yield of 96% was observed beginning with a 1.2 nmole yield
of the NH_2-terminus (1.8 nmoles was applied to the
spinning cup). By cycle 40 even peaks corresponding to
20-50 pmoles were clearly identified above background
levels.

An example of the use of this system for the
separation of hydrophobic peptides obtained from the
integral membrane protein, pig testes microsomal
cytochrome P450, is shown in Figure 4. Approximately 40
tryptic fragments (1-4 nmoles each) were separated in a
single 2 hr run. Chromatograms from the microsequence
analysis of 4 nmoles of one of the peptides (T34) are
shown in Figure 5. The hydrophobic peptide was treated
with a slight molar excess of SPITC in order to reduce its
solubility in the organic solvent washes. This treatment
did not affect the identification of glutamic acid at the
NH_2-terminus or lysine at cycle 7 (this lysine was not
hydrolyzed by trypsin because it is followed by a proline).

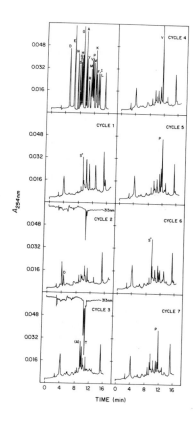

Figure 3.   Chromatograms from the microsequence analysis of a 24K dalton DNA-binding fragment from human plasma fibronectin. The upper left panel depicts the separation of 200 pmoles each of 4 standard PTH-derivatives. Approximately 40% of the sample from each cycle was injected. Diethyl phthalate (the last peak on the chromatograms) was used as an internal standard. Separations were achieved on an Ultrasphere ODS column, using stepwise elution from solvent I: 10% CH_3CN, 90% trifluoroacetate/acetate buffer (33 mM/2 mM, pH 5.6) to solvent II: 75% CH_3CN, 25% trifluoroacetate buffer (30 mM, pH 3.4), a flow rate of 1.3 mL/min, and temperature of 43°.

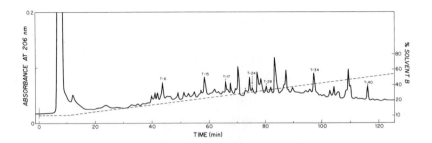

Figure 4. Reverse-phase HPLC separation of the TFA soluble tryptic fragments from microsomal cytochrome P450 from pig testes. The HPLC system is described in Figure 1A.

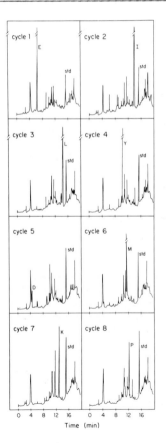

Figure 5. Chromatograms of peptide T34. The HPLC system is described in Figure 1A.

Based on our experience we feel that it is now reasonable to routinely separate complex mixtures of peptides in the low µg range by reverse-phase HPLC, and proceed to perform direct microsequence analysis on them. The same methodology can often be applied to proteins, but usually requires consideration of problems such as precipitation, aggregation, and denaturation. We believe that most of these problems will be solved during the next few years.

References

1. Wittman-Liebold, B. Hoppe-Seyler's Z. Phys. Chem. 354 1415-1431 (1973).
2. Hunkapiller, M.W., and Hood, L.E. Biochemistry 17 2124-2133 (1978).
3. Hunkapiller, M.W., and Hood, L.E. Science 251 523-525 (1980).
4. Shively, J.E. Methods Enzymology 79 31-48 (1981).
5. Shively, J.E., Hawke, D., and Jones, B.N. Anal. Biochem, submitted.
6. Wittman-Liebold, B., Graffander, H., and Hohls, H. Anal. Biochem. 75 621-633 (1976).
7. Zimmerman, C.L., Apella, E., and Pisano, J.J. Anal. Biochem. 77 569-573 (1977).
8. Johnson, N.D., Hunkapiller, M.W., and Hood, L.E. Anal. Biochem. 100 335-338 (1979)
9. Hawke, D., Yuan, P.-M., and Shively, J.E. Anal. Biochem., submitted.
10. Tarr, G.E., Beecher, J.F., Bell, M., and McKean, D.J. Anal. Biochem. 85 126-131 (1978).
11. Levy, W.P., Shively, J., Rubinstein, M., Del Valle, U., and Pestka, S. Proc. Natl. Acad. Sci. USA 77 5102-5104 (1980)
12. Stein, S., Kenny, C., Friesen, H.-J., Shively, J., Del Valle, U., and Pestka, S. Proc. Natl. Acad. Sci. USA 77 5716-5719 (1980).
13. Feiesen, H.-J., Stein, S., Evinger, M., Familletti, P.C., Maschera, J., Meienhoffer, J., Shively, J., and Pestka, S. Arch. Bioch. Biophys. 20 432-450 (1981).
14. Wetzel, R., Perry, L.J., Estell, D.A., Lin, N., Levine, H.L., Slinker, B., Fields, F., Ross, M.J., and Shively, J. J. Interferon Res. 1 381-390 (1981).
15. Levy, W.P., Rubinstein, M., Shively, J., Del Valle, U., Lai, C.-Y., Moshera, J., Brink, L., Gerber, L., Stein, S., Pestka, S. Proc. Natl. Acad. Sci. USA, in press.

16. Stern, A.S., Jones, B.N., Shively, J.E., Stein, S.,
     and Udenfriend, S.   Proc. Natl. Acad. Sci. USA 78
     1962-1966 (1981).
17. Jones, B.N., Stern, A.S., Lewis, R.V., Kimura, S.,
     Stein, S., Udenfriend, S., and Shively, J.E.   Arch.
     Biochem. Biophys. 204 392-395 (1980).
18. Kilpatrick, D.L., Taniguchi, T., Jones, B.N., Stern,
     A.S., Shively, J.E., Hullihan, J., Kimura, S., Stein,
     S., and Udenfriend, S.   Proc. Natl. Acad. Sci. USA 78
     365-3268 (1981).
19. Hancock, W.S., Bishop, C.A., Meyer, L.J., Harding,
     D.R.K., and Hearn, M.E.W.   J. Chromatogr. 161 291-298
     (1978).
20. Yuan, P.-M., Pande, H., Clark, B.R., and Shively,
     J.E.   Anal. Biochem., submitted.
21. Bohlen, P., and Kleeman, G.   J. Chromatogr. 205 65-75
     (1981).
22. Bohlen, P., Stein, S., Stone, J., and Udenfriend, S.
     Anal. Biochem. 67 438-445 (1975).

Acknowledgements: This research was supported in part by
the following grants: NSF grant no. PCM-8022705, and NCI
grants CA 16434 and CA 19163.

ANALYTICAL AND PREPARATIVE APPLICATIONS OF HIGH-PERFORMANCE

LIQUID CHROMATOGRAPHY. COMPARISONS WITH OTHER METHODS.

Hedvig von Bahr-Lindström, Mats Carlquist,
Viktor Mutt and Hans Jörnvall
Departments of Physiological Chemistry I and
Biochemistry II, Karolinska Institutet
Box 60 400, S-104 01 Stockholm, Sweden

INTRODUCTION.

Reverse phase high-performance liquid chromatography (reverse phase HPLC) and gel permeation HPLC are frequently used separation methods for peptides and proteins. In this report, we show results from analytical and preparative use of both types of matrices, and from comparisons of HPLC with other separation methods.

MATERIALS AND METHODS.

Proteins and Peptides.

Aldehyde dehydrogenase was prepared from horse liver (1), using DEAE to separate the two isozymes. After reduction and ^{14}C-carboxymethylation as described (2), both isozymes were digested (4 h at 37°C) with TPCK-trypsin (1:100 by weight) in 0.1 M ammonium bicarbonate. Bovine (3) and chicken (4) secretins, prepared from the intestines, were treated with trypsin under the same conditions. The hexon viral capsid protein from cells infected with adenovirus type 2 was reduced and ^{14}C-carboxymethylated in guanidine-HCl (5).

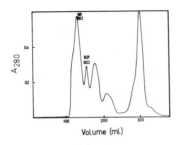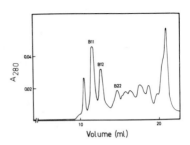

Fig. 1. Fractionations of CNBr-fragments of the hexon protein from adenovirus type 2. Left curve shows separation on
Sephadex G-50 fine (1.5 x 200 cm), right curve separation
on Varian TSK 2000 SW (7.5 x 500 mm).

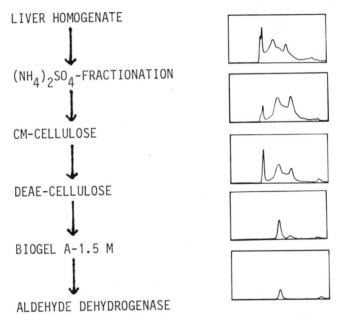

Fig. 2. Use of gel permeation HPLC to monitor the purification of the basic isozyme of horse liver aldehyde dehydrogenase. Separations were on LKB 2135 UltroPac TSK SW 3000
(7.5 x 600 mm) of samples from each step in the purification.
Horisontal: elution volume (27 ml for the entire length shown).
Vertical: absorbance at 206 nm (1.0 absorbance unit full scale).

Separations.

Gel permeation HPLC was performed in 0.1 M ammonium bi-carbonate on a Varian TSK 2000 SW column (7.5 x 500 mm) at a flow rate of 0.2 ml/min and on an LKB 2135 Ultropac TSK SW 3000 column (7.5 x 600 mm) at a flow rate of 1.0 ml/min.
Chromatography on Sephadex G-50 fine (1.5 x 200 cm) was carried out in 30% acetic acid while DEAE-Biogel A (100-200 mesh) was used in ammonium bicarbonate with a linear gradient of 0.01-0.2 M for elution.
Tryptic peptides, purified on paper by high voltage electrophoresis and by chromatography (2), were detected by fluorescamine staining (Fluram, Roche; 4 µg/ml in acetone) and eluted with water.
Reverse phase HPLC was carried out on a µBondapak C_{18} column (Waters; 3.9 x 300 mm) in 0.1% phosphoric acid with a linear gradient of 0-50% acetonitrile; or in 5mM ammonium acetate, 0.1% acetic acid, 1% ethanol (solvent A) with a 0-100% linear gradient of ethanol containing 10% solvent A. Flow rates were 0.5-1.5 ml/min and gradient times 30-60 minutes.

Structural analysis.

Amino acids in acid hydrolysates were determined with a Beckman 121 M analyzer. Liquid phase sequencer degradations were performed in a Beckman 890C instrument with a 0.1 M Quadrol peptide program using precycled polybrene (5). Phenylthiohydantion amino acids were analyzed by HPLC (6), while dansyl amino acids from manual dansyl-Edman degradations were separated on polyamide sheets in four solvent systems (7).

RESULTS.

Gel Permeation HPLC.

CNBr peptides from the hexon protein of adenovirus were fractionated by gel permeation HPLC (Varian TSK 2000 SW) and by exclusion chromatography (Sephadex G-50 Fine). Chromatograms from the two methods are compared in Fig. 1. Of 29 CNBr fragments from hexon (5), only the three longest,

ranging from 7-17 K, were recovered in the useful part of
the chromatogram after HPLC-fractionation. Smaller peptides
were not resolved, in agreement with the specified separation range of the matrix. Sephadex exclusion chromatography
gave no better resolution in either the low or the high
molecular weight region.

The same type of HPLC-gel (LKB 2135 UltroPac TSK SW
3000) was used to follow the purification of horse liver
aldehyde dehydrogenase. Samples were analyzed after each
purification step, as shown in Fig. 2.

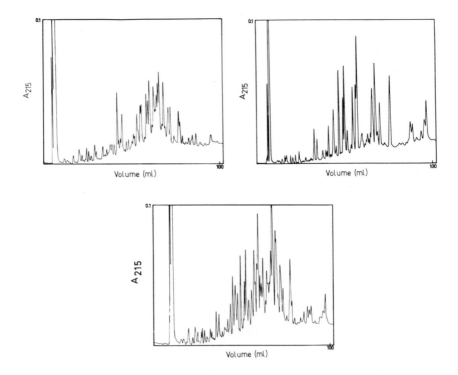

Fig. 3. Reverse phase HPLC fractionations of tryptic peptides
from acidic (upper left) and basic (upper right) isozymes of
horse liver aldehyde dehydrogenase. Lower curve shows a mixture of the two tryptic digests.

Peptide Mapping.

Tryptic digests of the basic and acidic isozymes of
horse liver aldehyde dehydrogenase were separately applied
to a reverse phase HPLC-column (μBondapak C_{18}) in 0.1% phosphoric acid and eluted with a gradient of acetonitrile
(Fig. 3). Fractionation of a mixture of the two isozyme
digests shows that only few peaks coincide (Fig. 3).

Preparative Fractionation.

A tryptic digest of aldehyde dehydrogenase was first

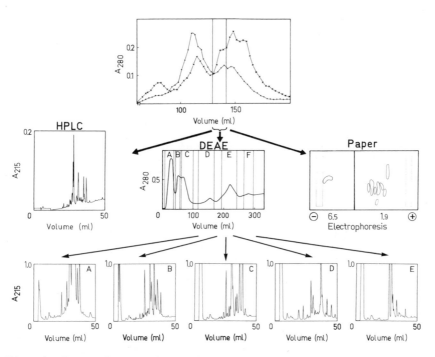

Fig. 4. Comparisons of purifications of tryptic peptides
from aldehyde dehydrogenase.
Pre-fractionation on Sephadex (top), followed by HPLC directly (middle left), and via an intervening DEAE step
(middle center + bottom), or by paper electrophoresis and
chromatography (middle right).

Table 1. Comparisons of yields in HPLC-based and paper-based purifications of peptides form aldehyde dehydrogenase.

Peptide	Recovery (%)	
	DEAE + HPLC	Paper
ILDLIESGK	40	12
IFINNEWHDSVSGK	37	9
EMGEYGFHEYTEVK	30	7
LFSNAYLMDLGGCLK	20	8
LILATMESK	13	4
ANNTTYKIFAGSFT(GSZG$_2$IL)K	11	0
KFPVFNPATZZK	8	6
VAFTGSTEVGK	7	5
IAKEEIFGPVQQI(PBSGM)K	8	0
YVLGNPLTPGVSQGPQIPKEQYK	7	0
SLDDVIK	7	5
LCQVQQGDKEDVNK	7	0
YCAGPADK	0	4
Average yield	15	5

fractionated on Sephadex G-50 (Fig. 4), and a pool containing peptides of suitable size (around 10-20 residues) were further purified by either of three methods: A) reverse phase HPLC, B) paper high voltage electrophoresis and chromatography, or C) DEAE-chromatography followed by reverse phase HPLC. The results (Fig. 4) show that HPLC-fractionation alone does not give as complete separation of the peptides as paper electrophoresis plus chromatography, or ion exchange chromatography plus HPLC do. The peptide yields from the paper-based purification and from the ion exchange-HPLC purification are compared in Table 1. The method including HPLC gives a higher yield (up to 4-fold) than the paper method. This is most evident for the longer peptides, of which some are not even recovered after paper separations.

Recoveries of tryptic peptides from chicken and bovine secretins were compared in a similar way. Peptides from the bovine hormone were purified by reverse phase HPLC (3) in ammonium acetate and those from chicken secretin by electrophoresis on paper. The recoveries with the two methods are compared in Table 2. HPLC-fractionation gave yields about twice those from the method utilizing paper electrophoresis.

Table 2. Comparisons of yields in HPLC-based or paper-based
purifications of peptides from secretins.

Peptide	Recovery (%)	
	HPLC	Paper
GNAQVQK		52
MR		50
HSDGLFTSEYSK		48
FIQNLM		35
DSAR	85	
LR	93	
HSDGTFTSELSR	87	
LLQGLV	89	
LQR	95	
Average yield	90	46

Reverse phase HPLC was also used as a final step for
purification of the native hormones. In this way, bovine secretin (3), chicken secretin (4) and bovine vasoactive intestinal peptide (8) were obtained in improved yields as
compared to earlier preparation methods.

DISCUSSION.

Gel Permeation HPLC.

The results from the CNBr-peptide fractionations (Fig. 1)
suggest that sensitivity, speed and peak resolution favor
the HPLC-method for peptides longer than about 60 residues.
The advantage of shorter separation time is, however, largely
lost when the matrix is used for preparative purposes because of its low capacity.
The use of HPLC in monitoring enzyme purifications is
demonstrated by the analytical separations after each step
in the purification of aldehyde dehydrogenase (Fig. 2).
Speed and sensitivity make gel permeation HPLC a good alternative to SDS polyacrylamide gel electrophoresis.

Reverse Phase HPLC.

The high resolving power of reverse phase HPLC is useful for single-step peptide mapping as demonstrated by the
comparison of the two isozymes of aldehyde dehydrogenase
(Fig. 3). Over 20 different peaks are resolved, and the low
coincidences of the peaks in mixtures of the two digests
demonstrate that the two isozymes are less similar than
earlier suggested (2).

For preparative purposes, the yields also favor HPLC-
based methods over paper methods (Tables 1 and 2). However,
in complex mixtures a pre-fractionation utilizing conventional exclusion and ion exchange chromatography is advantageous for purification of all fragments (Fig. 4). In
digests of large proteins therefore, the preferred method
is an initial Sephadex G-50 fractionation followed by DEAE-
chromatography and reverse phase HPLC. In this way, all
fragments (Fig. 4) are obtained pure in high yield (Table 1).

ACKNOWLEDGEMENTS.

Help with gel permeation chromatographies by Marianne
Bornefeldt is gratefully acknowledged. This work was
supported by grants from the Swedish Medical Research
Council, the Swedish Cancer Society and Magn. Bergvall's
Foundation.

REFERENCES.

1. Eckfeldt, J., Mope, L., Takio, K. & Yonetani, T. (1976).
    J. Biol. Chem. 251, 236-240.
2. von Bahr-Lindström, H., Sohn, S., Woenckhaus, C., Jeck,
    R. & Jörnvall, H. (1981). Eur. J. Biochem. 117, 521-526.
3. Carlquist, M., Jörnvall, H. & Mutt, V. (1981). FEBS
    Lett. 127, 71-74.
4. Nilsson, A., Carlquist, M., Jörnvall, H. & Mutt, V.
    (1980). Eur. J. Biochem. 112, 383-388.
5. Jörnvall, H., Akusjärvi, G., Aleström, P., von Bahr-
    Lindström, H., Pettersson, U., Appella, E., Fowler, A.V.
    & Philipson, L. (1981). J. Biol. Chem. 256, 6181-6186.
6. Zimmerman, C.L., Appella, E. & Pisano, J.J. (1977).
    Anal. Biochem. 77, 569-573.
7. Jörnvall, H. (1970). Eur. J. Biochem. 14, 521-534.
8. Carlquist, M., Mutt, V. & Jörnvall, H. (1979). FEBS
    Lett. 108, 457-460.

SEPARATION OF HINGE GLYCOPEPTIDES OF HUMAN IgD BY HPLC

NOBUHIRO TAKAHASHI, DANIEL TETAERT AND
FRANK W. PUTNAM
Department of Biology, Indiana University
Bloomington, Indiana 47405

Amino acid sequence analysis of myeloma proteins has established that all five classes of human immunoglobulins have the same tetrachain structural pattern. However, the five classes differ greatly in the structure of the hinge region connecting Fab and Fc (1). Previous studies (2) of the human IgD protein WAH showed that the hinge region has an unusual structure with two dissimilar sections; the C-terminal half, which is highly charged, is extremely sensitive to proteolysis, whereas the N-terminal half is rich in galactosamine (GalN), which may confer on IgD an important function as a cell receptor (3). Determination of the amino acid sequence of the IgD hinge region has been very difficult for several reasons: 1) the heterogeneity and steric hindrance of the carbohydrate chains, 2) difficulty in purification of a series of similar large glycopeptides, 3) technical problems of sequence analysis. To facilitate structural studies of IgD, we have developed reverse phase high pressure liquid chromatography (HPLC) for preparative isolation of glycopeptides.

MATERIALS AND METHODS

WAH IgD and a mixture of peptides containing GalN from the hinge region obtained after papain and tryptic digestion were prepared as described by LIN and PUTNAM (4). Chymotryptic digestion of the δ chain was also done, and the large glycopeptides were purified by gel filtration on a Sephadex

G-50 column (2.6 x 90 cm) eluted with 10% n-propanol containing 0.1% TFA. Further purification of the heterogeneous fractions rich in GalN was achieved by use of an HPLC
system consisting of a BECKMAN controller (model 421) and
pump (model 110A). The columns (SYNCHROPAK RP-P and ULTRA-
SPHERE ODS) (0.4 x 25 cm) were equilibrated with 0.1% TFA
or 0.1% HFBA as counter-ions and eluted with a flow rate of
0.7 ml/min with an isocratic or programmed gradient of n-
propanol. Methods for amino acid analysis and for sequence
determination with the BECKMAN model 890C sequencer have
been described (4). To determine the amino sugar content,
the sample was hydrolyzed in 4 M HCl at 110°C and was analyzed with the BECKMAN model 121M amino acid analyzer.

A large chymotryptic peptide (CP1) that was rich in
GalN was digested with trypsin and the digest (3.9 mg) was
separated first by gel filtration on a Sephadex G-50 column
(1.5 x 56 cm) eluted at a flow rate of 3.8 ml/hr with 10%
n-propanol containing 0.1% TFA. Further purification was
done by HPLC as described above. The GalN-rich hinge glycopeptide was also digested with proline-specific endopeptidase. The hinge glycopeptides were treated by three different approaches to remove the sugar: 1) with HF at 4°C for
6 hr or at room temperature for 3 hr or 24 hr, 2) with tri-
fluoromethane sulfonic acid (TMSA) at room temperature for
10 hr, 3) with neuraminidase and endo-α-galactosidase in
0.1 M cacodylate buffer pH 6.0 at 37°C for 18 hr.

RESULTS AND DISCUSSION

The chymotryptic digest of the IgD heavy chain was
fractionated by gel filtration on a Sephadex G-50 column,
and the profile revealed eight peaks (Fig. 1). Uncleaved
IgD and chymotrypsin were eluted in the first peak; peak 2
contained a large glycopeptide (CP1). This was purified by
HPLC (Fig. 2) and submitted to amino acid analysis (Table I)
and sequence analysis (Fig. 3A), which stopped at the threonine expected to be the sugar attachment site (Thr-28).
After trypsin and papain digestion of IgD, separation of the
GalN-containing peptides by gel filtration or ion-exchange
chromatography was always unsuccessful (4). However, these
glycopeptides were separated into three main fractions by
HPLC (Fig. 4a). Glycopeptide 1 (GP1) was pure according to
the amino acid composition (Table I) and N-terminal analysis.

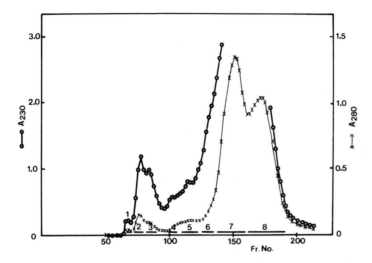

Fig. 1. Gel filtration of the chymotryptic digest of the δ chain (163 mg). Bars indicate the fractions pooled.

After the first 6 residues were identified, sequence analysis stopped probably because of the presence of sugar and/or the formation of dehydroalanine. Fraction II was heterogeneous but fraction III was a large pure peptide containing all the GalN (Table I). At high concentration (>200 nmoles), the glycopeptides eluted rapidly, and their hydrophilicity decreased the resolving power of the RP-P column (Fig. 4b).

In contrast to the highly charged section of CP1, which is easily cleaved by proteases (2), the GalN-rich segment (positions 1-41) (Fig. 3A) is very resistant to many proteolytic enzymes of different specificity (e.g. trypsin, papain, S. aureus protease, proline-specific endopeptidase, clostripain, and chymotrypsin). However, tryptic digestion of CP1 was somewhat effective, and fractionation of the tryptic glycopeptides was done on Sephadex G-50 (Fig. 5). The GalN-rich fractions (1 and 2) were further purified by HPLC (Fig. 4d,f) and denoted TP1, TP2, and TP3 (Table I). Tryptic cleavage occurred with a yield of 10% at Lys-26 and Arg-35 to give TP2 and TP3, respectively (Fig. 3D, 3E), but 90% of TP1 was recovered. TP4, purified by HPLC from fraction 4 did not possess GalN, showing that Thr-37 is not a sugar attachment site (Fig. 3F). Sequence analysis of TP1 and TP2

Table I.   Amino Acid Composition of Glycopeptides

	Glyco-peptides		Chymo-tryptic peptide 1		Subdigested Peptides						
					Tryptic peptides (TP)				Proline peptidase		
	1	3			1	2	3	4	a	b	c
Lys	2.1	1.8	7.7	8	0.7	1.0					
His			0.9	1							
Arg	2.0	1.7	4.1	4	2.1		1.9	0.9	1.7		
Asp	1.1	1.0	1.1	1	0.7		0.8	0.9	1.0		
Thr	5.0	6.6	9.3	8	4.9	1.2	4.5	0.9	2.7	0.8	1.7
Ser	0.9	2.9	4.7	5	2.6	2.5					
Glu	2.7	6.3	18.0	17	4.2	4.1				1.0	
Pro	1.0	3.4	7.7	6	4.0	1.9	1.1			0.9	1.0
Gly	3.1	4.0	4.4	4	2.0	1.0	1.0	1.0	1.2		
Ala	4.0	6.6	8.0	8	7.1	4.9	3.0		0.8	1.0	1.0
Cys/2			0.8	1							
Val		0.8	1.2	1	1.0	1.0					
Leu	0.8	1.0	2.3	2	1.0	0.9					
GalN	2.9	3.5	4.2		4.2	0.8	2.8	0	0.2	0	0
Sum	23	37	70		32	19	13	4	8	4	4
Yield (nm)	80	30	1040		440	83	55	20	4	13	8
End Group	Gly	Ala	Arg		Ala	Ala	Ala	Asn	-	-	-
Fig.3	B,H		A		C,G	D	E	F	K	I	J

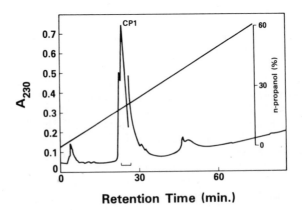

Fig. 2.   Purification of the largest chymotryptic peptide CP1 by HPLC on an RP-P column (2 mg of peak 2 of Fig. 1).

```
                                                    GalN  GalN              GalN
        1              10 GalN                30                      40
Arg-Trp-Pro-Glu-Ser-Pro-Lys-Ala-Gln-Ala-Ser-Ser-Val-Pro-Thr-Ala-Gln-Pro-Gln-Ala-Glu-Gly-Ser-Leu-Ala-Lys-Ala-Thr-Thr-Ala-Pro-Ala-Thr-Thr-Arg-Asn-Thr-Gly-Arg-Gly-Glu-Glu-Lys
                                                         20

Arg-Trp-Pro-Glu-Ser-Pro-Lys-Ala-Gln-Ala( )Ser-Val( )( )Ala( )( )Ala-Glu-Gly( )Leu-Ala-Lys-Ala--------------------------------->
A

                         Gly-Ser-Leu-Ala-Lys-Ala--------------------------------------------
                         B

Ala-Gln-Ala-----------------------------------------------------------------
C

Ala-----------------------------------◄Ala( )( )Ala-Pro-Ala( )( )Arg-Asn( )Gly◄
D                                      E

                                                         Asn-Thr-Gly-Arg
                                                         F

Ala-Gln-Ala( )Ser-Val-Pro-Thr-Ala-Gln-Pro-Gln-Ala-Glu-------------------------------------------
G

                         Gly-Ser-Leu-Ala-Lys-Ala-Thr-Thr-Ala-Thr-Ala-Pro-Ala( )( )Arg-Asn( )Gly-Arg-Gly-Gly-----------◄
                         H

(Thr,Ala,Gln,Pro)
I

                                (Thr,Thr,Ala,Pro)(Ala,Thr,Thr,Arg,Asn,Thr,Gly,Arg)
                                J                                              K
```

Fig. 3. Sequences of the different hinge peptides are shown: A) the largest chymotryptic peptide CP1, B) glycopeptide GP1, C) tryptic peptide TP1, D) tryptic peptide TP2, E) tryptic peptide TP3, F) tryptic peptide TP4, G) TP1 after treatment with HF, H) GP1 after treatment with HF, I), J) and K) proline-endopeptidase peptides from TP1 after treatment with HF; these were placed by amino acid composition. For the complete sequence of the hinge region including the high-charge area see reference 2.

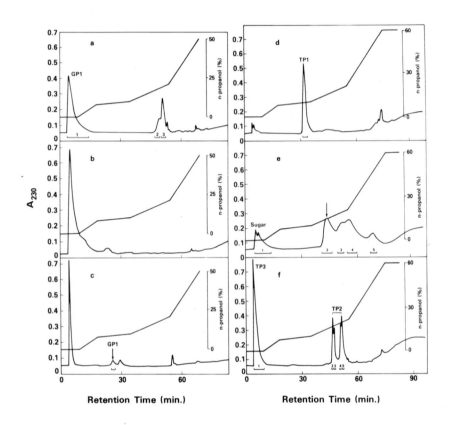

Fig. 4.   Separation on an RP-P column of the glycopeptide
mixture from a partial digest of IgD with papain and trypsin (Figs. a, b, c) and of tryptic subpeptides of CP1 (Figs.
d, e, f).   Amount applied: Fig. 4a, 0.5 mg; 4b, 1.0 mg;
4c, 0.1 mg after enzymatic removal of sugar; 4d, 0.6 mg of
fraction 1 of Fig. 5; 4e, 0.6 mg of TP1 from Fig. 4d after
HF treatment; 4f, 0.2 mg of fraction 2 of Fig. 5.   Peak 1
of Fig. 4f contained tryptic peptide TP3; TP2 was obtained
in multiple forms from peaks 2, 3, 4 and 5.

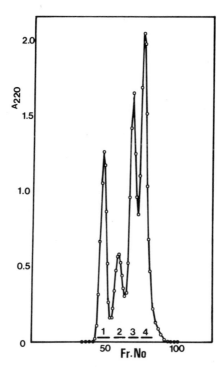

Fig. 5. Gel filtration of subpeptides of CP1.

was not successful, and only 7 positions of TP3 were identified (Fig. 3C-E).

        Despite their purity, the GalN-rich glycopeptides were
difficult to sequence by automatic Edman degradation. Therefore, they were submitted to enzymatic and chemical reactions
to remove sugar to facilitate sequence analysis. After action of neuraminidase and endo-α-galactosaminidase, separation of sugar and peptides was performed by HPLC (2). The
elution pattern (Fig. 4c) differed from that of Fig. 4a;
suggesting that sialic acid was removed. In this case, sequence analysis seemed to be prevented by the remaining GalN
(3.0 residues). Although TMSA was relatively effective for
the removal of GalN, the yield was very low and it was difficult to desalt the samples.

        Treatment with HF was first carried out under weak con-
ditions (4°C, 6 hr). Separation of the sugar and peptides
was done by HPLC (Fig. 4e), and 45% of the deglycosylated

peptide was obtained without destruction of serine and threonine. Under these conditions, only about 40 to 50% of the
GalN was removed, and sequence analysis was incomplete (Fig.
3G). When the glycopeptides were submitted to stronger HF
treatment, i.e., at room temperature, the removal of GalN
took place with destruction of serine and threonine, but the
partially deglycosylated peptides were more amenable to sequence analysis (Fig. 3H). (After 3 hr at room temperature,
60% of the GalN was removed but only 75% after 24 hr.)

In order to complete the sequence of the hinge region
TP1 was treated with HF (4°C, 6 hr), the sugar and peptides
separated by HPLC (Fig. 4e), the peptide digested with proline-specific endopeptidase, and the subpeptides purified by
gel filtration and HPLC. The amino acid composition of the
subpeptides confirmed the sequence (Table I). The tentative
location of 4 to 5 oligosaccharides was assigned by difference in the GalN content of the various subpeptides and by
the failure to detect serine or threonine at a given step in
the sequencing of the hinge region. Thus, the use of HPLC
allowed the determination of the complete sequence of the
IgD hinge region and the probable assignment of the oligosaccharides. This was the most difficult part of the IgD
structure, and the complete amino acid sequence of the entire IgD molecule has now been completed in our laboratory.

Acknowledgments: We thank J. Madison, S. Dorwin, P. H.
Davidson, J. Dwulet, and Y. Takahashi for valuable assistance and Dr. Jacques Baenziger for the enzymatic removal of
sugar. Supported by Grants IM-2G from the American Cancer
Society and CA08497 from the National Cancer Institute.

REFERENCES

(1) Putnam, F. W. (1977) in The Plasma Proteins, ed. Putnam,
     F. W. (Academic, New York), 2nd ed., Vol. III, p.1-153.
(2) Putnam, F. W., Takahashi, N., Tetaert, D., Debuire, B.
     and Lin, L.-C. (1981) Proc. Natl. Acad. Sci. (USA) 78,
     (in press).
(3) Kettman, J. R., Cambier, J. C., Uhr, J. W., Ligler, F.
     and Vitetta, R. S. (1979) Immunol. Rev. 43, 69-95.
(4) Lin, L.-C. and Putnam, F. W. (1979) Proc. Natl. Acad.
     Sci. (USA) 76, 6572-6576.

USE OF HIGH PERFORMANCE LIQUID CHROMATOGRAPHY IN CHARACTER-

IZING NUCLEOTIDE BINDING SITES AND ANTIGENIC DETERMINANTS IN

cAMP-DEPENDENT PROTEIN KINASE

Susan Taylor, Anthony Kerlavage, Norman Nelson,
Sharon Weldon, and Mark Zoller
Dept. of Chemistry, University of California
at San Diego
La Jolla, California   92093

cAMP-dependent protein kinase is a multisubunit protein whose activity and quaternary structure are both modulated by cAMP (1,2). In its native holoenzyme form, the enzyme is an inactive tetramer containing two regulatory (R) and two catalytic (C) subunits. In the presence of cAMP, the holo-enzyme dissociates into a regulatory subunit dimer (R_2) and two monomeric catalytic subunits with each R_2 binding 4 mol-ecules of cAMP (3,4). Because of the multiple aggregation states that serve a regulatory role and because of several substrate and modulator binding sites, this enzyme provides an excellent model for investigating protein-protein inter-actions, nucleotide-protein interactions, and enzyme regu-lation. We have utilized high performance liquid chromato-graphy (HPLC) in conjunction with solid phase sequencing to characterize a variety of specific sites on both subunits of the kinase purified from porcine cardiac muscle.

Catalytic Subunit

Although there are two major classes of cAMP-dependent protein kinases, the differences are associated primarily with the R subunits. The C subunits are virtually indis-tinguishable (5,6). The C subunit (M_r=40,000) has a variety of functional sites, including an ATP-binding site, two magnesium-binding sites (7), a peptide or protein substrate-binding site, and a recognition site for the R subunit.

471

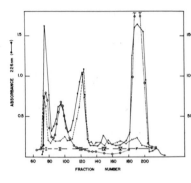

Figure 1. *Separation of the CNBr Peptides of the C Subunit on Sephadex G-50*. Peptides were eluted from Sephadex G-50 (2 x 188cm) in 1% HOAc (flow rate = 30 ml/hr). FSO$_2$BzAdo-labeled peptides were monitored for absorbance at 226 nm and [^{14}C]cpm (●---●). IAA-labeled peptides from an essentially identical elution were monitored for [^{14}C] cpm (O---O) and superimposed.

As an initial step to characterizing at least some of these functional sites, we have developed a rapid and convenient procedure for separating the 7 cyanogen bromide (CNBr) peptides of the C subunit. This separation based on gel filtration followed by HPLC is indicated in Figures 1 and 2. Three of the CNBr peptides (I, II, and IV) are pure following gel filtration alone (Fig. 1) whereas peptides III and V are each resolved into two peptides by reverse-phase HPLC (Fig. 2).

ATP-binding Site. Affinity labeling was utilized in an effort to identify some of the specific amino acid residues that are associated with or contribute to nucleotide binding. For this purpose, the ATP analog, p-fluorosulfonylbenzoyl 5'-

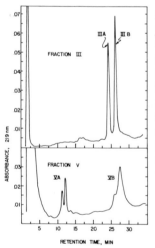

Figure 2. *HPLC separation of CNBr fraction III and V*. Fractions III and V from Fig. 1 were applied to a C$_{18}$ µBondapak column (0.39x30cm). Buffer A=0.1% H$_3$PO$_4$; Buffer B=CH$_3$CN. Peptides were eluted with a linear gradient from 0 to 50% B in 30 min. at a flow rate of 2 ml/min. The doublets seen in V correspond to the homoserine and homoserine lactone form of the same peptide. Figure taken from (10).

adenosine (FSO_2BzAdo), was used. This analog contains a reactive fluorosulfonyl group in a position analogous to the γ-phosphate of ATP (8). Reaction of the C subunit with FSO_2-BzAdo resulted in complete irreversible inhibition of enzymatic activity with a stoichiometry of 1 mole of FSO_2BzAdo incorporated per mole of enzyme (9). Following inhibition with $FSO_2[^{14}C]BzAdo$, the protein was oxidized with performic acid, digested with trypsin and eluted with a pH gradient from DEAE-Sephadex A25. At this point, the modified peptide was selectively purified by taking advantage of the relatively labile ester bond in FSO_2BzAdo. When the radioactive peak was pooled, subjected to mild alkaline hydrolysis sufficient to cleave the ester bond (see inset, Fig. 3), and rechromatographed under identical conditions on DEAE-Sephadex A25, the elution of the modified peptide was selectively shifted to a lower pH, yielding a pure peptide (Fig. 3). Alternatively, when the $FSO_2[^{14}C]BzAdo$-modified protein was cleaved with CNBr, radioactivity was associated only with peak III following gel filtration (Fig. 1). When this region was pooled and subjected to reverse-phase HPLC, only CNBr IIIA (Fig. 2A) contained radioactivity. The composite sequence from this region of the molecule, *Met-Leu-Val-Lys-His-Lys-Glu-Thr-Gly-Asn-His-Phe-Ala-Met-LYS*-Ile-Leu-Asp-Lys-Gln-Lys-Val-Val-Lys-Leu-Lys-Gln-Ile*, revealed that covalent modification was associated with a single unique lysine residue (*LYS**)(10). Based on homology with the bovine

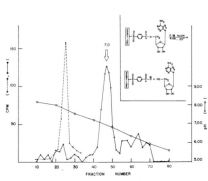

Figure 3. *DEAE-Sephadex A25 elution of $FSO_2[^{14}C]BzAdo$-modified tryptic peptides*. Radioactive peak from the first DEAE Sephadex elution (see text) was subjected to mild alkaline hydrolysis (inset) and rechromatographed on the same column. Elution was carried out with a 500 ml linear gradient of 1% pyridine, 1% collidine from pH 8.0 to 5.0. (-----) Indicates the position of the labeled peptide from the first column prior to NaOH treatment. Taken from (10).

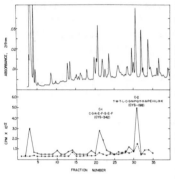

Figure 4. *HPLC separation of the cysteine-containing tryptic peptides of native C subunit in the presence and absence of MgATP (12).* Native enzyme was modified with [^{14}C]IAA, digested with trypsin, and chromatographed as described in Fig. 2. Upper panel shows absorbance; lower panel shows radioactivity incorporated in the absence (•——•) and presence (•----•) of 0.67 m*M* MgATP. Sequences of C1 and C2 appear above the respective peaks.

catalytic subunit (11), this residue has been identified as LYS 71.

  Cysteine Residues. The C-subunit contains 2 cysteine residues, both of which can be readily alkylated in the native enzyme with the sulfhydryl specific reagent, iodoacetic acid (IAA). Following tryptic digestion, the two cysteine-containing peptides were well-resolved by HPLC (Fig. 4). At pH 7.8, loss of enzymatic activity was concommitant with alkylation of both cysteine residues. On the other hand, under identical conditions but in the presence of Mg-ATP not only was C completely protected against inactivation, but, in addition, the alkylation of both cysteine residues was blocked (Fig. 5). These data indicate that both cysteine residues are in close proximity to the ATP-binding site, or that ATP binding promotes conformational changes that significantly decrease the reactivity of both groups (12).

  When the [^{14}C]IAA alkylated enzyme was cleaved with CNBr and chromatographed on Sephadex G-50, CNBr I and II were covalently modified (Fig. 1). The [^{14}C]IAA modified

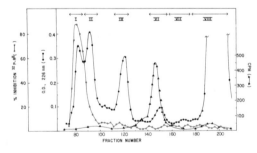

Figure 5. *Elution of the CNBr peptides of the R subunit from Sephadex G50.* Elution conditions are as described in Fig. 1.

tryptic peptides were isolated using the reverse-phase HPLC
shown in Figure 4. The sequences of these peptides were
determined using a combination of manual dansyl-Edman degradation and automated solid phase techniques (12). Comparison with the bovine catalytic subunit (11) identified
the sites of modification as CYS 198 and CYS 342.

Regulatory Subunit

     Although the R subunit of cAMP-dependent protein kinase
has no catalytic activity, it does have several functional
sites including binding sites for cAMP, a site of autophosphorylation, and a recognition site for the C subunit. In
addition, monoclonal antibodies have been generated against
the R subunit and their recognition sites provide another
probe for regions on the surface of the molecule. Using a
procedure analogous to that described for the C subunit, the
CNBr peptides of the R subunit can be separated by a combination of gel filtration (Fig. 5) and HPLC.

Limited Proteolysis. Limited proteolysis of the R subunit (monomeric M_r=55,000) with chymotrypsin yielded two
fragments with the larger fragment (M_r=37,000) corresponding
to the carboxy-terminal region of the molecule. The amino-
terminal sequence of this domain determined by dansyl-Edman
degradation as well as solid phase and liquid phase sequence
analysis is as follows: Asp-Arg-Arg-Val-*SER(P)**-Val-Cys-
Ala-Glu-Thr-Tyr-Asn-Pro-Asp-Glu-Glu-Glu-Glu-Asp-Thr-Asp-Pro-
Arg-Val-. In contrast to the native dimeric R subunit, the
carboxy-terminal domain is monomeric. Nevertheless, the
carboxy-terminal domain does retain several functional sites
such as the site of autophosphorylation, *Ser(P)**, which is in
close proximity to the site of proteolytic cleavage. In
addition, if the native R subunit was photolabeled with
$[^{32}P]8-N_3cAMP$ either before or after limited proteolysis,
 the radioactivity was associated exclusively with the carboxy-terminal domain, indicating that this domain also retained binding sites for cAMP (13).

cAMP Binding Sites. Having established the general
location of cAMP binding, the site was further localized by
cleaving $[^{32}P]N_3cAMP$-labeled R with CNBr. Following gel
filtration, all of the radioactivity was associated with
peak VI (Fig. 5). When this peak was further resolved by
HPLC, three major doublets (A', A, & B) were found, and both
peaks of each doublet had identical amino acid compositions
and amino-terminal residues. The doublets were subsequently

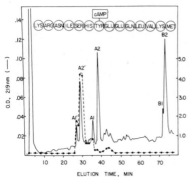

Figure 6. *HPLC separation of CNBr VI from [^{32}P]8-N$_3$cAMP-labeled R.* HPLC conditions are as described in Fig. 2, except that elution time was 50 min. Inset indicates the sequence of A' determined by automated solid phase sequencing of 18 nmol. Taken from (14).

shown to be the homoserine and homoserine lactone form of the same peptide. Two of the peptides (A' and A) had lysine amino terminals, and 90% of the radioactivity was associated with A' (Fig. 6). Sequence analysis showed A' and A to be the 8-N$_3$cAMP-modified and unmodified forms of the same peptide. When a mixture of the two peptides was coupled to glass beads and sequenced (Fig. 6, inset), radioactivity was released at step 7, identifying the tyrosine residue as the site of N$_3$cAMP modification (14). Using purified R subunit, a maximum stoichiometry of 0.5 mole 8-N$_3$-cAMP/mole R monomer was achieved, whereas if photolabeling was carried out using holoenzyme, a stoichiometry of 1 mole 8-N$_3$cAMP/mole R monomer was obtained. Only the peptide shown in Figure 6 was modified whether isolated R or holoenzyme was photolabeled, indicating that 8-N$_3$cAMP was specific for only one of the two cAMP binding sites on each protomer (15).

Antigenic Sites. A final approach for characterizing functional sites on the R subunit has been to identify specific antigenic sites that are recognized by monoclonal antibodies. The monoclonal antibody indicated in Fig. 7 was generated against bovine R; however, as indicated, it cross-reacts equally well with porcine R (16). To further localize the antigenic site, the immunoreactivity of the CNBr fragments was investigated. Immunoreactivity was associated exclusively with CNBr I (Fig. 5) which cross-reacted with an affinity similar to native R. CNBr I (M_r=26,000) corresponds to the amino-terminal CNBr fragment (Fig. 8). The cAMP-binding domain generated with chymotrypsin also cross-reacted (Fig. 7), and the amino-terminal CNBr fragment of this domain (CNBr CI) likewise retained the antigenic site. Both CNBr I and CNBr CI were further fragmented with various proteases in order to identify a smaller immunoreactive peptide. Proteolysis of either fragment with trypsin destroyed all antigenicity. On the other hand, antigenicity was not lost

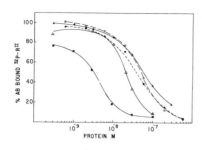

Figure 7. *Relative Affinitives of Bovine R, Porcine R and CNBr Fragments.* Binding affinities for bovine R (●), dephospho (○) and phospho (▲) porcine R, CNBr I (△) and CNBr CI (■) were established by competitive displacement radioimmunoassay using *Staphylococcus aureus* (Cowan I) and 1 pmol ^{32}P-R_2 (porcine). Protein concentration was determined by amino acid analysis.

following proteolysis of either fragment with chymotrypsin. Two immunoreactive chymotryptic peptides (Fig. 8) were isolated from CNBr CI and CNBr I respectively.

In summary, several functionally important sites on cAMP-dependent protein kinase have been localized using an approach that combines limited proteolysis, affinity labeling, and antibody recognition with HPLC and solid phase sequencing. The use of HPLC has been particularily significant in that relatively small quantities of protein (5-10 mg) were required when eventual sequence analysis was carried out and even less (0.1-0.5 mg) for analytical purposes.

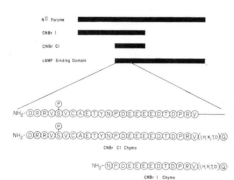

Figure 8. *Localization of Antigenic Site in R Subunit.* CNBr CI (Chymo) was isolated from a chymotryptic digest of CNBr CI purified on C_8-Syn Chrom (75 min gradient of 0.1% TFA to 50% n-propanol). The digest was chromatographed with a 2 h linear gradient of 0.1% TFA to 50% CH_3CN on C_{18} µBondapak. CNBr I (Chymo) was derived from chymotryptic cleavage of CNBr I. The peptides were chromatographed on Sephadex G-50 in 50 mM NH_4OH followed by HPLC on C_{18} µBondapak.

References

1. Beavo, J.A., Bechtel, P.J., and Krebs, E. G. (1975) *Adv. Cyclic Nuc. Res.* *5*,241-251.
2. Rosen, O.M., Erlichman, J., and Rubin, C.S. (1975) *Adv. Cyclic Nuc. Res.* *5*,253-263.
3. Corbin, J.D., Sugden, P.H., West, L., Flockhart, D.A., Lincoln, T.M., and McCarthy, D. (1978) *J. Biol. Chem.* *253*,3997-4003.
4. Builder, S.E., Beavo, J.A., and Krebs, E.G. (1980) *J. Biol. Chem.* *255*,2350-2354.
5. Hofmann, F., Beavo, J.A., Bechtel, P.J., and Krebs, E.G. (1975) *J. Biol. Chem.* *250*,7795-7801.
6. Zoller, M.J., Kerlavage, A.R., and Taylor, S.S. (1979) *J. Biol. Chem.* *254*,2408-2412.
7. Granot, J., Kondo, H., Armstrong, R.N., Mildvan, A.S., and Kaiser, E.T. (1979) *Biochemistry* *18*,2339-2345.
8. Pal, K., Wechter, W.J., and Colman, R.F. (1975) *J. Biol. Chem.* *250*,8140-8147.
9. Zoller, M.J. and Taylor, S. S. (1979) *J. Biol. Chem.* *254*,8363-8368.
10. Zoller, M.J., Nelson, N.C., and Taylor, S.S. (1981) *J. Biol. Chem.* *256*,in press.
11. Titani, K., Shoji, S., Ericsson, L.H., Demaille, J.G., Walsh, K., Neurath, H., Fischer, E.H., Takio, K., Smith, S.B., and Krebs, E.G. (1981) *in Cold Spring Harbor Symposium on Protein Phosphorylation, Vol.8*, Book A, 19-32.
12. Nelson, N.C. and Taylor, S.S. (1981) *J. Biol. Chem.* *256*, 3743-3750.
13. Potter, R.L. and Taylor, S.S. (1979) *J. Biol. Chem.* *254*, 9000-9005.
14. Kerlavage, A.R. and Taylor, S.S. (1980) *J. Biol. Chem.* *255*,8483-8488.
15. Kerlavage, A.R. and Taylor, S.S. (1981) *J. Biol. Chem.* *in press.*
16. Mumby, M. and Beavo, J.A. (1981) *in Cold Spring Harbor Symposium on Protein Phosphorylation, Vol.8*, Book A, 105-124.

THE USE OF RADIALLY-COMPRESSED REVERSED PHASE HPLC FOR THE SEPARATION OF TRYPTIC PEPTIDES OF APOLIPOPROTEINS

W.A. Bradley, A.M. Gotto, Jr. and J.T. Sparrow

Department of Medicine, Baylor College of Medicine, The Methodist Hospital, Houston, TX, USA

INTRODUCTION

The plasma apolipoproteins are an interesting family of proteins[1] which are responsible for the transport of apolar lipids in the blood. These unique detergents accomplish this role by forming structures known as amphipathic helices, in which, upon folding into α-helices, one face of the structure contains the hydrophobic residues (lipid-binding) while the opposite face contains the hydrophilic residues. In concert with polar lipids (phospholipids and cholesterol), these proteins form a surface monolayer which surrounds a core of apolar lipids (i.e. cholesteryl ester, triglyceride). In addition, several of these proteins (C-II, A-I, C-I) serve as cofactors in the activation of enzymes (lipoprotein lipase, LCAT) involved in lipid metabolism (For recent reviews, see references 2-4).

Our laboratory has been involved in structure-function analyses of apoproteins for several years[5] and in solid-phase peptide synthesis of various regions of these proteins[6]. It became increasingly necessary that we develop sensitive, rapid, and highly versatile methods for the isolation and characterization of both native and synthetic peptides from the apoproteins.

The use of high performance liquid chromatography (HPLC) and reversed-phase columns has begun to fill these

479

needs. Present-day HPLC equipment is reliable and accessible to most protein chemists. Its increased acceptance and utilization is documented by the current number of publications seen throughout the literature. It is soon obvious to most HPLC users that the heart of the method lies with the quality of the columns, both the packing material and the packing method. Good quality steel columns, if not personally prepared, are relatively expensive. The recent introduction of radially-compressed columns gives the protein chemist a versatile, moderately-priced system for the separation and isolation of peptide mixtures. This manuscript deals with our recent use of radial-compression HPLC for the separation of native, enzymatically-produced peptides of apolipoproteins and their subsequent uses for sequence analysis.

METHODS

Reagents. Orthophosphoric acid was obtained from Fisher (Fair Lawn, NJ). Water was glass-distilled, while 2-propanol, methanol, pyridine and acetonitrile was purchased from Burdick and Jackson Laboratories (Muskegon, MI). Trifluoroacetic acid was obtained from Halocarbon Product Corporation (Hackensack, NJ) and distilled, collecting only the constant boiling fraction. Triethylamine was obtained from Aldrich (Milwaukee, WI) and used without further purification. 4-N,N-dimethylamino azobenzene 4'-isothiocyanate (DABITC) was purchased from Pierce (Rockford, IL). Triethylammonium phosphate (TEAP) was prepared from a 1% aqueous solution of phosphoric acid, titrated to pH 3.2 with triethylamine (final concentration ca. 0.17 $\underline{M}$). This solution was then chromatographed on a Waters Preparative 500 LC equipped with a C-18 reverse phase column which had been extensively washed with 100% 2-propanol. Trifluoroacetic acid (0.1% v/v) solutions were prepared from freshly distilled TFA dissolved in glass distilled water.

Apoproteins (C-I, C-II, C-III, A-I, and A-II) were isolated as described previously[10]. Purity was determined based on SDS-PAGE, isoelectric focusing, amino acid composition, N-terminal analysis, and retention time on reversed-phase C-18 as reported by Hancock et al.[7]

Tryptic digestions were carried out generally in 0.1 $\underline{M}$ NH_4HCO_3 at r.t. or 37° C for various time periods at 1-4%

trypsin.   Reactions were quenched by the addition of acid
(either H_3PO_3 or 0.1% TFA) just prior to analysis. Simultaneous trypsin blanks were used as controls for background
in HPLC traces.

Samples for chromatography in the TEAP system were
mixed with an equal volume of 1% TEAP, pH 3.2, containing
6 $\underline{M}$ guanidine hydrochloride and 5 μl of 50% orthophosphoric
acid.   Samples for the TFA chromatographic system were either dissolved directly in 0.1% TFA aqueous buffer, or
mixed with, at least, an equal volume of this buffer.   On
occasion, samples were dissolved in 6 $\underline{M}$ Gdn·HCl, 0.1% TFA,
pH 3.

Radial Compression Module.   The radial compression
module (RCM-100) was purchased from Waters Associates.   A
detailed description of this unit is found in Waters Associates Users Manual (Manual No. 1M82154) and it is used in
accordance to the supplier's suggestions.

Column Use and Maintenance.   Radial-PAK liquid chromatographic cartridges were obtained from Waters Associates,
Inc. (Milford, MA).   The packing materials used were C-18
(10 microns) or C-8 silica housed in a 10 x 0.8 cm polyethylene cartridges.   New columns were extensively washed with
methanol and 2-propanol prior to equilibration with the organic solvent used in the chromatographic system, then cy-
cled with a linear gradient to 100% of the aqueous phase
solution before any analyses were performed.   Cartridges
were reversed when back pressures exceeded 2500 psi under
the usual flow conditions and washed with 50% aqueous - 50%
organic - overnight at .3-.5 ml/min flow rates.   The rever-
sal of flow extended the use of the column with no loss of
resolution.

High Performance Liquid Chromatography.   Separations
were carried out on either a Spectra Physics 8000B equipped
with a SP 8400 UV/VIS detector or a SP 8000A equipped with a
Schoeffel 720 UV spectrophotometer which monitored the effluent at 220 nm.   Injector loops from 0.1 to 2.8 ml were
utilized.   Solvents were degassed by helium entrainment, before and throughout the analyses.   Bubble formation by solvent outgassing in the flow detector is prevented by locat-
ing the effluent exit approximately three feet above the
cell.

RESULTS

The separation and detection (at 220 nm) of the tryptic peptides of apolipoprotein C-II are shown in Figure 1.
In the 1% TEAP system, peptides T5 and T8, Leu-Arg and
Gly-Glu-Glu are found in the initially eluting peak. T7 is
insoluble and is isolated after precipitation. Five major
peptide peaks are seen in the chromatogram, eluting in the
order T6, T3, T4, T2 and T1. In addition, a deamidation
fragment of T1 elutes just prior to T2 and is designated
$T1_d$. Also a peptide, T7-8, is isolated and is a partial
digestion product containing the carboxyterminal tryptic
tripeptide, Gly-Glu-Glu. As seen in Figure 1, the resolution and peak shapes are excellent in the 1% (0.17 $\underline{M}$) TEAP-2
propanol gradient. Amino acid compositions, retention times
and recoveries of the apoC-II tryptic peptides are given in
Table I. The compositions are in good agreement with those
obtained by more conventional chromatograpy, demonstrating
the nondestructiveness of the ion pair methodology.

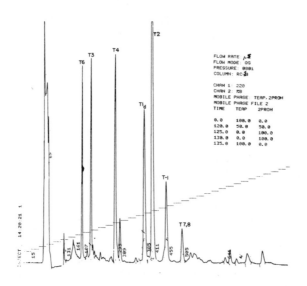

FIGURE 1:   Elution profile of the soluble tryptic peptides
of apoC-II on Radial-PAK (C-18, 10 microns) developed with
a 1% TEAP-2-propanol gradient.   70 nmoles was dissolved in
100 µl 6 M Gdn·HCl, 1% TEAP.

TABLE I

AMINO ACID COMPOSITIONS OF TRYPTIC PEPTIDE ApoC-II ISOLATED FROM RADIALLY-COMPRESSED REVERSED PHASE C-18 HPLC, 1% TEAP-2-PROPANOL (0-50% 2-PrOH, 2 HRS)

	T1	T2	T3	T4	T5[2]	T6
Retention Time (min)	45.5	41.1	21.7	29.5	6.9	18.7
Aspartic acid	1.0		1.1	1.1		1.1
Threonine	3.1		0.8	0.9		
Serine	1.1	3.3				0.9
Glutamic acid	6.3	2.0	2.1	1.4		
Proline	3.1			1.1		
Alanine		1.0	2.2	1.3		
Valine	1.1			1.2		
Methionine	1.0					
Leucine	1.0	1.0	1.0	1.1	1.0	1.0
Tyrosine		0.9	1.0	1.1		1.1
Phenylalanine	1.0					
Lysine	1.0	1.0	1.0	1.0		1.0
Arginine					1.0	
Tryptophan		0.3				
Total residues	19	11	9	9	2	5
Sequence position	1-19	20-30	31-39	40-48	49-50	51-55
% Recovery[1]	51	50	54	70	N.D.[2]	60

[1] 70 NMOLES LOADED.
[2] NOT ISOLATED.

Figure 2 illustrates the equivalent separation of the tryptic digestion of apoC-II in a 0.1% TFA-2-propanol gradient. The elution order of the peptides is identical to the TEAP system. The peak shapes are less symmetrical in the TFA system but the resolution of the tryptic peptides easily allows their complete separation. (The small peak just to the right of T6 is a chymotryptic peptide, STAAMSTY, derived from T7, the insoluble tryptic fragment.) A comparison of retention times indicates that each peptide in the 0.1% TFA gradient is eluted at a later time (higher organic solvent concentration). The advantage of the TFA system, however, is the ease of isolation of the peptides by simple lyophilization of the collected peak.

Analysis, by the DABITC microsequence method[8], of the tryptic peptides (Table II), isolated from the elution profile seen in Figure 2, confirmed the identification of the peptides[9]. The purity of the peptides was apparent from N-terminal analyses. The high yield of glutamine and

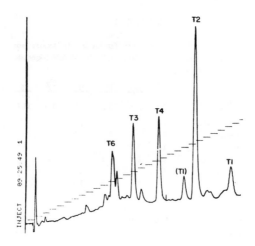

FIGURE 2:   Elution profile of the soluble tryptic peptides
of apoC-II on Radial-PAK (C-18, 10 microns) developed with
a 0.1% TFA-2-propanol gradient.   The run parameters were
identical to those described in Figure 1.   The sample was
dissolved in 0.1% TFA, 100 µl.

TABLE II

DABITC MICROSEQUENCE ANALYSIS OF THE TRYPTIC PEPTIDES

OF APOLIPOPROTEIN CII ISOLATED FROM A RADIAL-PAK (10 MICRON)

C18 COLUMN, 0.1% TFA-2-PROPANOL GRADIENT

PEPTIDE	RETENTION TIME (MIN)	SEQUENCE
T-1	60.3	TQQPQQNEMPSPTFLTEVK
T-2	49.1	E(S)LSSY(W)ESAK
T-3	30.0	TAAQNLYEK
T-4	37.0	TYLPAVDEK
T-6	23.6	DLYSK

asparagine (data not shown) during the Edman degradation suggests very little deamidation occurred in the TFA buffer system.

In our experience, it was not possible to sequence peptides directly from the TEAP system. Conventional de-salting methods, although applicable, were tedious, and hy-drophobic peptides were often lost to the column matrix. A quick and convenient desalting method was developed using the radial-compression system with 0.1% TFA as the aqueous buffer. The peptide in TEAP-2-propanol (up to 0.9 ml) was mixed with an equal volume of the 0.1% TFA buffer (0.9 ml), and loaded into a 1.8 ml injection loop. Elution gradients were developed (i.e. Fig. 3) which allowed the rapid remov-al of the TEAP salt. The peptide itself eluted at approxi-mately the 2-propanol composition that it would in a linear 0.1% TFA-2-propanol gradient. Upon analytical analysis, peptides exposed to TEAP had identical retention times as the same peptide chromatographed in .1% TFA. This suggest-ed that the TEAP was completely removed and had no delete-rious effect on the peptides.

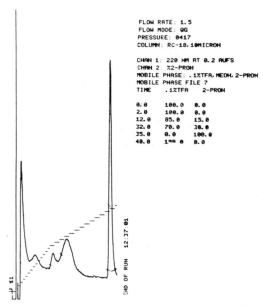

FIGURE 3: Elution profile of the removal of TEAP from tryptic fragment T2. Details are described in the text.

DISCUSSION

The separation and purification of peptides by radial-compression columns is rapid, sensitive, and extremely convenient. The method can be used in both the analytical as well as the preparative mode under identical conditions (only the detector range need be changed). Microgram to milligram quantities are readily separated and detected. Samples may be injected in the presence of denaturants (i.e. 6 M Gdn·HCl, 8 M urea) and at relatively high sample volumes, up to 1.8 ml in this study.

The reproducibility of peptide retention times is good to within ±0.2 min when proper precautions are taken concerning the recycling of the column, and the equilibration time allowed prior to the start of subsequent analyses. Although there are probably many alternatives, we allow the gradient to reach 100% organic, at a higher rate of solvent change (% organic/min) in about 5 min, from the end of the gradient, and then to return to 100% aqueous phase. Equilibration is allowed to continue until the detector response reaches the starting baseline. Careful control of these parameters has yielded consistent results and allowed hundreds of analyses from a single column.

Recoveries of the peptides are comparable to or better than most other chromatographic methods. The use of 2-propanol has been superior to other commonly used organic solvents (i.e. methanol, acetonitrile) in the isolation of these lipid-binding peptides.

One problem, which has not been resolved, concerns the amino acid analysis of peptides at low levels (1.0 nmole or less) directly from the TEAP system without desalting. The TEAP interferes with the resolution of the acidics and neutral amino acids. However, this is not a problem when analyzing at the 10 nmole level on the .9 x 60 cm style columns. As described above, the peptides may be desalted prior to analysis. The advantages of the TEAP system, however, include excellent peak shape, allowing good resolution, high recoveries, and depending upon the peptides, a change in selectivity for the reverse-phase coating. Even though desalting is necessary, these advantages often make TEAP the system of choice.

Although only results from apoC-II are reported here, we have utilized this methodology on other apoproteins (C-I, C-III, A-I, A-II and E) with essentially equivalent results. This methodology should prove useful for many other lipid-interacting peptides and proteins.

The introduction of the radial compression module and polyethylene cartridges adds a useful tool to the separation techniques of the protein chemist, both for analytical and preparative purposes. The uses for which it might be designed seem only to be limited by the adroitness of the user.

ACKNOWLEDGEMENTS

We thank Alice Lin and Ellen Gilliam for skillful technical assistance, Susan McNeeley for the artwork, and Debbie Mason for typing the manuscript. This work was supported, in part, by NIH grant HL 17269, and a grant from the American Heart Association, 78-875. J.T.S. is an Established Investigator of the American Heart Association.

REFERENCES

1. Dayhoff, M. O. (1978) in Atlas of Protein Sequence and Structure, Vol. 5 (Suppl. 3) p. 21.

2. Osborne, J. C., Jr., and Brewer, H. B., Jr. (1977) Adv. Protein Chem. 31, 253-337.

3. Smith, L. C., Pownall, H. J., and Gotto, A. M., Jr. (1978) Ann. Rev. Biochem. 47, 751-777.

4. Bradley, W. A., and Gotto, A. M., Jr. (1978) in Disturbances in Lipid and Lipoprotein Metabolism (Dietschy, J. M., Gotto, A. M., and Onko, J. A., eds.) Waverly Press, Inc., Baltimore, Md., p. 111-137.

5. Smith, L. C., Voyta, J. C., Catapano, A. L., Kinnunen, P.K.J., Gotto, A. M., Jr., and Sparrow, J. T. (1980) Ann. N. Y. Acad. Sci. 348, 213-223.

6. Sparrow, J. T., and Gotto, A. M., Jr. (1980) Ann. N. Y. Acad. Sci. 348, 187-211.

7. Hancock, W. S., Capra, J. D., Bradley, W. A., and
    Sparrow, J. T. (1981) J. Chromatogr. 206, 59-70.

8. Chang, J. Y., Brauer, D., and Wittmann-Liebold, B.
    (1978) FEBS Lett. 93, 205-214.

9. Jackson, R. L., Baker, H. N., Gilliam, E. B., and
    Gotto, A. M., Jr. (1977) Proc. Natl. Acad. Sci. USA
    74, 1942-1945.

10. Baker, H. N., Gotto, A. M., and Jackson, R. L. (1975)
    J. Biol. Chem. 250, 2725-2738.

ROUTINE PEPTIDE MAPPING BY HIGH PERFORMANCE LIQUID CHROMATOGRAPHY

C. S. Fullmer and R. H. Wasserman

N.Y.S. College of Veterinary Medicine

Cornell University, Ithaca, NY

INTRODUCTION

The routine application of high performance liquid chromatography (HPLC) to many phases of protein primary structural determination is currently underway and all aspects cannot thoroughly be detailed herein. Examples of such applications include: 1) subnanomolar analytical mapping of protein digests for comparative purposes such as comparison of digestion conditions and comparison of proteins from different species or tissues; 2) nanomolar analytical mapping for amino acid compositional and end group studies on isolated peptides; 3) preparative mapping of milligram quantities of protein digests for sequence analysis, immunological and biological activity measurements, etc. Intestinal calcium-binding proteins (CaBP) from the cow (75 residues, M_r = 8,500) and chick (242 residues, M_r = 28,000), the primary structures of which have been under investigation in this laboratory, will serve as examples of these applications.

The routine mapping procedure is simple, rapid, nondestructive and offers high recoveries of even very hydrophobic peptides. The entire sequence of the bovine CaBP could be defined by only 7 tryptic or 5 chymotryptic peptides and these were all recovered in high yield (60-100%)[1].

For detailed information regarding procedures and results, the reader is referred to references 1-3.

METHODS

Peptide mapping of enzymatic digests is accomplished by HPLC (Waters Associates ALC/GPC Model 244 liquid chromatograph) on a reverse-phase support medium (C_{18} μBondapak) as previously described[2]. Routine separation of peptides is effected by generation of a linear gradient from 0.1% orthophosphoric acid (solvent A) to 60% acetonitrile (solvent B) in 1 hr. at a constant flow rate of 2 ml/min, at room temperature. Effluents are monitored at 210, 254 and 280 nm, simultaneously. As required for individual separations or applications, solvent composition, run time or gradient curve shape may be varied.

RESULTS AND DISCUSSION

The sensitivity and speed of HPLC mapping provide an opportunity to monitor the progress of enzymatic digestions as a means of optimizing reaction conditions, obtaining suitable products, or as an approach to kinetic studies. The bovine CaBP serves as a straightforward example of such an application (Figs. 1 and 2) in which an entire timed series of tryptic digests was completed in less then 7 hr. requiring a total of 140 μg protein (2 nmol each map). All peaks were collected individually for compositional and end group analyses. Where such analyses are not required, high quality maps can be performed on as little as 4 μg protein, without supression of baseline drift.

Figure 3 shows the similar separation of chymotryptic peptides from the same protein. Peptides represented in Figs 2 and 3 contain all the sequence and overlap data required to construct the primary structure of this protein [1].

The application of HPLC to semipreparative peptide mapping, on an analytical scale column, is demonstrated in Figure 4. One mg quantities of the bovine CaBP, subjected to tryptic digestion, were mapped repeatedly (3 times) and the individual peptide peaks collected and pooled for sequencer analysis (Fig. 4a). In this instance, the orthophosphoric acid concentration of solvent A was altered to 0.01% in order to reduce the amount of acid applied to the sequencer cup. The total volume of collected effluent was 6 ml or less for each peak, and no further sample preparation was undertaken. In all cases, collected peptides were shown to be at least 95% pure. The isolated peptide peak 8 (30 residues), shown to be homogeneous by analytical mapping (Fig. 4b), was successfully sequenced for 29 cycles via the peptide volatile buffer (N,N-dimethyl-allylamine) program.

Figure 5 demonstrates the use of HPLC mapping for compari-

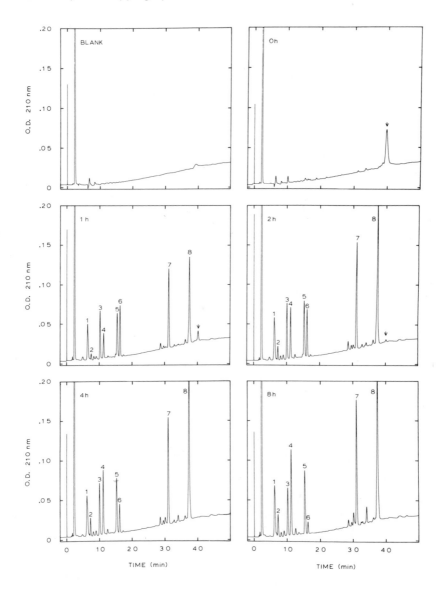

Fig. 1. Time course of tryptic digestion of bovine intestinal CaBP. Twenty μg (2 nmol) digest removed and mapped at indicated times. Arrow denotes undigested protein. Note disappearance of Peak 6 and concomitant increase in Peaks 2 and 4, with time. From Fullmer and Wasserman, ref. 2.

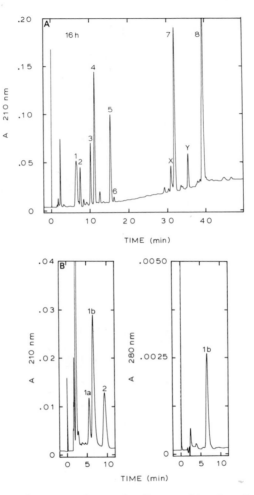

Fig. 2. Sixteen hr. map of tryptic digest of bovine CaBP (20 µg).
a) gradient elution of all peptides and b) isocratic separation of
Peak 1 components with solvent A adjusted to pH 3 with NaOH.
From Fullmer and Wasserman, ref. 1.

son of similar proteins following tryptic digestion. Three variant
forms of the bovine CaBP are isolated, differing only slightly at
their NH_2-termini. Peaks 3 and 3a represent peptides with sequences K-S-P-E-E-L-K and S-P-E-E-L-K, respectively (see ref. 3, for
details).

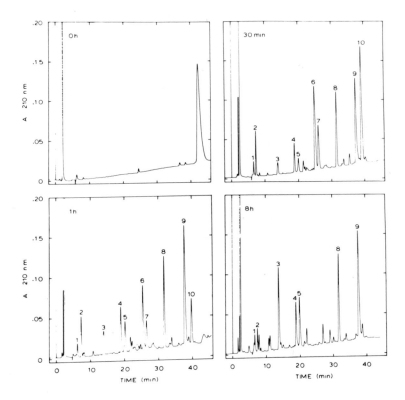

Fig. 3. Selected HPLC maps of timed chymotryptic digestion of bovine CaBP (25 µg, each map). From Fullmer and Wasserman, ref. 1.

The advantages of simultaneous multiple wavelength monitoring are demonstrated by the tryptic digest of chick CaBP (Fig. 6). Peptides containing tryptophan (and tyrosine) are evident, eliminating the need for routine tryptophan determination and providing clear reporter groups for a variety of uses.

Peptides isolated by this procedure have been shown to be entirely suitable for amino acid compositional and end group determinations, provided the total collected volume of effluent is small (less than 2 ml). For liquid phase sequence analysis of

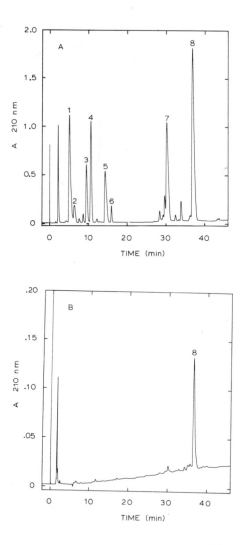

Fig. 4. a) Semi-preparative isolation of tryptic peptides from bovine CaBP (1 mg) and b) analytical map (0.5 nmol) of isolated Peak 8 peptide prior to sequence analysis. From Fullmer and Wasserman, ref. 1.

peptides, it becomes necessary to reduce the total amount of orthophosphoric acid. Reduction of the acid concentration in solvent A is a satisfactory method when employed in conjunction with larger peptides and the volatile buffer (DMAA) program[1]. However, when sequence analysis is via the dilute Quadrol (or dilute Quadrol with Polybrene) program, removal of acid becomes critical due to the limited buffering capacity of this coupling reagent. Acid-soluble peptides may be desalted by gel filtration in 1-10% acetic or formic acid. Hydrophobic peptides, eluting at higher concentrations of acetonitrile and not soluble in water or dilute acid, may be dried directly in the sequencer cup, and the residual H_3PO_4 extracted with the appropriate solvent prior to the final sample application subroutine. These relatively simple sample preparation techniques have provided excellent results for sequence analysis of peptides 3-40 residues in length (unpublished observations). Replacement of orthophosphoric acid with volatile ion-pairing reagents such as trifluoroacetic and heptafluorobutyric acid eliminates the necessity for sample preparation. These solvents, however, limit the practicability of low nanomolar and subnanomolar mapping due to poor U.V. transparency in the region of 206 nm and difficulties in balancing solvent absorbance during gradient runs.

ACKNOWLEDGEMENTS

This work was supported by Grant AM-04652 from the National Institutes of Health and conducted at the Cornell University Amino Acid Analysis and Protein Sequencing Facility, Ithaca, New York.

REFERENCES

1.   Fullmer, C. S. and Wasserman, R. H. (1981) J. Biol. Chem. 256, 5669-5674.
2.   Fullmer, C. S. and Wasserman, R. H. (1979) J. Biol. Chem. 254, 7208-7212.
3.   Fullmer, C. S. and Wasserman, R. H. (1980) In: Calcium-Binding Proteins: Structure and Function (F.L. Siegel et al, eds.) Elsevier North Holland, N.Y., pp 363-370.

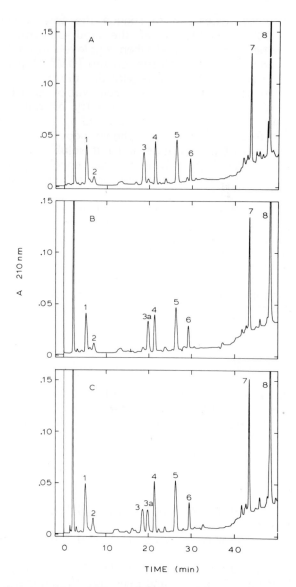

Fig. 5. Comparative HPLC tryptic maps of bovine CaBPs (30 μg each). Conditions as described except a slightly concave gradient (curve #7) was employed to separate the difference peptides (Peaks 3 and 3a. See ref. 3, for details.

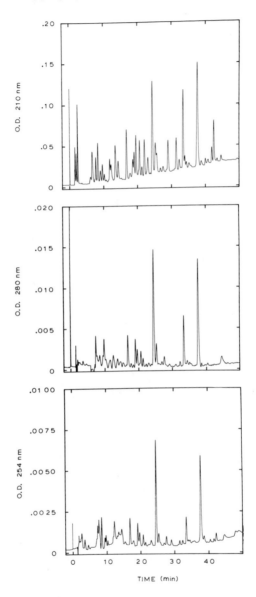

Fig. 6. HPLC map of 16 hr. tryptic digest of chick CaBP (30 μg) monitored at 3 wavelengths, simultaneously. Peptide peaks containing the 2 Trp residues are evident at 254 and 280 nm.

COMPARATIVE PEPTIDE MAPPING BY HPLC: IDENTIFICATION OF SINGLE AMINO ACID SUBSTITUTIONS IN TEMPERATURE SENSITIVE MUTANTS

K.R. WILLIAMS[*], J.J. L'ITALIEN[*], R.A. GUGGENHEIMER[+], L. SILLERUD[*], E. SPICER[*], J. CHASE[+], W. KONIGSBERG[*],

[*]YALE UNIV., NEW HAVEN, CT. 06510; and

[+]ALBERT EINSTEIN COLLEGE, BRONX, N.Y. 10461

INTRODUCTION

The study of mutant proteins containing single amino acid substitutions offers a particularly powerful approach for exploring the relationship between protein structure and function. While proteins that have been subjected to chemical modification, partial proteolysis, and covalent protein:ligand crosslinking may contain multiple alterations in their primary structure, proteins containing missense mutations usually have only a single amino acid substitution. The value of this approach has been well demonstrated by the elegant studies of Perutz et al (1,2) on naturally occurring variants of human hemoglobin and by Miller (3) on altered lac repressor molecules that were generated in vitro. In order to realize the full potential of this approach, a method is required to rapidly define the particular amino acid replacement that has occurred. The reproducibility, sensitivity, and resolving power of peptide mapping and isolation by HPLC makes it the current method of choice for comparing the primary structures of wild-type and mutant proteins. This technique allows the identification of single amino acid substitutions at the 1-10 nanomole level with the limiting factor being amino acid analysis. HPLC has previously been used to separate tryptic peptides from 25 hemoglobin variants containing

known amino acid substitutions (4). In this study we have
used HPLC to identify the mutations that have occurred in
three temperature-sensitive ssDNA binding proteins and show
how this information can be correlated with other physicochemical data to examine the interrelationship between the
structure and function of a protein.

MATERIALS AND METHODS

The bacteriophage T4 ssDNA binding protein was
purified (5) from E. coli BW46 infected with either T4 (am
N134, am BL292), which overproduces the wild-type protein,
or T4 (tsP7, am N134, am BL292), which overproduces a
temperature sensitive ssDNA binding protein containing the
P7 mutation (6). The E. coli single stranded DNA binding
protein (SSB) was purified from KLC647 carrying the pDR1996
plasmid as described previously (7,8). SSB from a strain
containing the lexC113 mutation (9) was purified by a
procedure similar to that previously described by
Molineux et al (10). The purification of single strand
binding protein from an ssb-1 mutant strain (11) will be
described later (12). The bacteriophage T4 ssDNA binding
protein was carboxamidomethylated (5) and then digested
with trypsin (protein:enzyme weight ratio of 25:1) for 6
hrs at 37⁰ in 50 mM NH_4HCO_3. The various E. coli
ssb proteins were dialyzed versus 1 mM HCl, precipitated
with 10% (v/v) trichloroacetic acid and then digested
with trypsin as above. Chymotrypsin digestion of T-14,15
from SSB and lexC was for 4 hrs at 37⁰ in 25 mM
NH_4HCO_3 at a protein:enzyme weight ratio of 30:1.

The 25% acetic-acid soluble peptides from all digests
were applied to a μBondapack C-18 column (0.39 x 30 cm,
Water Assoc.) which had been equilibrated with 20 mM
potassium phosphate (pH 2.55). The column was then
developed using linear gradients of acetonitrile and the
eluent was monitored by direct UV detection at both 214 and
254 nm. The elution profile for each peptide mixture was
optimized by screening various gradient conditions for separation at the 1 nanomole level. Once suitable elution
conditions were obtained, the peptides were separated at
the 10-20 nanomole level and collected for amino acid
analysis. Differential scanning microcalorimetry and
fluorescence quenching experiments were carried out as
described in references 13 and 14 respectively.

RESULTS AND DISCUSSION

The P7 Temperature-Sensitive Mutation in Bacteriophage T4 Gene 32

The protein encoded by gene 32 of bacteriophage T4 (32P) has served as a prototype for a class of proteins which bind preferentially to ssDNA and, therefore, destabilize dsDNA (see reference 15 for a review). Because of its tight, cooperative binding (16) and relative abundance in vivo, 32P (and similar DNA binding proteins from other organisms) would be expected to completely cover the transient single stranded regions of DNA that normally arise in vivo as a result of DNA replication, repair, and recombination. The crucial function that 32P plays in DNA metabolism in vivo can be clearly shown by studies on bacteriophage T4 containing the temperature-sensitive mutation in gene 32 termed P7. When cells infected with the P7 mutant are shifted from 25° to 42° T4 DNA synthesis (17) and repair (18) ceases and the intracellular T4 DNA is rapidly degraded (19). Detailed genetic studies suggest that while the P7 gene 32 protein is unable to bind ssDNA above 37°, it nonetheless retains its ability to interact with other T4 proteins (6,20). Hence it was proposed that 32P contained several independent domains and presumably the P7 mutation results in an unfavorable amino acid substitution at the DNA binding site which would account for the inability of this protein to bind ssDNA at temperatures above 37°.

In an attempt to localize this DNA binding domain we used HPLC to determine the location of the P7 mutation. After trypsin digestion of the wild-type protein and HPLC we were able to identify 28 of the 30 expected acid-soluble peptides (Fig. 1). With the exception of the peak corresponding to T-13 and T-27, which frequently did not separate from the nearby T-4,5 peak, the overall pattern shown in Fig. 1 was quite reproducible. While the yield varied from 12% for T-22 to 100% for several peptides, most were recovered in yields of at least 70%. Six peptides (T-1,9,15,16,17,20) were recovered in pure form that had previously been difficult to isolate (21). The HPLC profile of the P7 tryptic digest (Fig. 1) revealed that T-7 was missing and subsequent amino acid analysis demonstrated that T-8 no longer co-chromatographed with T-12 and T-19. Since T-8 contains several amino acids (Ser,Pro,Phe) not found in T-12 or T-19, its absence from this pool was readily detected. In addition, there was a noticeable

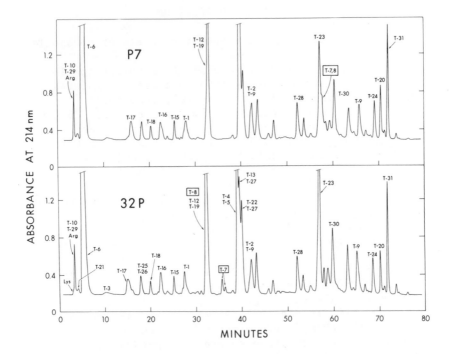

Fig. 1. HPLC separation of tryptic peptides from 14.9
nanomoles of wild-type (lower curve) or 11.4 nanomoles of
the P7 mutant (upper curve) bacteriophage T4 gene 32
protein. The column was eluted with linear gradients of
acetonitrile (solvent B) into 20mM KH_2PO_4, pH 2.2: 0-60
min (0-30%B), 60-90 min (30-60%B).

shoulder on the peak for T-23 that was not present in the
chromatogram of the wild-type digest. Amino acid analysis
of this shoulder revealed that its composition exactly
matched that expected for T-7,8 (residues 35-51) except
that there is a cysteine in place of arginine 46.

   This single amino acid replacement in the P7 protein
has a profound effect on a number of physical properties of
the protein. Although both the wild-type and P7 protein
require 2M NaCl for elution from ssDNA-cellulose,
fluorescence quenching measurements indicate that the P7
mutation decreases the affinity of the protein for ssDNA
from $2 \cdot 10^8$ M^{-1} to $0.4 \cdot 10^8$ M^{-1} at 25°. While
increasing the temperature to 42° has very little effect

on the wild-type protein, fluorescence quenching indicates
that at least 75% of the P7 protein is denatured by
incubating for a few minutes at 42°. In contrast to in
vivo studies (6,20) suggesting that only the DNA binding
domain of P7 denatures at nonpermissive temperatures,
differential scanning microcalorimetry reveals that the
entire P7 protein unfolds at 37°. The Arg to Cys
substitution at position 46 shifts the single thermal
transition from 56.3° for the wild-type protein to
37.5° for the P7 mutant. Unlike that for 32P (13), the
P7 transition is not altered by ssDNA. Arginine 46
therefore appears to be crucial for maintaining the native
conformation of 32P above 37°.

The ssb-1 and lexC Temperature Sensitive Mutations in the E. coli Single Strand Binding Protein

The E. coli ssDNA binding protein (SSB) seems to be
functionally homologous to the T4 gene 32 protein. Thus
DNA synthesis stops when E. coli containing the ssb-1
temperature sensitive mutation are shifted from 30° to
42° (22). Similarly, even at 37° the ssb-1 mutant is
extremely sensitive to ultraviolet irradiation and only
about one-fifth as active as wild-type E. coli in recombination (23). While bacteria containing the lexC mutation
in the ssb gene are also temperature-sensitive for growth
and DNA replication (24), there are important differences
between the phenotypic expression of these two allelic
mutations. Hence on T-broth at 42° the colony forming
ability of the ssb-1 mutant is only about 0.13% that of
the lexC mutant or wild type bacteria (25). Also, there
is more extensive DNA degradation after ultraviolet
irradiation at 42° of the ssb-1 mutant than the lexC
mutant or wild type E. coli (26). In contrast, cells
containing the lexC mutation are considerably more sensitive to methyl methane sulfonate than those containing the
ssb-1 mutation (25). These in vivo studies are similar to
those done on 32P (6,20) in that they suggest that SSB may
have discrete functional domains and that the ssb-1 and
lexC mutations occur in different domains. To test this
idea we used HPLC of tryptic digests to determine the amino
acid substitutions that have occurred in each.

As shown in Fig. 2 we were able to identify all 14 of
the acid-soluble tryptic peptides from SSB. About 50%
cleavage occurred at the Arg-Pro bond in between T-14 and
T-15 giving rise to some T-15, which co-chromatographs with
uncleaved T-14,15, and a low yield of T-14, which elutes

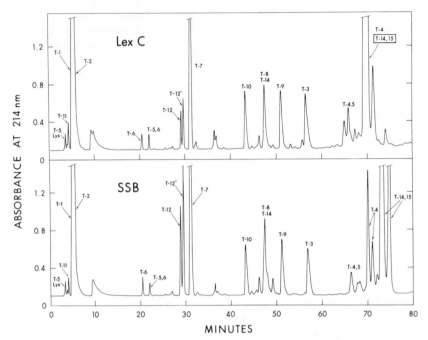

Fig. 2. HPLC separation of tryptic peptides from 20 nanomoles of wild-type (lower curve) or lexC mutant (upper
curve) E. coli single strand binding protein (SSB). The
column was eluted with linear gradients of acetonitrile
(solvent B) into 20mM KH_2PO_4, pH 2.2: 0–40 min
(0–20%B), 40–70 min (20–30%B), 70–90 min (30–60%B).

very close to T–8. The multiple peaks for T–4 and T–14,15
appear to be due to methionine oxidation. In each case,
amino acid analysis revealed that the first peak in each
doublet contained some methionine sulfoxide. When a
similar tryptic digest was run from the lexC mutant protein
(Fig. 2) it was observed that the peak corresponding to the
carboxy terminal T–14,15 peptide eluted earlier and now co-
chromatographed with T–4. Because this was too complex a
mixture to analyze directly, this pool as well as the pool
corresponding to T–14,15 from wild-type SSB was digested
with chymotrypsin and then rechromatographed. Although no
chymotryptic cleavage occurred within T–15, both T–4 (which
was a contaminant in the lexC T–14,15 peak) and T–14 were
cleaved at several sites. Hence the last peak in both the

wild type and lexC digests was the carboxy-terminal T-15 and it eluted significantly earlier in the lexC as compared to the wild-type digest (data not shown). Amino acid analysis of the lexC T-15 (residues 155-177) revealed that it contained a serine residue in place of one of the 6 prolines normally in the wild-type T-15. Chymotryptic digestion of the wild-type T-14,15 resulted in the isolation of a peptide spanning residues 116-135. No serine was found in the amino acid analysis of this peptide hence the published SSB sequence (8) contains an error at "serine" 133.

Our analysis of the ssb-1 mutation was complicated by the insolubility of this protein after trichloroacetic acid precipitation. As a result only partial cleavage occurred at several sites. Nonetheless, Fig. 3 demonstrates that T-7 elutes about 15 minutes later in the ssb-1 digest compared to the wild type digest. Amino acid analysis of this peak (as well as T-5,6,7 and T-6,7) revealed that it contains a tyrosine residue in place of histidine 55. As expected, this missense mutation requires only a single base change (codon for His 55 is CAC (8) and that for tyrosine is UAC).

Our preliminary physicochemical studies on SSB-1 and lexC indicate that the lexC protein binds equally as well as the wild type protein to ssDNA. In addition, lexC binding to ssDNA is not temperature sensitive so there must be another explanation for the inability (at least in some media) of bacteria containing the lexC mutation to grow at temperatures above 37° (24,25). In contrast, even at 5°C the ssb-1 protein elutes from ssDNA cellulose with less than 0.5 M NaCl (compared to 1-2 M NaCl for wild type SSB) and so it does not appear to bind tightly to ssDNA even at permissive temperatures. These results are consistent with our previous partial proteolysis studies which showed that residues 116-177 in SSB are not essential for DNA binding (27). The ssb-1 mutation occurrs in the middle of the longest predicted stretch (residues 45-70, (8)) of α-helix in SSB. Based on a Chou and Fasman (28) conformational analysis, the tyrosine substitution in SSB-1 would be expected to prematurely terminate this α-helical segment at tryptophan 54. The inability of the ssb-1 protein to bind ssDNA strongly suggests this region of SSB may play a direct role in DNA binding.

Our results clearly demonstrate the utility of HPLC for rapidly confirming protein sequences and for comparing the primary structures of mutant proteins. By combining

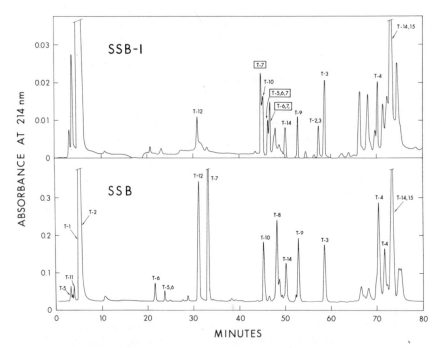

Fig. 3. HPLC separation of tryptic peptides from 23 nanomoles of wild-type (lower curve) or 15 nanomoles of ssb-1
mutant (upper curve) E. coli single strand binding protein.
The column was eluted as in Fig. 3.

in vivo and in vitro data on proteins containing single
missense mutations considerable insight can be gained
concerning the relationships between a protein's structure
and function. In addition, the ability to rapidly detect
single amino acid substitutions in proteins may result in
more precise diagnosis of inherited diseases as well as
providing a possible approach for determining the impact of
environmental agents on mutational frequencies in higher
organisms.

                    REFERENCES
1. Perutz, M.F. (1970) Nature 228, 726.
2. Perutz, M.F. and Lehmann, H. (1968) Nature 219, 902.
3. Miller, J. (1979) J. Mol. Biol. 131, 249.
4. Wilson, J., Lam, H., Pravatmuang, P., and Huisman, T.
    (1979) J. Chromatography 179, 271.

5. Williams, K., LoPresti, M., Setoguchi, M., and Konigsberg, W. (1980) Proc. Natl. Acad. Sci. 77, 4614.

6. Mosig, G., Berquist, W., and Bock, S. (1977) Genetics 86, 5.

7. Chase, J., Whittier, R., Auerbach, J., Sancar, A., and Rupp, W. (1980) Nucleic Acids Res. 8, 3215.

8. Sancar, A., Williams, K., Chase, J., and Rupp, W., (1981) Proc. Natl. Acad. Sci. 78, 4274.

9. Johnson, B. (1977) Molec. Gen. Genet. 157, 91.

10. Molineux, I., Friedman, S., and Gefter, M. (1974) J. Biol. Chem. 249, 6090.

11. Glassberg, J., Meyer, R., and Kornberg, A. (1979) J. Bacteriol. 140, 14.

12. Chase, J., and Williams, K.R., ms. in preparation.

13. Williams, K., Sillerud, L., Schafer, D., and Konigsberg, W. (1979) J. Biol. Chem. 254, 6426.

14. Spicer, E., Williams, K.R., and Konigsberg, W. (1979) J. Biol. Chem. 254, 6433.

15. Williams, K.R., and Konigsberg, W.H. (1981) DNA Helix-Destabilizing Proteins in "Gene Amplification and Analysis," Vol. II: (Eds. J.G. Chrikjian and F.S. Papas) Elsevier Press, 475.

16. Alberts, B. and Frey, L. (1970) Nature 227, 1313.

17. Riva, S., Cascino, A., and Geiduschek, E. (1970) J. Mol. Biol. 54, 85.

18. Wu, J. and Yeh, Y. (1973) J. Virol. 12, 758.

19. Curtis, M., and Alberts, B. (1976) J. Mol. Biol. 102, 793.

20. Breschkin, A., and Mosig, G. (1977) J. Mol. Biol. 112, 279.

21. Williams, K., LoPresti, M., and Setoguchi, M. (1981) J. Biol. Chem. 256, 1754.

22. Meyer, R., Glassberg, J. and Kornberg, A. (1979) Proc. Natl. Acad. Sci. 76, 1702.

23. Glassberg, J., Meyer, R., and Kornberg, A. (1979) J. Bacteriology 140, 14.

24. Greenberg, J., and Donch, J. (1974) Mutation Research 25, 403.

25. Vales, L., Chase, J., and Murphy, J. (1980) J. Bacteriology 143, 887.

26. Baluch, J., Chase, J., and Sussman, R. (1980) J. Bacteriology 144, 489.

27. Williams, K.R., Guggenheimer, R., Chase, J.W., and Konigsberg, W.H. (1981) Fed. Proc. 40, 1731.

28. Chou, P., and Fasman, G. (1978) Adv. in Enzymol. 47, 45.

A New Low Cost, Fully Automated Amino Acid Analyzer

Using a Gradient HPLC

David G. Klapper

Department of Bacteriology and Immunology

University of North Carolina at Chapel Hill, N. C.

Since the first generally useful, dependable amino acid analyzer was developed by Spackman, Stein, and Moore (1), a large number of investigators have modified and improved the original scheme for separation of amino acids. Resins, buffers, detection reagents, and hardware have all undergone major refinements in the past 24 years, although the mechanics have basically remained the same. In a sense, the only amino acid analyzer configuration which has deviated in a fundamental way from the original design was the system described by Piez and Morris (2) in 1960. Rather than "switch" from one buffer to another, a complex buffer gradient was utilized to elute various amino acids. That system, requiring a multichambered gradient device (3) never really became as popular as the original system since the gradient maker had to be cleared and refilled after each run. The gradient system did not lend itself as readily to automation as the "step" buffer system where a simple valve switches from one buffer to another to another, etc., until the run is over and initial buffer conditions are reestablished.

As early as 1963, Hamilton (4) developed a procedure utilizing a step gradient, but only a single column. Mechanistically, this is the procedure which has evolved and is considered most useful today since it requires only a single aliquot of sample. Other improvements in buffers (pH "steps" vs ionic strength "steps"), detection reagents

509

(ninhydrin vs orthophthalaldehyde), ninhydrin reaction conditions (boiling water bath vs 130°C coil), automatic
sample injection, pulse-free and reliable pumping systems,
and microbore columns are all variations of the same methodology.

In an attempt to take advantage of the most reliable
pumping systems currently available, a high pressure liquid
chromatograph (HPLC) has been adapted to perform amino acid
analysis. During these trials, it became clear that a
multiple buffer switching valve was not a necessary addition to commercially available HPLC systems, but, in
fact, a simple two buffer linear gradient gave excellent
separation of all amino acids in a standard hydrolysate.

MATERIALS AND METHODS

Chemicals and resins: The two buffers used in this report are standard amino acid analyzer buffers. Buffer "A"
is pH 3.25 sodium citrate (final sodium molarity = 0.2)
purchased from Pierce Chemical Company, Rockford, IL. For
best threonine-serine resolution, this buffer contains 1.5%
n-propanol (Pierce). Buffer "B" is a sodium borate buffer,
pH 9.8, also with a final sodium molarity of 0.2. This
buffer has always been made in the laboratory, but is now
commerically available (Beckman Bioproducts, Palo Alto,
CA). As a slight modification of the reagent described by
Moore (5), ninhydrin (Pierce) is dissolved in dimethyl-
sulfoxide (DMSO) and stored as a 3.0% (w/v) solution in the
dark in an amber bottle. To make up analyzer ninhydrin,
125 mg hydrindantin is dissolved in 200 ml of DMSO-
ninhydrin at room temperature with bubbling nitrogen. When
the hydrindantin is dissolved, 100 ml ice cold 4M lithium
acetate, pH 5.5 is added. The reagent becomes dark
reddish-purple in color and slightly warm to the touch.
Nitrogen bubbling is continued for 3-4 minutes and the reagent transferred to the ninhydrin storage vessel of the
analyzer. Orthophthalaldehyde (Fischer Chemical Co.) is
stored at -20°C as a 0.4% solution in methanol. The
potassium borate buffer (pH 10.4) is made as the recommended modification from Pierce Chemical Co., except that the
boric acid content is doubled (i.e., 50 grams/liter) and
the orthophthalaldehyde concentration is halved.

Several resins have been tested and found to be adequate for this procedure. These include HPC from Biorad

Laboratories, Richmond, CA., (4 x 250 mm), A9 from Biorad Laboratories (4 x 250 mm), and a test column from Waters Associates, Medford, MA., (4 x 250 mm) with excellent threonine-serine resolution. We have packed our own glass columns (3 x 250 mm, 3 x 500 mm, 6 x 200 mm) with bulk A9 resin also with excellent results.

Instrumentation: Ninhydrin and orthophthalaldehyde (OPA) solutions are stored at room temperature in amber 1 liter bottles under N_2 (<3 p.s.i.) using the Glenco N_2 supply console (Glenco Instrument Co., Houston, TX). Buffers also are stored at room temperature under N_2 in the nitrogen supply console. The teflon caps on the buffer bottles were drilled to accommodate 1/8 inch tubing. Two Waters M6000 pumps and a model 660 gradient maker are the heart of the system. The gradient device is automatically reset to initial conditions (100% pH 3.25) at the end of a run by the WISP automatic sample injector (Waters Associates). The column is water jacketed at 62°C by means of a small circulating water bath. A Milton-Roy mini pump is used to pump ninhydrin or OPA through 10 feet of 0.02 i.d. teflon tubing to a teflon "T" where the reagent mixes with column eluate. A low pressure 3-way slider valve (Rheodyne, Berkeley, CA) directs the flow of reagent-column eluate to either a low temperature, 4 foot reaction coil (0.02 i.d. teflon) in the case of OPA or a high temperature (125°C) 35 foot reaction coil of the same type of tubing for ninhydrin reaction. The low temperature coil is immersed in an oil bath consisting of a 250 ml water jacketed wide mouth reaction vessel. The circulating water bath provides 62°C water to this vessel as well as to the column. At typical flow rates, the OPA-column eluate mixture spends 10-15 seconds in this bath and most peak heights are approximately doubled. The high temperature bath is a heating mantel controlled by a variable transformer. The heating mantel holds a 500 ml round bottom flask. For convenience, the top was cut off the flask to make it easier to change reaction coils during development of this instrument. Three glass "knobs" were melted around the top half of the flask and matching "knobs" were placed on the lower part of the vessel. Three small steel springs hook over these knobs and serve to clamp the flask together during use. The flask is filled with 200 ml of high temperature heating oil (Fischer Chemical) which is adequate to cover the coil when it is lowered into the bath. At typical flow rates, the ninhydrin-column eluate mixture spends 2.5 minutes in the reaction coil.

The low temperature (OPA detection) coil flows to an Aminco filter fluorometer (American Instrument Co., Silver Springs, MD) which is connected to an integrator and re-corder. The integrator is started by the automatic sample injector in this particular instrument arrangement. The high temperature coil flows to a 590/440 nm photometer (Glenco Instruments) which is also connected to the inte-grator and a two pen recorder. When not in use, the coil is lifted out of the high temperature oil bath and sus-pended above the hot oil. To eliminate "boiling" at 125°C in the coil, the teflon effluent tube from the photometer has a back pressure valve set for 40.0 p.s.i. (Rainin Instrument Co., Woburn, MA.).

RESULTS

Figures 1 and 2 depict chromatograms obtained with the instrument described in this communication. These parti-cular chromatograms were run with a linear gradient from 100% pH 3.25 buffer to 100% pH 9.8 buffer over a period of 48 minutes. The pH 9.8 buffer is continued for 21 minutes and the gradient device resets at 100% pH 3.25 buffer. The column is reequilibrated and ready for the next injection 11 minutes after the switch from pH 9.8 to pH 3.25. Super-

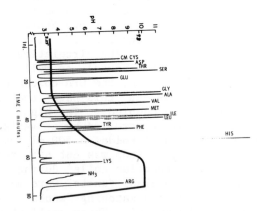

Figure 1: Elution profile of a standard amino acid hydrolysate (see text for details). Sample contained 1 nanomole of each amino acid and detection was with OPA.

imposed on the elution profile in Figure 1 is the pH gradient measured at the column outlet in a separate run without reagent or sample injection. It is possible to shorten
the gradient by as much as 10 minutes (and therefore the
total run time by 10 minutes) and still maintain quite good
resolution. The isoleucine/leucine doublet is most affected by this time change. It should also be noted that the
glycine / alanine doublet is well separated even though the
pH 3.25 buffer contains 1.5% propanol. This is presumably
due to the fact that by the time glycine and alanine are
eluted, the propanol concentration is less than 0.7% and
therefore does not adversely affect that separation.

The pH of the borate buffer is critical for the elution of arginine, but does not appreciably affect elution
times of the neutral amino acids unless the pH is significantly above or below pH 9.8. Even a difference of 0.1 pH
units will result in arginine coeluting with ammonia or
eluting 4 or 5 minutes after ammonia. A linear gradient
has been found to be best as either convex or concave gradients negatively affect elution of acidic or neutral amino
acids respectively.

The particular column used to generate the figures in
this report was supplied by Waters Associates and is a sul-

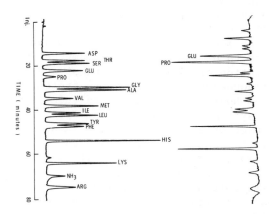

Figure 2: Elution profile of a standard amino acid
hydrolysate (see text for details). Sample contained 10
nanomoles of each amino acid and detection was with ninhydrin. Both the 590nm and 440nm detectors were set on 1.0
A.U.F.S.

fonated polystyrene resin packed in a steel 0.4 x 25 cm tube. Optimal buffer flow for this column is 0.4 ml / minute, but that will vary (as will gradient and total run time) depending upon column configuration. At this flow rate with this column, back pressure is 400-450 p.s.i. Reagent flow rate is 0.2 ml/minute. Higher reagent flow rates require longer reaction coils and begin to compromise peak resolution. Because buffer is flowing at twice the rate of reagent, it is necessary to increase the molarity of the orthophthalaldehyde borate buffer or it will not have sufficient buffering capacity to "break" the pH 3.25 running buffer. DMSO -ninhydrin has sufficient buffering capacity so that no increase in lithium acetate molarity is necessary.

DISCUSSION

It is clear from Figures 1 and 2 of this manuscript that a simple gradient high pressure liquid chromatograph can be easily adapted to perform amino acid analysis. Using only 2 buffers, a linear gradient can now replace the more traditional multiple buffer system and step valve. The system is fast (injection to injection times of 70 minutes at present), resolves all amino acids found in normal protein hydrolysates, and can utilize virtually any gradient HPLC. Back pressure depends upon column configuration and is under 500 p.s.i. for the 0.4cm x 25cm column reported here. Some concern has been expressed by HPLC manufacturers regarding the effects of salt (NaCl) on pump heads. We have taken the precaution of flushing the pumps and auto sampler with 0.1% trifluoroacetic acid when the system is idle for more than a few hours. This is done quickly and efficiently in this system by installing a "prime/purge" valve between the injector and the column. This is opened and the buffer inlets on both pumps switched from "Buffer 1" (analyzer buffer) to "Buffer 2" (0.1% TFA). The flow rate is run up to 2.4 ml/minute, the gradient device set on 50% and the sampler switched to manual and purged. In less than 2 minutes, all stainless steel flow lines are salt free. 0.1% TFA has been flushed through the column and is a superb storage buffer. It seems clear that high speed, automated amino acid analysis is an option available to virtually any laboratory with even a simple gradient HPLC. The ability to use a linear, two buffer elution system to elute amino acids immeasurably extends the usefulness of HPLC systems and makes amino acid analysis available at a quite modest cost.

REFERENCES

1. Spackman, D.H., Stein, W.H., Moore, S. (1958) Anal. Chem., $\underline{30}$, 1190.

2. Piez, K.A., Morris, L. (1960) Anal. Biochem. $\underline{1}$, 187.

3. Peterson, E.A., Sober, H.A. (1959) Anal. Chem. $\underline{31}$, 857.

4. Hamilton, P.B. (1963) Anal. Chem. $\underline{35}$, 2055.

5. Moore, S. (1968) J. Biol. Chem. $\underline{243}$, 6281.

Identification of Phenylthiohydantoins

Fused Silica-Capillary-Gas Chromatography.
A Novel Procedure for the Identification of PTH
in the Analysis of Protein.

D. Tripier

HOECHST AG

6230 Frankfurt|M 80, Germany

It is necessary to evaluate the purity of peptides with a high precision. Capillary gas chromatography is an ideal solution for quantitative
PTH-analysis.

Until now, the capillary chromatography of PTH-
amino acids has only been used infrequently. In
1973 Eyem and Sjöquist (1) analysed the volatile
Methylthiohydantoins and various PTH-amino acids
on a short (4,5 m) glass column at 260° C. A
special coating mixture for the stationary phase
on a noncommercial capillary limited the usefulness of the procℓdure.

Fused silica columns have recently become available. The older glass columns contained metal
ions which, at elevated temperatures necessary
for the resolution of some polar PTH-amino acids
caused catalytic destruction and thereby uncon-
trolled absorption. As a result, good analyses
were impossible. Fused silica columns, on the
other hands have virtually no catalytic properties, The result is a very high thermal stability
combined with low absorption phenomena.

A new gas chromatographic technique by which the
PTH-derivative of all naturally occuring amino
acids (except Arg) can be identified and quanti-

fied is reported here:
The PTH derivatives resulting from the sequencing of a protein are isolated and transformed
into their Trimethylsilyl-derivatives. The mixture is analysed on a capillary of fused silica.
One analysis lasts forty minutes. The method of
internal standard is used to calculate yields
and to check proper performance of the system.
The splitless technique yields a sensitivity of
about fifty picomoles for standard PTH derivatives.

Methods and Reagents:

A 10 µl aliquot of stock solution (2 mg of each
PTH-derivative/ml) was dried in a 300 µl vial,
derivatized with 10 µl of a 50 % solution of
MSTFA in dry acetonitrile and 0,15 µl were injected after 5 minutes.

        Fused Silica Capillary Chromatography:
Apparate: Hewlett-Packard 5880 A.
Column   : H-P OV-1. Silanox deactivated or
           Chrompack. Cp-TM-Sil-5
25 m x 0,25 mm. TZ = 21 at K = 6 for $C_{15}-C_{16}$
at 140° C. He = 30 cm/min. 17 psi.
Injection Port 250° C. FID detector 250° C.
Attenuation 1 p A|cm. Split = 1|20. Chart Speed
0,5 and after 10 minutes 1 cm|min.

        Stepwise temperature Program:
            a) Split mode
3 minutes at 140° C. 4°C/minute increase to 190°C.
3 minutes isotherm. 3°C/minute increase to 205° C.
1 minute isotherm. 8°C/minute increase to 295° C.

            b) Splitless mode
5 minutes at 90° C. 30°C/minute increase to 140°C.
4 minutes isotherm. 4°C/minute increase to 190° C.
3 minutes isotherm. 3°C/minute increase to 200° C.
1 minute isotherm. 7°C/minute increase to 295° C.

Results and Discussion:

            a) Separation of PTH
Chromatogram 1 shows the separation of the PTH-
derivatives. The first isothermic plateau

(190° C) facilitates the separation of Proline
and Valine. The next low temperature increase of
3° C/min causes Isoleucine and Leucine to dissolve. The complex mixture of the two Phenylalanine PTH-derivatives as well as Glutamic Acid
and the three Asparagine compounds are separated
following the second isothermic plateau (205° C).
Norleucine is used as the internal standard. The
twin-peak of Isoleucine demonstrates the race-
misation at the alpha-carbon yielding allo-
isoleucine. Limitations of the method have to be
mentioned: Histidine and Lysine have poor
response factors. As a result, the method can
only be used as an identification tool. In
addition, Arginine can not be identified. -
In the course of derivatisation, only one product normally results in which the PTH-ring
system and functional groups of the amino acid
side chain are silylated. Unfortunately, some
PTH derivates (Glycine, Phenylalanine, Asparagine) react in a more complex manner and give
rise to several products.
Glycine, for instance, quickly forms one product
with a GC-retention time of fifteen minutes. This
first peak disappears within one hour resulting
in a second peak with higher retention time.
This and other stepwise derivatisations have been
elucidated by mass spectroscopy. The three peaks
resulting from silylation of PTH-Asparagine
proved to be the mono-, di-, and thisilyl derivatives according to mass spectroscopy . We suspect
that these compounds are in equilibrium with
MSTFA and trifluoromethylacetamide. Glutamine
behaves similarly exept that the resulting peaks
can only be resolved if a column overload is
avoided.

b) Quantification

Quantification is accomplished with Norleucine as
internal standard. Standard solutions at concentrations of 0.5, 1.0, and 2.0 mg/ml are made.
A 10 µl aliquot of standard solution is mixed
with internal standard, dried, and treated with
a 50 % solution of MSTFA in acetonitrile. This
procedure is performed three times for each of
the three concentrations of standard solution

PTH	a	+	b	x	[m]	r	Coefficient of Variation	n
Gly =	0,009	+	2,567	x	[m]	0,999	1,57 %	9
Ala =	0,026	+	2,040	x	[m]	0,989	5,67 %	9
CM-Cys =	0,054	+	1,289	x	[m]	0,984	7,30 %	8
Val =	0,004	+	2,664	x	[m]	0,991	4,90 %	9
Pro =	0,045	+	2,146	x	[m]	0,984	6,83 %	9
Ile =	-0,041	+	2,572	x	[m]	0,990	5,39 %	9
Leu =	0,035	+	2,132	x	[m]	0,993	4,25 %	9
Ser =	0,024	+	2,210	x	[m]	0,985	6,54 %	9
Thr =	0,029	+	2,030	x	[m]	0,984	6,88 %	9
Asp =	0,036	+	2,215	x	[m]	0,990	5,22 %	9
Glu =	0,003	+	2,110	x	[m]	0,976	8,40 %	9
Met =	0,026	+	2,097	x	[m]	0,989	5,96 %	9
Trp =	-0,004	+	1,872	x	[m]	0,996	3,35 %	9
Gln:	0,037	+	0,642	x	[m]	0,868	23,10 %	9
Tyr:	0,114	+	1,553	x	[m]	0,868	21,59 %	9

Table 1

so that nine values are obtained and then correlated. Normally (see Table 1) a very good linear
regression of the formula 1 is obtained, the
intercept is approximately zero and the regression near one.

$$\frac{\text{Area PTH-amino acid}}{\text{Area PTH-Nle}} = a + b \times [m]$$

$$m \text{ in } \mu g$$

Formula 1

But some amino acids do not show this excellent
behaviour. The regression of Glutamine for instance is 0,868, the coefficient of variation
approximately 23 %. This finding is due most
probably to the high reactivity of the corresponding PTH-derivative. The silica columns are
very inert but nevertheless not completely so.
There are still a few reactive sites where absorption can take place. This is the case with
Asparagine, Glutamine, and surprisingly Phenylalanine too. As for Asparagine and Glutamine it
was already mentioned that the di- and trisilyl
derivatives are unstable because they are in
equilibrium with the strong alkylating agent
MSTFA.
As for PTH-Phenylalanine however no explanation
for the occurence of a second peak is yet available. Mass spectroscopy revealed only the mono-
lilylated species.
The mischievious properties of Asparagine and
Phenylalanine are shown very clearly when calcu-
lating the calibration curves. Instead of a
linear equation a square equation results
(Formula 2) which represents a considerable loss
in sensitivity.

$$\frac{\text{Area PTH-Asp}}{\text{Area PTH-Nle}} = 0,038+0,040\text{x}[m]+0,801\text{x}[m]^2$$
$$n = 8 \qquad\qquad r = 0,984$$

$$\frac{\text{Area PTH-Phe}}{\text{Area PTH-Nle}} = 0,015+1,072[m]+3,068\text{x}[m]^2$$
$$n = 9 \qquad\qquad r = 0,989$$

Formula 2

c) Splitless technique
In normal capillary chromatography only approximately 5 % of the injected sample is used for
detection, i.e. 95 % of the analytical sample
is lost as a result of splitting. For very
dilute samples this loss has to be avoided. It
can be done by sealing the split so that the
entire injected sample enters the column. Consequently, the initial column temperature has
to be kept low in order to displace the large
amount of solvent before substantial migration
of the compounds of interest takes place.
A comparison of chromatogram 1 and 2 clearly
shows the increase in sensitivity. The stock
solution was diluted 1 : 20 for chromatogram 2.
Each peak corresponds to about 15 ng.
The standard solution was diluted 1 :200 for
chromatogram 3. Each peak represents only 1,5 ng
silylated PTH. In the case of PTH-Aspartic acid
that is only 7,2 p.mole. However it can be seen
that a certain limitation has been reached since
Gln and Lys escape detection.

In Summary:
The splitless capillary GC, of silylated PTH's
on fused silica columns proves to be a valuable
tool for accurate analyses of small amounts of
proteins.

Reference:

1. Eyem, J and Sjöquist, J, Anal. Biochem. 52
   (1973) 255 - 271

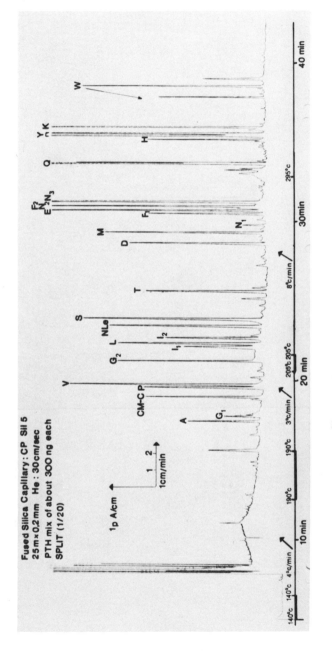

Chromatogram 1

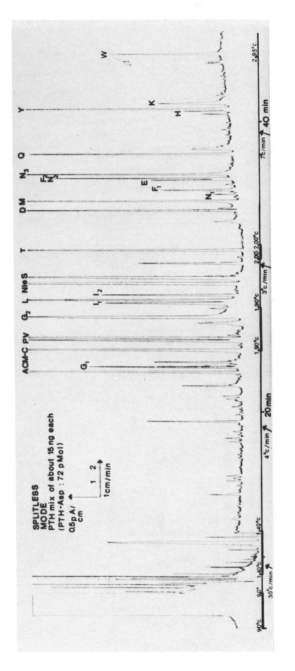

Chromatogram 2

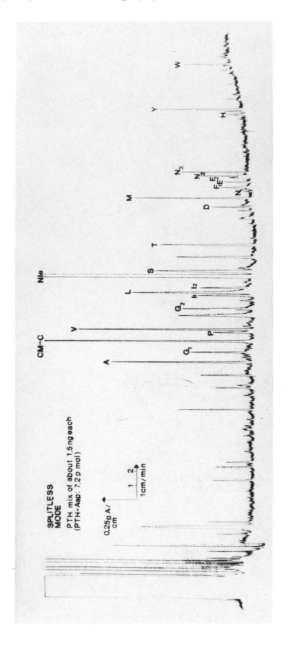

Chromatogram 3

Communications: Sequence Methods

Strategies of Sequence Analysis of Regulatory Subunit of Bovine Cardiac Muscle cAMP-Dependent Protein Kinase Type II

K. Takio, S. B. Smith, E.G. Krebs & K. Titani

Howard Hughes Medical Institute Laboratories

Univ. of Wash., Seattle, WA 98195

The regulatory subunit of bovine cardiac muscle cAMP-dependent protein kinase has a molecular weight of 45,000 and is composed of a single chain with blocked amino terminus and 400 amino acid residues including 9 Met, 6 CySH and one Trp.

Limited digestion with various proteolytic enzymes including trypsin, mast cell protease type II and staphylococcal protease resulted in generation of two complementary segments in each case. The region of the cleavage site included the "autophosphorylation" site and is assumed to represent a "hinge" region between domains (Takio et al. (1980) FEBS Lett. 114, 83). Extended tryptic digestion yielded five fragments by additional cleavages. These fragments, together with eleven cyanogen bromide fragments (10 major and one minor), served as the starting materials for the sequence determination of the whole protein. Cleavage at the single tryptophanyl residue and at four aspartylproline bonds also provided useful information.

Reversed phase high pressure liquid chromatography proved to be a powerful tool for the separation of subdigest peptides. Most of the methionine overlaps were established during analyses of large fragments. One tryptic fragment placed five cyanogen bromide fragments in sequence while two other methionyl overlaps were provided from analyses of fragments generated by cleavage of the single tryptophanyl bond or of an aspartyl-proline bond. The 9 methionyl residues were radiolabeled (Shoji et al. (1981) Proc. Natl. Acad. Sci. 78, 848) in a large segment (derived by limited proteolysis), which was then cleaved at arginyl bonds to obtain a complete set of methionyl overlaps.

Limited proteolysis generates far less fragments because the native conformation of proteins serve as barriers to digestive enzymes. Apparently, only the specific residues existing in a "hinge" or "fringe" region are available as cleavage sites.

Similar strategies are now being applied to a structural study of cGMP-dependent protein kinase.

ENHANCEMENT OF SENSITIVITY AND RELIABILITY OF AUTOMATED SEQUENCE ANALYSIS.

I.V. Nazimov, N.B. Levina, L.G. Snezhkova,

O.S. Reshetova

Shemyakin Institute of Bioorganic Chemistry,
USSR Academy of Sciences, Moscow, USSR.

In order to increase the speed, reliability and sensitivity of amino acid sequence determination, we have investigated the combined use of automated cleavage, HPLC separation and mass spectrometric identification of amino acid and peptide derivatives. We have also developed a reliable procedure for identifying PTH amino acids using HPLC (for the majority of PTHs) followed by silica gel HPLC or MS for unresolved pairs of PTHs. The total amount of sample required is 50-150 picomoles.

The use of diphenylindenonylthiohydantoin (ITH) amino acids allows and increase in sensitivity down to the 5-picomole level.

We have developed a scheme of sequence determination based on the HPLC separation of methyl esters of DNS-peptides followed by MS analysis of eluted fractions. This technique is particularly convenient when dealing with hydrophobic peptides. A comparative study of different techniques (trifluoroacetylation, dansylation, trifluoroacetylation plus dansylation) was undertaken for sequence determination of short peptides possessing a C-terminal α-amide. It was shown that the MS of DNS-peptides is the most reliable and sensitive method for α-amidated fragments.

The latter technique was used for the structure determination of two native analogs of mastoparan (14-membered peptides which stimulate the release of histamine from mammalian mast cells). The primary structure of these peptides from the venom of the hornet Vespa orientalis was determined by solid phase sequencing in combination with HPLC separation and MS identification of DNS-peptides, obtained by tryptic or restricted acid hydrolysis of the native peptide.

Manual Sequence Analysis of Retinoid-Binding Proteins

John W. Crabb and John C. Saari

University of Washington School of Medicine

Seattle, Washington 98195

Rapid manual microsequencing and isocratic HPLC methodology have been employed to demonstrate sequence homology among retinoid-binding proteins. Manual Edman degradation according to Tarr was performed using a large peptide/protein strategy. Coupling was carried out under N_2 in 10% PITC, 10% triethylamine, 65% pyridine, 15% H_2O for 3 min at 50°. The coupled sample was washed and precipitated as a film with heptane/ethyl acetate (1:1) then washed 2-3 times with ethyl acetate and vacuum dried. Cleavage was routinely carried out under N_2 in trifluoroacetic acid for 4 min at 50°. Extraction of the thiazolinones from a dry film was effected with heptane/ethyl acetate (1:2). Total cycle time averaged about 15 min. PTH-amino acids were identified and quantified according to Tarr using an isocratic HPLC system which separates all common PTHs in 6 min [Anal. Biochem. 111, 27 (1981)]. Using from 10-30 nmoles of protein, 29-30 positions of the amino-terminal sequences of three retinoid-binding proteins were obtained with an average repetitive yield of about 90%. The amino-terminal sequences of the cellular retinol-binding protein (CRBP) and cellular retinoic acid-binding protein (CRABP) from bovine retina are identical at 16 of the 30 positions which were determined. The partial sequence of CRBP from bovine retina shows 90% identity to that reported for rat liver CRBP. The amino-terminal sequence of 29 residues of bovine serum retinol-binding protein (SRBP)

535

is identical to that reported for human SRBP except for
an Ala/Ser interchange. The cellular 11-cis-reti-
nal-binding protein (CRALBP) from bovine retina contains a blocked amino-terminus. The results provide the
first evidence that CRBP and CRABP belong to a family
of structurally related proteins which share a common
genetic origin. Supported by USPHS Grants EY-02317,
00343, 01730 & 07013.

THE USE OF H.P.L.C. FOR STRUCTURAL STUDIES ON COLLAGEN.

M. van der Rest[+] and P.P. Fietzek.
Department of Biochemistry, CMDNJ-Rutgers Medical School, Piscataway, New Jersey, and [+]Genetics Unit, Shriners Hospital, Montreal, Canada.

The establishment of collagen primary structure poses specific challenging problems that are linked to the size of collagen polypeptide chains, their unique amino acid compositions and repetitive sequences. The purification of the fragments needed for sequencing is particularly difficult with standard chromatographic techniques. We have thus developed a comprehensive H.P.L.C. approach for the purification of collagen chains and their CNBr and other proteolytic fragments.

Our approach is based on the use of a large pore (33 nm) C_{18} reversed-phase column eluted with an acetonitrile gradient in the presence of eitehr 0.01 M heptafluorobutyric acid (HFBA) Or 0.01 M trifluoroacetic acid (TFA) as counterions. This solvent system is completely volatile and permits a highly sensitive detection at 210 nm. Human pro $\alpha 1(I)$ and pro $\alpha 2(I)$ procollagen chains (approximately 1450 amino acid residues), intact α chains (1,050 residues), CNBr derived peptides (36-650 residues) and tryptic fragments (3-250 residues) have been purified by one or two chromatographic steps. Consecutive chromatographies using TFA or HFBA as counterions helped in the purification of several peptides since the elution time of the peptides is not the same in the two systems. Yields were better than 80%. Amino acid sequencing of several peptides was carried out in a Beckman automatic sequence in order to demonstrate the usefulness of this approach for collagen primary structure analysis.

NH$_2$-TERMINAL SEQUENCE OF THE CLOSTRIPAIN AND NBS PEPTIDES OF STREPTOCOCCAL M5 PROTEIN PURIFIED IN ONE STEP BY HPLC

B.N. Manjula, S.M. Mische and V.A. Fischetti

The Rockefeller University

1230 York Ave., New York, NY   10021

M protein, the antiphagocytic surface molecule of the Group A streptococcus is a type specific, immunologically diverse molecule. We have shown that the molecular characteristics of M proteins and the seven residue periodicity within their partial sequences are similar to those found in tropomyosin suggesting an alpha-helical coiled-coil structure for these molecules (1, 2). To determine if the seven residue periodicity extends throughout the molecule, sequence studies have been undertaken on a biologically active 19,000 dalton M protein fragment, namely Pep M5, isolated by limited proteolysis of the type 5 streptococci with pepsin (3).

Since there are six arginines and two tyrosines in Pep M5, peptides for sequencing were obtained by two selective cleavage methods: a) Enzymic cleavage at the arginyl peptide bonds with clostripain (4);   b) Chemical cleavage at the tyrosyl peptide bonds by N-bromosuccinimide (NBS) (5). The clostripain and the NBS peptides were fractionated by high performance liquid chromatography (HPLC) on a system assembled in the laboratory utilizing a Whatman ODS 2 reverse phase column and a high pressure pump. Peptides were eluted with a linear gradient of the volatile solvent, acetonitrile containing 0.05% trifluoroacetic acid. The column effluent was monitored directly by measuring the peptide bond absorbance at 210 nm.

The clostripain peptides of Pep M5 were essentially completely resolved with one gradient elution step by HPLC with recoveries ranging from 30-100%. Each clostripain peptide thus isolated revealed a single amino terminal residue and hence was suitable for sequencing without further purification. Chemical cleavage at the two tyrosyl residues of Pep M5 by NBS and subsequent fractionation by HPLC resulted in one small and two large peptides, the small peptide corresponding to residues 1-20 of the native molecule.

The partial sequences of the clostripain and NBS peptides were determined by automated sequence analysis in the presence of polybrene with or without glycylglycine. The NBS peptide derived from the carboxyl terminal region of the Pep M5 molecule provided an overlap between two of the clostripain peptides. Thus based on these partial sequences and the sequence of the amino terminal region of the uncleaved Pep M5 molecule, it was possible to align the clostripain and the NBS peptides. The resulting sequence accounted for nearly 2/3rds of the Pep M5 molecule. A significant part of the sequenced segments contained the seven residue periodicity of non polar amino acid residues thus strengthening the presence of coiled-coil interactions in the Pep M5 molecule.

Supported in part by USPHS grants AI11822 and HL25219 and AHA grant 80793. BNM is an AHA established investigator and VAF is a recipient of USPHS RCDA.

REFERENCES

1. Phillips, G.N., Jr., Flicker, P.F., Cohen, C., Manjula, B.N. and Fischetti, V.A. (1981) Proc. Natl. Acad. Sci. USA 78:4689-4693.

2. Manjula, B.N. and Fischetti, V.A. (1980) J. Immunol. 151:695-708.

3. Manjula, B.N. and Fischetti, V.A. (1980) J. Exp. Med. 124:261-267.

4. Mitchell, W.M. (1977) Meth. Enzymol. 47:283-299.

5. Ramachandran, L.K. and Witkop, B. (1969) Meth. Enzymol. 11:283-299.

HIGH-SPEED LIQUID CHROMATOGRAPHY OF PTH-AMINO ACIDS

J. Van Beeumen, J. Van Damme and J. De Ley

Laboratory of Microbiology and Microbial
Genetics, State University of Gent,
Ledeganckstraat, 35 ,   9000 Gent, Belgium

It is now generally accepted that the quickest and most
sensitive method for the determination of PTH-amino acids is
'reversed phase' liquid chromatography on columns of either
CN- or C_{18}-bonded phase packing (1,2). Isocratic separations
giving the best resolution on the routinely used 0.46 x 25
cm columns with 10 μ packing material can be achieved in
30 min (3) when carried out at elevated temperature (e.g.
64°). Gradient elution using a buffered aqueous solvent and
increasing amounts of either methanol and/or acetonitrile
reduce the analysis time to 20 min, also at ca 60°.
We show in this paper that the technique of high-speed liquid chromatography, recently introduced (4), allows the
gradient type of analysis to be carried out in 6 min, including the reequilibration of the column.
    The equipment used was a Perkin Elmer Series 3B chroma-
tograph, a model LC 85 variable wavelength detector with 2.4
μl cell and a Rheodyne model 7125 injector with 6 μl loop.
All tubing from the injector to the detector had an internal
diameter of 0.007 inch. The column, a Perkin Elmer HS3 C_{18}
with 3 μ particles, was 2.5 times shorter than the one used
in the 'conventional' analyses. It was eluted at 3 ml/min
giving an initial column pressure of 270 bar. The gradient
(curve 4 of the Series 3B gradient profiles) was made using
0.01 M sodium acetate,pH 4.5 (solvent A), and acetonitrile
(solvent B). Analyses were carried out at room temperature.
    As seen in Fig.1, all the PTH's except those of Met and
Val were separated in about 4 minutes. (The latter compounds
can be differentiated using a linear gradient; Van Beeumen,

541

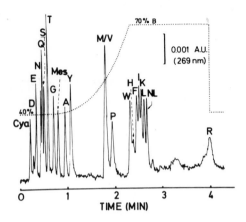

Fig. 1. Separation of PTH-amino acids (16 picomoles each)
on a 0.46x10 cm HS3 C_{18} column using high-speed liquid chromatography with gradient elution (········). Baseline noise: 1.5
x 10^{-4} AU. Other experimental details: see text.

unpublished). Since only 2 min of reequilibration was required, a new analysis could be started every 6 minutes.
Compared to the 'conventional' separations at 64°, the analysis time is thus shortened about 5 times.

A concomittant feature we consider to be even more important is the increase in sensitivity of the PTH-analysis
by a similar factor. Taking a S/N ratio of 3/1 as the ultimate limit of detection, it was possible to determine unambiguously 2 picomoles of PTH-cysteic acid and 3 picomoles of
PTH-Leu. In our view, this high-speed method offers a lower
cost alternative to the picomole level determination of PTH-
amino acids using 'conventional' reversed phase HPLC and
computer-aided baseline correction (2).

We thank the Etn. Van der Heyden for their cooperation.

1. Zimmerman,C.L., Appella,E. and Pisano,J.J. (1977)
   Anal. Biochem. 77, 569-573.
2. Hunkapiller,M. and Hood,L.R. (1980) Science 207, 523-525.
3. Van Beeumen,J., Van Damme,J., Tempst,P. and De Ley,J.
   (1980) in Methods on Peptide and Protein Sequence Analysis (Birr,Chr.,ed.), Elsevier, Amsterdam, 503-506.
4. DiCesare,J.L., Dong,M.W. and Ettre,L.S. (1981) Chroma-
   tographia, 14, 257-268.

Communications: New Sequences

NADP-DEPENDENT 6-PHOSPHOGLUCONATE DEHYDROGENASE

STUCTURE OF THE ENZYME FROM SHEEP LIVER.

ALAN CARNE and JOHN E. WALKER

Laboratory of Molecular Biology,

The MRC Centre, Hills Rd.,

Cambridge CB2 2QH, U.K..

INTRODUCTION

The NADP-dependent 6-phosphogluconate dehydrogenase (6PGDH) (E.C. 1.1.1.44.) converts 6-phosphogluconate into ribulose-5-phosphate forming NADPH in the pentose phosphate pathway. Determination of the primary amino acid sequence of the enzyme was undertaken in conjunction with the 3-D X-ray crystallographic structure analysis by Dr M.J. Adams et al (1). The enzyme is dimeric with identical monomers each of molecular weight 50,000 (2).

METHODS AND RESULTS

The enzyme was isolated from sheep liver (2) with modifications including triazine dye affinity chromatography on Matrex gel Red A as a final step purification enabling production of 250 mg enzyme in 10 days.

The strategy employed to determine the sequence of 6PGDH (fig. 1) was to generate large peptide fragments by cyanogen bromide and by succinyl tryptic digestion, purification of these peptides by gel filtration, ion exchange chromatography and R.P. HPLC, and sequence analysis using either liquid or solid phase sequencer. To determine other overlaps and complete the C-termini of some peptide fragments, digestion with either cyanogen bromide, trypsin, S. aureus protease, limited chymotryptic and asn-gly

545

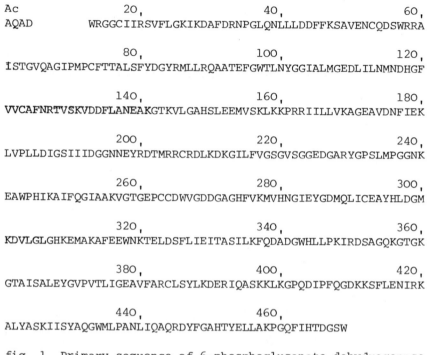

Ac   20,   40,   60,
AQAD   WRGGCIIRSVFLGKIKDAFDRNPGLQNLLLDDFFKSAVENCQDSWRRA

                      80,   100,   120,
ISTGVQAGIPMPCFTTALSFYDGYRMLLRQAATEFGWTLNYGGIALMGEDLILNMNDHGF

                     140,   160,   180,
VVCAFNRTVSKVDDFLANEAKGTKVLGAHSLEEMVSKLKKPRRIILLVKAGEAVDNFIEK

                     200,   220,   240,
LVPLLDIGSIIIDGGNNEYRDTMRRCRDLKDKGILFVGSGVSGGEDGARYGPSLMPGGNK

                     260,   280,   300,
EAWPHIKAIFQGIAAKVGTGEPCCDWVGDDGAGHFVKMVHNGIEYGDMQLICEAYHLDGM

                     320,   340,   360,
KDVLGLGHKEMAKAFEEWNKTELDSFLIEITASILKFQDADGWHLLPKIRDSAGQKGTGK

                     380,   400,   420,
GTAISALEYGVPVTLIGEAVFARCLSYLKDERIQASKKLKGPQDIPFQGDKKSFLENIRK

                     440,   460,
ALYASKIISYAQGWMLPANLIQAQRDYFGAHTYELLAKPGQFIHTDGSW

fig. 1. Primary sequence of 6-phosphogluconate dehydrogenase.

hydroxylamine cleavage were employed with purification of
the peptides as above and sequence analysis by automated
sequencer or manual DABITC sequencing (3). The blocked
N-terminus of 6PGDH was determined by mass spectrometry of
the tetrapeptide obtained by dowex chromatography and R.P.
HPLC of a thermolysin digest of succinyl 6PGDH. The primary
sequence is complete apart from a short region near the
N-terminus and one overlap to be confirmed (fig. 1).
     The sequence has not yet been fitted to the X-ray
electron density map but the X-ray data indicate a high
proportion of α-helix, little β-sheet and when compared
with other NADP-enzymes that have been studied (dihydrofolate
reductase, glutathione reductase, and p-hydroxybenzoate
hydrolase)(4,5,6), they show greater variation in structure
than do the known NAD-enzyme structures (7).
     Homologous sequences are found in sheep 6PGDH to the
pyridoxal 5'-phosphate binding site sequences determined in
Candida utilis 6PGDH (8):

C. utilis:	TVSKVAHFIZBZAK	ILBZAGGKGZTK
Sheep:	TVSKVDDFLANEAK	IRDSAGQKGTGK

Homology is also found between <u>Bacillus stearothermophilus</u> and sheep 6PGDH cysteine sequences (9):

B. stear.:	GGCIIR	ICSYAQGFAQM
Sheep:	GGCIIR	ICEAYHLDGMK

ACKNOWLEDGEMENTS

To the late Dr J.I. Harris for instigating the project. Dr M.J. Adams and the Enzyme Unit, Zoology Dept., Oxford for discussions and assistance with purification of 6PGDH. Dr H. Morris for mass spectral analysis. This work was supported by an MRC grant.

REFERENCES

1. Adams, M.J. et al J. Mol. Biol. (1977) 112 183-197, and personal communication.
2. Silverberg, M. and Dalziel, K. Eur. J. Biochem. (1973) 38 229-238.
3. Chang, J.Y. et al FEBS Lett. (1978) 93 205-214.
4. Matthews, D.A. et al J. Biol. Chem. (1978) 253 6946-6954.
5. Schulz, G.E. et al Nature (1978) 273 120-124.
6. Hofsteenge, J. et al Eur. J. Biochem. (1980) 113 141-150.
7. Rossman, M.G. et al The Enzymes Vol. XI 3rd. Edtn. (Boyer, P.D. ed.) (1975) 61-102.
8. Minchiotti, L. et al Biochim. Biophys. Acta (1981) 657 232-242.
9. Pearse, B.M.F. personal communication.

SEQUENCE STUDIES ON Na,K-ATPase

J.H. Collins, W.J. Ball, L.K. Lane, A.S. Zot

Dept. of Pharmacology and Cell Biophysics

Univ. of Cinti., Col. of Med. Cinti., OH 45267

Na,K-ATPase is the enzyme responsible for the active transport of Na^+ and K^+ across cell membranes. Cardiac glycosides, such as digitalis and ouabain, specifically bind to and inhibit Na,K-ATPase, and it is thought that this enzyme is the pharmacological receptor for these drugs. For our sequence studies we have purified gm quantities of Na,K-ATPase from the outer medulla of lamb kidney. The highly purified enzyme is solubilized in 1% SDS, and chromatographed on a Sepharose CL-6B column run in 0.1% SDS. This procedure separates the three proteins present in the preparation: α (catalytic subunit, $M_r \simeq 100,000$), β (glycoprotein, $M_r \simeq 50,000$) and γ (proteolipid, $M_r \simeq 12,000$). The γ component is separated from salts, lipids and other low molecular weight material by chromatography on Sephadex LH-60 in an organic solvent.

The α subunit was reduced and carboxymethylated at SH groups with iodoacetic acid, and its N-terminal sequence was determined to be Gly-Arg-Asn-Lys-Tyr-Glu- with the use of the Beckman Sequencer. The carboxymethylated α was passed through and AGIX-2 column in order to remove SDS, and then digested with trypsin. The digestion was very incomplete, presumably due to the inaccessability of the lysine and arginine residues. A rapid and complete clevage at the 40 arginines of α could be obtained, however, if the protein were succinylated prior to tryptic digestion. When a tryptic digest of carboxymethylated, succinylated α is chromatographed on a Sephadex G-50 column, four major, well-

defined fractions (A to D) are reproducibly obtained.
Fraction A, which elutes at the void volume of the column,
contains aggregated, very hydrophobic peptides presumably
derived from regions of α that are buried within the membrane lipid bilayer in the native enzyme. Fractions B to D
contain water-soluble peptides and together account for
about 75% of the α polypeptide chain. Seventeen small
(2 to 15 residues) peptides have been isolated from fraction
D by conventional chromatographic procedures and these are
being sequenced. The peptides of fractions A to C are being
separated by HPLC.

In addition, preliminary studies were carried out to
determine whether antibodies raised to the native Na,K-
ATPase could be used to identify and purify specific tryptic
peptides of α. We found that carboxymethylation and
succinylation do not significantly decrease total antibody
binding to α, although the affinity of the anti-Na,K-ATPase
antibodies for α is reduced by about 50%. Tryptic digestion
of the carboxymethylated and succinylated α causes little
further decrease in antibody binding. The fractions A to C
(see above) of the tryptic digest all bind antibodies, and
competition binding studies suggest that at least some of
the tryptic peptides are present in more than one fraction.

We have developed the first preparative-scale procedure
(see above) for the isolation of the proteolipid associated
with Na,K-ATPase, and have called this very hydrophobic
protein the γ component of the enzyme. Sephadex LH-60
chromatography yields two very similar proteolipid fractions:
γ_1, which emerges at the void volume of the column, and γ_2.
On rechromatography on the same LH-60 column γ_1 and γ_2 are
interchangeable, suggesting that γ_1 is an aggregated form of
γ_2. Structural studies on γ are in progress, and the results to date suggest that γ is similar to proteolipids
that are present in sarcoplasmic reticulum from cardiac and
skeletal muscle (Supported by a grant from the Muscular
Dystrophy Association and by NIH grants R01-AM-20875, K04-
HL-00555, P01-HL-22619 and R23-HL-24941.)

N-TERMINAL AMINO ACID SEQUENCES OF γ-GLIADINS FROM BREAD WHEATS

Mary D. Dietler[*] and Donald D. Kasarda

Western Regional Research Center; USDA

Albany, CA 94710

Present-day bread wheats are hexaploid in nature (genomes designated AABBDD), having evolved from crosses between different diploid grass species, followed by chromosome doubling. γ-Gliadin protein components have been purified from the endosperm of a bread wheat cultivar (Triticum aestivum) and characterized according to electrophoretic patterns and N-terminal amino acid sequences. In addition, proteins with electrophoretic mobilities comparable to γ-gliadins were purified from three diploid species, T. monococcum, Aegilops speltoides, and Ae. squarrosa, that are likely candidates for being the progenitors of the A, B, and D genomes of T. aestivum. N-terminal amino acid sequences of these proteins were determined and found to be comparable to either γ_2- or γ_3-type gliadins from T. aestivum. The protein from T. monococcum differed slightly in sequence from the proteins from Ae. speltoides and Ae. squarrosa, which were identical in sequence to their equivalents in the hexaploid. This difference may be an example of divergence between the genera Triticum and Aegilops, especially if both the γ_2- and γ_3-gliadins of the hexaploid were contributed by species of Aegilops.

RECOGNITION OF PATTERNS IN PROTEIN SEQUENCES

BY METRIC ANALYSIS

Bruce W. Erickson, Janet M. Sekulski, and
Peter H. Sellers

The Rockefeller University
New York, New York 10021

Metric analysis is a series of mathematical procedures useful for assigning the order of fragments generated during protein sequence determination and for measuring the similarity of the final amino acid sequence to other sequences. This analysis uses theoretically proven combinatorial algorithms that are implemented by digital computer programs. For a given sequence, the entire *specified* sequence is denoted as S and the set of all *unspecified* intervals present in S is designated as U. Algorithm SS calculates the evolutionary distance and all metric alignments between two specified sequences (1). Evolutionary distance is a *metric* function defined as the minimum number of nucleotide mutations, insertions, and deletions needed to interconvert two nucleic acid sequences that could code for the amino acid sequences. Algorithm SU finds all of the locally best ways to align a specified sequence (a pattern) with unspecified intervals of a longer sequence (2,3). Algorithm UU identifies the locally best ways to align unspecified intervals of one sequence with those of another (3). Problems of protein structure previously explored by metric analysis are evolution of the eight α-helices of calmodulin from a single α-helical segment (4), comparison of the heavy chain variable regions of two human cryoglobulins with those of normal immunoglobulins (5), alignment of a 45-residue segment of a primitive shark protein with segments of the first constant domains of human immunoglobulins (6), and alignment of the circularly permuted sequences of concanavalin A and favin (3,7).

Two other procedures can save siqnificant time and effort
during the sequencing of a target protein if the sequence
of a related model protein is available. They require not
the sequence but only the *amino acid composition* (denoted
as C) of a target protein fragment produced by enzymatic or
chemical means. Algorithm CU finds all intervals of the
model protein sequence whose compositions are close to the
compositions of the target protein fragment. When combined
with the specificity of the cleavage that generated the
fragment, this analysis usually allows assignment of the
fragment to one or a few specific intervals in the sequence
of the model protein. For each fragment and close model
interval, algorithm CS finds a series of *sequences* that are
consistent with the composition of the target protein
fragment and are close to the sequence of the model protein
interval. These predicted sequences and their alignments
with intervals of the model protein sequence provide a
useful picture of the unknown target protein sequence. Even
before any sequencing is begun, this picture can focus
attention on the fragments that are relevant to a given
structural study, such as the complementarity determining
regions in a study of antibody specificity. These
predictions of the relative positions and sequences of the
target protein fragments are also useful for selecting
enzymes or reagents to generate peptide fragments that
overlap the initial set of fragments. When the sequence of
a fragment is determined, the best ways to align it with the
model protein sequence can be found using algorithm SU.

Acknowledgment. This work was supported in part by U.S.
Public Health Service grants AI 15301, RR 07065, and
CA 24435.

1. Sellers, P.H. (1974) SIAM J. Appl. Math 26, 787-793.
2. Sellers, P.H. (1980) J. Algorithms 1, 359-373.
3. Erickson, B.W. & Sellers, P.H. (1982) In "Time Warps,
   String Edits, and Macromolecules" (Sankoff, D. &
   Kruskal, J.B., eds.), in press.
4. Erickson, B.W., Watterson, D.M. & Marshak, D.R. (1980)
   Ann. N. Y. Acad. Sci. 356, 378-379.
5. Erickson, B.W., Gerber-Jenson, B., Wang, A.C. &
   Litman, G.W. Mol. Immunol., in press.
6. Litman, G.W., Erickson, B.W., Lederman, L. &
   Mäkelä, O. (1982) Mol. Cell Biochem., in press
7. Cunningham, B.A., Hemperly, J.J., Hopp, T. & Edelman,
   G.M. (1979) Proc. Natl. Acad. Sci. U.S.A. 76, 3218-3222.

Amino Acid Sequence of Bovine β_2-Microglobulin

Rae Greenberg and Merton L. Groves

Eastern Regional Research Center, USDA, ARS

600 East Mermaid Lane, Phila., PA 19118

Crystalline bovine β_2-microglobulin (β_2-m) was pre-pared from colostrum and its complete primary structure elucidated. β_2-m is a low molecular weight protein (11,600) widely found in various body fluids and bound to cell surface proteins. It is structurally related to the immunoglobulins and is non-covalently associated with histocompatibility antigens. Bovine β_2-m was isolated from colostrum casein by DEAE- and CM-cellulose chroma-tography, gel filtration and finally by crystallization. The bovine protein is the only reported crystalline homo-logue.

Bovine β_2-microglobulin was sequenced completely using only S. aureus V8 protease (Miles) under two sets of conditions to catalyze limited cleavage at glutamyl residues. Cleavage at pH 4 in 0.1 M ammonium acetate, 37°C, for one hour at an enzyme to protein ratio of 1/50 yielded only two fragments resulting from a single split at a glu-ile bond between positions 36-37. Catalysis mediated by the same enzyme at pH 8 in 0.5% ammonium bi-carbonate, 37°C, for two hours at a ratio of 1/100 pro-duced several peptides with the major additional cleavage of a glu-phe link at residues 68-69. Three sequences pro-vided the primary structure data: intact carboxymethylated β_2-m, sequenced 1-41; one pH 4 fragment, sequenced 37-72; and a pH 8 peptide, 69-98.

Bovine β₂-m contains 98 amino acid residues as compared with 99 for the other species whose complete sequence has been reported. This difference represents a deletion at position 49 where the other β₂-m molecules contain valine. The bovine protein is also unique in that it contains three di-prolyl sequences all located within the amino terminal third of the molecule. In a comparison of absolute differences among species, bovine β2-m ranges from a low of 24 substitutions as compared with the rabbit homologue and 26 for human to 32 for mouse and guinea pig. β₂-microglobulin is certainly a highly conserved protein; of the 26 differences between bovine and human, 19 represent single base changes.

In single letter notation, the sequence of bovine β₂-microglobulin is:

```
            10                    20
    I Q R P P K I Q V Y S R H P P E N G K P


            30                    40
    N Y L N C Y V V G F H P P Q I E I D L L


            50                    60
    K N G E K I K S E Q S D L S F S K D W S


            70                    80
    F Y L L S H A E F T P D S K D E Y S C R


            90                  98
    V K H V T L E Q P R I V K W D R D L
```

SEQUENCE ANALYSES IN CHARACTERIZATIONS OF FUNCTIONAL PROPERTIES

Hans Jörnvall, Jonathan Jeffery, Anders Carlström, Mats Carlquist and Viktor Mutt
Departments of Physiological Chemistry I and Biochemistry II, Karolinska Institutet, S-104 01 Stockholm, Sweden, Department of Clinical Chemistry, Karolinska Institutet, Danderyd's Hospital, S-182 88 Danderyd, Sweden, and Department of Biochemistry, University of Aberdeen, Marischal College, Aberdeen, AB9 1AS, UK.

Sequence analyses have shown new functional properties for dehydrogenases and for gastrointestinal polypeptide hormones.

DEHYDROGENASES

Sorbitol dehydrogenase has a reactive cysteine residue (1), a sensitivity to metal chelators, and a subunit size close to 350 residues, like "long" alcohol dehydrogenases (characterized from yeast, bacteria, birds and mammals). Sequence studies reveal extensive homologies around the active site. Similarly, sequence comparisons (2) show the same type of relationship between "short" alcohol dehydrogenases (characterized from an insect) and another sugar polyol dehydrogenase (from a bacterium). Finally, they show a different and limited homology between coenzyme binding regions of the "short" and "long" dehydrogenases. Sequence studies therefore define different active site regions and mechanisms of action of these enzymes, uniting crystallographic and sequential support in showing building units in the structures. These results suggest new functional properties, especially for the uncharacterized "short" dehydrogenases, and may lead to alternative methods of investigation.

557

GASTROINTESTINAL HORMONES

Seven new forms or species variants of gastrointestinal polypeptides have been determined, showing that the various types of peptide differ in extent of divergence. In
spite of partial cross-activities in biological properties,
each polypeptide type is a structurally unique entity. Vasoactive intestinal peptides are, like glucagons, comparative-
ly constant between species, while secretins (3) and gastrin-
releasing peptides (4) appear to differ more. Preparations
of gastric inhibitory peptide have two components, showing
N-terminal proteolysis (removal of two residues in a minor
component) and explaining observations on the heterogeneity (5).

All these structures may be compared and correlated
with functional properties. The comparisons suggest that
the most constant part among widely different types of hormone appears to involve the C-terminal halves. This might
indicate the position of a common and important function in
this region. Further comparisons with plasma proteins suggest
a new superfamily of proteins related to prohormones (6).
The results coordinate well with known properties of receptor binding and other interactions.

REFERENCES

1. Jeffery, J., Cummins, L., Carlquist, M. & Jörnvall, H.
   (1981) Eur. J. Biochem., in press.
2. Jörnvall, H., Jeffery, J. & Persson, M. (1981) Proc.
   Natl. Acad. Sci. USA 78, 4226-4230.
3. Nilsson, A., Carlquist, M., Jörnvall, H. & Mutt, V.
   (1980) Eur. J. Biochem. 112, 383-388.
4. McDonald, T.J., Jörnvall, H., Ghatei, M., Bloom, S.R. &
   Mutt, V. (1980) FEBS Lett. 122, 45-48.
5. Jörnvall, H., Carlquist, M., Kwauk, S., Otte, S.C.,
   McIntosh, C.H.S., Brown, J.C. & Mutt, V. (1981) FEBS
   Lett. 123, 205-210.
6. Jörnvall, H., Carlström, A., Pettersson, T., Jacobsson,
   B., Persson, M. & Mutt, V. (1981) Nature, 291, 261-263.

THE PRIMARY STRUCTURE DETERMINATION OF RIBOSOMAL PROTEINS

BY COMBINED AMINO ACID AND NUCLEOTIDE SEQUENCING

Makoto Kimura

Max-Planck-Institut für Molekulare Genetik,
Abteilung Wittmann,
D-1000 Berlin 33 (Dahlem), Germany

The primary structures of almost all of the 53 proteins present in the E. coli ribosome have now been completely determined (1). It was especially difficult to establish the structures of the proteins with molecular weight
between 20,000 and 60,000 because of the small amounts of
protein available and the complexity of the peptide mixtures.

Recently the amino acid sequence of protein S1 which
has a molecular weight of 61,000 and is the largest protein
of the E. coli ribosome has been determined (M. Kimura,
K. Foulaki, A.R. Subramanian and B. Wittmann-Liebold, manuscript submitted). Simultaneously the nucleotide sequence of
the gene for protein S1 has been elucidated by modern DNA
sequencing techniques (J. Schnier and K. Isono, manuscript
in preparation). The results from both approaches were
compared.

The experience gained by simultaneously establishing
both the protein and the nucleotide sequence can be discussed as follows: the advantages are   1) the time necessary
for isolation and sequencing of the protein is reduced since
smaller amounts of protein are required;   2) valuable information about the protein gene and codon occurrence fre-
quency are obtained;   3) questions about intervening sequences can also be answered. Furthermore, use of the two
techniques allows us to establish the sequences more reliable, especially if only very small amounts of material are
available.

Recently, sequence studies on Bacillus stearothermo-
philus proteins have been undertaken, making simultaneously
use of protein and nucleotide sequencing methods. The aim
of this approach is to reduce the amount of protein by sequencing only the N-terminal region of the protein and
those peptides which can be obtained in good yields. The
alignment of peptides is then deduced from the nucleotide
sequence. By this means, the primary structures of proteins
from ribosomes of other organisms than E. coli are obtained
much faster and easier than by amino acid sequence analysis
only. So far 17 ribosomal proteins from other organisms
have been completely sequenced (see Table 1). Comparison of
the primary structures of ribosomal proteins from different
organisms yields interesting information about the evolution of the ribosome.

TABLE 1

Completely Sequenced Ribosomal (and Related) Proteins
from Organisms Other than E. coli

Organisms	Pro-tein	aa-No.	Ref.	corresponds to E. coli
Bacillus stearothermophilus	BL10	177	2	EL6
" "	BL17	147	3	
" "	BL29	104	4	EL24
" "	BL34	62	5	EL30
" "	IF-3	183	5	IF-3
" "	BSb	90	5	NS2
Bacillus subtilis	BL9	122	6	L7/L12
Micrococcus lysodeikticus	MA1	118	7	L7/L12
" "	MA3	128	8	
Streptomyces griseus	A	126	8	L7/L12
MRCC 11227	A	122	9	L7/L12
Halobacterium cutrirubrum	HL20	112	10	L7/L12
Saccharomyces cerevisiae	YP-A1	110	8	L7/L12
" "	YP44	103	11	
" "	YP55	88	12	
Artemia salina	eL12	111	13	L7/L12
Rat (liver)	P3	104	14	L7/L12

REFERENCES

1. Wittmann, H.G., Littlechild, J. and Wittmann-Liebold, B. (1980) In "Ribosomes" (J.E. Davies et al., eds.), University Park Press, Baltimore, pp. 51-88

2. Kimura, M., Rawlings, N. and Appelt, K. (1981) FEBS Lett. (submitted)

3. Kimura, M., Dijk, J. and Heiland, I. (1980) FEBS Lett. 121, 323-326

4. Ashman, K. and Kimura, M., unpublished

5. Kimura, M. and Appelt, K., unpublished

6. Itoh, T. and Wittmann-Liebold, B. (1978) FEBS Lett. 96, 392-394

7. Itoh, T. (1981) FEBS Lett. 127, 67-70

8. Itoh, T., unpublished

9. Falkenberg, P., Yaguchi, M., Roy, C. and Matheson, A.T., unpublished

10. Oda, G., Yaguchi, M., Roy, C., Visentin, L.P. and Matheson, A.T., unpublished

11. Itoh, T. and Wittmann-Liebold, B. (1978) FEBS Lett. 96, 399-402

12. Itoh, T. and Wittmann-Liebold, B., unpublished

13. Amons, R., Pluijms, W. and Möller, W. (1979) FEBS Lett. 104, 85-89

14. Lin, A., Wittmann-Liebold, B. and Wool, I.G., unpublished

TWO DIFFERENT SPECIES OF CYTOCHROME b_5 IN ONE CELL

Florence Lederer, Rachid Ghrir, Bernard Guiard
and Akio Ito. Centre de Génétique Moléculaire
du CNRS, Gif-sur-Yvette (France) and Department
of Biology, Faculty of Sciences, Kyushu University, Fukuoka, Japan.

Cytochrome b_5 is traditionnaly extracted from liver microsomes. It is however known to be present in a number of
other cell membrane fractions : nuclear membrane, Golgi membrane, plasma membrane, outer mitochondrial membrane. In the
latter, it has been described to be part of a so-called ro-
tenone insensitive NADH-cytochrome c reductase system.

Rat liver cytochrome b_5 has been solubilized by proteolysis and purified from microsomes and outer mitochondrial
membrane. The two species were shown to be different according to a number of criteria (chromatographic and electrophoretic mobility, amino acid composition and immunological
reactivity) (A. Ito (1980),J.Biochem. Tokyo, 87, 63-71).

Partial sequences have now been determined for the two
molecules. The microsomal species contains one tryptophan
(invariant at position 22 in all other species studied thus
far), no sulfur amino acid and has a blocked amino-terminal
end. 0.1 µmole of the protein was cleaved by cyanogen bromide in the presence of heptafluorobutyric acid and the two
resulting fragments were separated by sephadex chromatography. The amino-terminal peptide was studied by proteolysis
and thin-layer peptide mapping ; the carboxyl-terminal fragment was submitted to a liquid-phase automatic degradation.

The mitochondrial species also contains two methionines
and was found to present a frayed amino-terminal end which
varied somewhat with the preparation. A sequenator run was

carried out on a sample containing about 5 nmoles of a major
component plus a number of species beginning at successive
positions, each one in amount smaller than 10-15%. About
30 nmoles of a less homogeneous preparation was succinylated, cleaved with BNPS skatole and submitted to an automatic
degradation without separation of the fragments.

The results show the two cytochromes b_5 to be indeed
different molecular species, but homologous ones. They raise
the question of the recognition by each species of its specific target membrane. Is it due only to differences in
membrane composition and protein structure, and/or to other
factors ?

```
            -5          -1 1       5          10          15
OM b5     D G Q G S D P A V T Y Y R L E E V A K ? N ? ? ?

MC b5     (XE',E',A,S,D')K D' V K(Y,Y,T,L)E(E',I,E')K(H D')D' S K
          |-------------->  |--->  |----------------->       |--->
               T1           T2            T3                  T4

          |------------------------------------->
                            S1

          20          25         30          35          40
OM b5     ? ?(W)M V I H G R V Y D I T R F L ? E H P G G E E

MC b5     (S,T,W)V I L H H K V Y D L T K F L E E H P G G E E
          |---->|->

          45          50         55          60          65
OM b5     V L L E Q A G . . .

MC b5     V L R E Q A G G D A T E ? F E D V G H ? T D A R . . .
```

The numbering adopted aligns residue 1 with residue 1 of
trypsin solubilized bovine MC b_5 (Ozols & Strittmatter,
J. Biol. Chem. 243, 3376 (1968)).

A HYDROPHOBIC TRYPTIC PEPTIDE FROM BOVINE BRAIN

M.B. Lees[1], B. Chao[1], R. Laursen[2],

J. L'Italian[2] and J. Evans[1].

1. E. K. Shriver Center, Waltham, MA. 02254

2. Boston University, Boston, MA. 02215

The major protein of central nervous system myelin is a hydrophobic intrinsic membrane protein designated proteolipid protein. The proteolipid apoprotein, devoid of complex lipids is completely soluble in chloroform-methanol mixtures but can be converted to a water soluble form. The primary structure of this protein has been difficult to obtain because of problems related to solubility and aggregation. In the present study we have used a combination of solid phase Edman degradation and mass spectrometry to obtain the primary structure of a major, chloroform-soluble tryptic peptide which is located in the COOH-terminal region of the protein.

A bovine brain white matter proteolipid preparation (40% protein) was digested with trypsin in the presence of 0.6% Triton X-100 for 18 hrs. The digest was partitioned between the two phases of a chloroform-methanol-water system. The lower, chloroform phase contained peptides accounting for approximately 10% of the starting material, along with Triton and lipids. Peptides were separated from other components by chromatography on Sephadex LH 60 using 2:1 chloroform-methanol as the eluant. The major peptide, purified by rechromatography on the same column, was enriched in apolar amino acids with Val, Ala, Phe and Leu accounting for 53% of the total amino acid residues. Based on amino acid composition and position on SDS gels, it had a molecular weight of approximately 4000. The peptide,

dissolved in chloroform-methanol, was coupled to p-phenylene
diisothiocyanate activated glass and the sequence of the
first 21 residues was obtained by solid phase Edman degradation. A combination of CNBr cleavage and chymotryptic
digestion led to the identification of 10 residues at the
COOH-terminus of the peptide. However, several residues
which should have been cleaved by chymotrypsin were
inaccessible and their sequence was obtained by mass spectrometry. The tryptic peptide was digested with pepsin
and partitioned in a chloroform-methanol-water system. HPLC
showed a large number (>50) of peptides in the methanol-
water phase. The peptide mixture was converted to trimethylsilyl trifluoroethyl amino alcohols and separated by
gas chromatography-mass spectrometry. Analysis of the data
provided sequence information on several peptides and led
to completion of the sequence. The sequence of the tryptic
peptide is shown in Fig. 1. Underlined sequences were
determined or confirmed by mass spectrometry. This sequence
spans the region CN_2, CN_3 and CN_4 of Jolles et al (BBRC
87:619, 1979). Analysis of this entire region (CN_2-CN_4)
by the method of Chou and Fasman indicates a segregation of
the molecule into domains with the tryptic peptide corresponding to a major highly ordered hydrophobic domain with
an equal probability of α-helical or β-structure. The
conformational flexibility of the original protein may be
dependent, in part, on properties of this region.

Thr-Ala-Glu-Phe-Gln-Val-Thr-Phe-X-Leu-Phe-Ile-Ala-Ala-Phe-

Val-Gly-Ala-Ala-Ala-Thr-Leu-Val-Ser-Leu-Thr-Phe-Met-Ile-Ala-

Ala-Thr-Tyr-Asn-Phe-Ala-Val-Leu-Lys-Leu-Met-Gly-Arg

Fig. 1 Sequence of tryptic lower phase peptide

COMPARISON OF N-TERMINAL AMINO ACID SEQUENCES OF ω-GLIADINS AND A RELATED PROLAMIN

D.D. Kasarda, J-C Autran, C.C. Nimmo, E.J-L Lew

USDA Western Regional Research Center

Berkeley, California 94710

The gliadins (prolamins) of wheat constitute the major storage protein fraction in the kernel endosperm. They contain large amounts of glutamine (36–56 mole percent) and proline (15–30 mole percent), presumably because these amino acids are a readily transformable source of nitrogen for the developing embryo. Prolamins are complex mixture of proteins which have similar amino acid compositions and properties. Two dimensional methods of electrophoresis separate the gliadin mixture from a single wheat variety into about 40 components. This complexity may be the result of multiple copies of ancestral genes which have diverged through mutation to produce distinguishable protein components. An understanding of the nature of this complexity may suggest how genes coding for gliadins and other prolamins originated.

We report two new types of N-terminal amino acid sequences for ω-gliadins from a hexaploid bread wheat variety 'Justin'. The sequences show partial homology with one another and also with sequences of an ω-gliadin from a diploid wheat Triticum monococcum (ω-Tm) and a related prolamin component from barley (C-hordein). The sequence -Pro-Gln-Gln-Pro-Tyr- and related sequences were repeated in these proteins; duplications of short DNA segments coding for such sequences may have given rise to genes for ω-gliadins and related prolamins.

567

Amino Acid Sequence of a Monomeric Hemoglobin in Heart

Muscle of Bullfrog, <u>Rana catesbeiana</u>.

Nobuyo Maeda and Walter M. Fitch

Department of Physiological Chemistry

University of Wisconsin-Madison

In order to use the amino acid sequence of orthologous proteins for the systematic study of evolutionary relationships, especially for the study of tetrapod origin, we've
been studying the structure of myoglobins from lower verte-
brates. Although myoglobins have been isolated and
sequenced from mammals, birds, reptiles and fish, no
amphibian myoglobin have been structurally studied so far.
Our effort to isolate myoglobin from skeletal muscles of
amphibians such as bullfrogs, a conger eel, and mud puppies
was unsuccessful because of the very low content of monomeric heme proteins in their muscles. The consideration
that the intercellular hemo-protein is important to facilitate the uptake and transport of oxygen in heart muscles
led us to study the heart muscles of bullfrog. <u>Rana</u>
<u>catesbeiana</u>. A monomeric heme protein isolated and
sequenced, however, is substantially different in size and
in sequence from other myoglobins and from hemoglobins.
The protein is composed of 132 amino acid residues and has
a molecular weight of about 14,000. This is the shortest
heme globin so far known.

In order to align its amino acid sequence with other proteins in the globin superfamily, three gaps common to
α-hemoglobin are required plus three more gaps unique in
this protein. Of the latter, one is at the end of the
EF region, the second near the beginning of the H-region,
and the third is at the c-terminus. With 62 amino acid
residues in common the amino acid sequence of this monomer

is more homologous to α-hemoglobin of tadpole of the bull-frog, <u>Rana catesbeiana</u>, than to any other globin. A phylogenetic study of it and other globins clearly reveals that it arose via a gene duplication of hemoglobin near the time of the duplication that gave rise to the alpha and beta genes. But residues in contact with the heme group are rather conserved while the residues in the $\alpha_1\beta_1$, $\alpha_1\beta_2$ subunit contact regions are significantly substituted, frequently reverting to a myoglobin-like residue. The absence of this monomeric protein from the blood, and the absence of myoglobin in heart muscle may indicate the protein functions as a myoglobin.

PURIFICATION AND MICRO SEQUENCE ANALYSIS OF ACTIVE PEPTIDES

FROM AMPHIBIAN SKINS

Pier Carlo Montecucchi and Luigia Gozzini

Farmitalia Carlo Erba S.p.A.- Chemical Research Department

via dei Gracchi,35 - 20146 Milan (Italy)

PEPTIDE PURIFICATION. In addition to the normally routine methods, we have applied the following techniques to solve some structural problems observed during the isolation of active peptides from amphibian skins: (i) isoelectric focusing in combination with electrophoresis to separate sauvagine I and II; (ii) reverse-phase HPLC to purify dermorphins and triptokinins; in particular, HPLC appears to be a versatile technique to isolate isocratically (on a μ Bondapak C-18 column) caerulein from the analogue beta Asp^3-caerulein (using as eluent 30% MeOH-10% MeCN in 0.02 M NH4OAc at pH 3.5), probably because of a different protonation of their aspartyl residues at the 3rd position; these results are in accordance with those obtained on HVPE at pH 2.7 (pyridine/glacial acetic acid/water 1/100/899 by vol.).

ENZYMATIC FRAGMENTATION. Subtilisin (protease type VIII, Sigma-0.5M NH4OAc, pH 8.2, 37°C,4h, E/S = 1/50)cleaves the tryptophyl bonds in caerulein and analogues (Table 1); in addition it is able to split the linkage Tyr*-Xxx and Thr-Xxx, exclusively when they occupy the 4th position from N-terminus. Beta-Asp^3-caerulein possesses a cross-link between two peptidic segments (segment 1: Z Q D; segment 2: Y* T G W M D F■) consequently the bond Tyr*-Thr occupies an abnormal first position and it is not split. The same results are obtained on the desulphated peptides.

APPLICATION OF FIELD DESORPTION(FD) MASS SPECTROMETRY. The sensitivity of the method is estimated approximately 10^{-6} to 10^{-7} g. As reported in Table 2, only the peak corresponding to the molecular weight is obtained for the peptides with-out any free amino groups.

Table 1 - Degradation by subtilisin of caerulein and analogues. The arrows
indicate the cleaved bonds. Tyr* or Y* = tyrosine-O-sulfate.

```
Pyr-Gln-Asp-Tyr*↓Thr-Gly-Trp↓Met-Asp-Phe amide caerulein
Pyr-Asn-Asp-Tyr*↓Leu-Gly-Trp↓Met-Asp-Phe amide Asn², Leu⁵-caerulein
Pyr-Glu-----Tyr*-Thr↓Gly-Trp↓Met-Asp-Phe amide phyllocaerulein
Pyr-Glu-----Thr-Tyr*↓Gly-Trp↓Met-Asp-Phe amide Thr³, Tyr*⁴-phyllocaerulein
Pyr-----Asp-Tyr*-Thr↓Gly-Trp↓Met-Asp-Phe amide des-Gln²-caerulein
Pyr---------Tyr*-Thr-Gly-Trp↓Met-Asp-Phe amide des Gln², des-Asp³-caerulein
Pyr-Gln-Asp⌐Tyr*-Thr-Gly-Trp↓Met-Asp-Phe amide beta-Asp³-caerulein
```

Table 2 - FD mass spectrometry (peaks at m/z corresponding to M^+, MH^+, MNa^+).

				M^+			MH^+			MNa^+	
dermorphin	MW	802	§	–	§		803	§		825
Hyp⁶-dermorphin		818		–			819			841
deamidated dermorphin		803		–			804			826
cyclo(-L-Phe - L-Leu-)		260		260			–			–
cyclo(-L-Tyr - L-Pro-)		260		260			–			–
cyclo(-L-Lys - L-Trp-)		314		314			315			–
Pyr-Pro-Trp-Val amide		510		510			–			–
Met-Asp-Phe amide		410		–			411			–
Pro-Val amide		213		213			214			–
Pro-Ser amide		201		201			202			–

EDMAN MICROSEQUENCING. The degradation has been performed in a manual version, using Reacti-Vials (1ml) with mininert valves (from Pierce) conveniently modified. The application of this method has shown satisfying repetitive yields, also in the presence of particular peptide modules (. .
L P R . . ; . . . P P P I . . .).

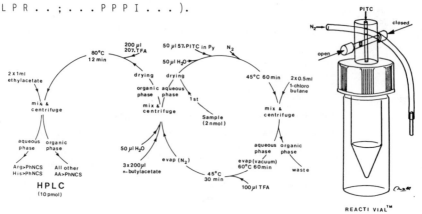

THE AMINO ACID SEQUENCE OF A NOVEL DECAPEPTIDE FROM THE SKIN OF THE

NEOTROPICAL FROG PHYLLOMEDUSA SAUVAGEI

Pier Carlo Montecucchi and Luigia Gozzini

Farmitalia Carlo Erba S.p.A. - Chemical Research Department

via dei Gracchi, 35 - 20146 Milan (Italy)

A novel decapeptide with the following structure L M Y Y T L P R P V ■ has
been recently isolated from the methanol extracts of the fresh skins of
Phyll. sauvagei (Tucuman, Argentina). The purification procedure involves:
(i) chromatography on alkaline alumina column (the peptide is eluted with
90% ethanol); gel filtration on Bio-Gel P-4 and (iii) RP-HPLC [column,Li-
Chrosorb RP-2 (10 μu) Merck; eluent, 34% ethanol-66% 0.1M HCOOH at pH 2.5
with NH₄OH; flow rate, 0.7ml/min; detection UV at 215 nm; room temperature].
The peptide appears then homogeneous as judged by chromatography, including RP-HPLC, and several kinds of electrophoresis. The molecule has been
analyzed for primary structure and the strategy employed is summarized in
the scheme. The following considerations can be carried out: (i) trypsin
(typeXI, DPCC treated - Sigma - E/S=1/50) does not split the bond Arg-Pro,
but it shows an aspecific action on the linkage Tyr-Thr; clostripain (Boeh-
ringer) hydrolyzes the bond Arg-Pro (0.025M phosphate buffer, pH 7.6, containing 0.04 mM $CaCl_2$ and 7.8 mM dithiothreitol,25°C, 1 h, E/S=1/50 approximately), producing two fragments which are separated by paper electrophoresis; (iii) on C-terminal amino acid analysis by digestion with CP-A
or by hydrazinolysis no amino acid is released, indicating a blocked C-terminus; however, with CP-Y Val and Pro are found in approximately equal amounts; (iv) when the dipeptide Pro-Val-X is examined by Fd mass spectral
technique two peaks are obtained at m/z 213 (M^+) and 214 (MH^+) in agreement
with the postulated presence of an amide group blocking the COOH of valine;
(v) the two CNBr fragments have been separated by gel filtration on Bio-Gel
P-4 and curiously the segment CB_1 elutes from the column faster than the

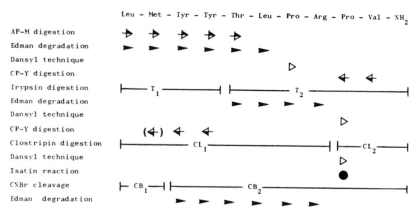

C-terminal octapeptide. The pharmacological properties of this molecule as
neuromodulator/neurotransmitter are in progress. The decapeptide shows similarities with tuftsin (T K P R); in addition, it contains the segment
P R P which has been demonstrated to possess a central antalgic activity.
The same tripeptide is present also in other biologically active molecules,
as reported below: 1, decapeptide from Phyll. sauvagei; 2, snake bradyki-
nin-potentiating peptide B; 3 to 6 angiotensin-converting enzyme inhibitors from South American vipers (V-6-I,V-6-II,V-2 and V-9 respectively); 7,
viscotoxin (17-31) from European mistletoe; 8, human plasminogen (603-617);
9, human complement C1Q subcomponent A chain (85-97) .

1   L M Y Y T L P R P V ■

2   Z - G L P R P - - K I P P

3   Z - - W P R P - - Q I P P

4   Z - N W P R P - - Q I P P

5   Z - - W P R P T P Q I P P

6   Z G G W P R P G P E I P P

7   • • R L T G A P R P T C A K L S G • •

8   • • C L E K S P R P S S Y K V I L • •

9   • • N I K D Q P R P A F S A I • •

Structural and Functional Studies on Nucleolar Protein C23

Using Limited Chemical Cleavage and HPLC

Mark O. J. Olson, S. V. V. Rao, O. Bouwsma
and Z. Rivers
Department of Biochemistry
The Univ. of MS Medical Center
Jackson, MS   39216

Protein C23 (Mr 110,000; pI 5.5) is a nucleolus localized DNA binding phosphorylated protein which contains several highly acidic regions; e.g., -AAPAAPA-SEDEDEEDDDDEDDDDDSQEESEEEDEEVMEITPAK- (all three serines may be phosphorylated.) In contrast, the amino terminal sequence is basic: VKLAKAGKT-. To determine the linear distribution of acidic and basic regions, protein C23 was partially cleaved by N-bromosuccininimide (NBS). The presence of only 8 residues of tyrosine and no tryptophan makes it possible to generate large, overlapping fragments for sequence analyses and functional studies. At high NBS: Tyr ratios (25:1) a 60 K fragment and several lower molecular weight fragments are produced. The 60 K fragment is highly acidic (pI~4) and contains all of the phosphoryl groups in the molecule. The 60 K and several smaller fragments were purified by Sephadex G-75 followed by HPLC on a column of Synchopak RP-P using gradients formed with 0.1% trifluoroacetic acid (TFA) in water and 0.1% TFA in acetonitrile. The 60 K fragment had the same NH_2-terminal sequence as the parent protein, thereby placing it at the amino terminal end of the molecule. At lower NBS ratios (3:1) a series of larger fragments, 70 K, 78 K, and 88 K, are produced. These are extensions of the 60 K fragment and provide overlaps with it and C-terminal portions of the molecule. The 60 K fragment retains DNA binding activity as does a basic 13 K fragment and several inter-

mediate sized fragments. Thus, by carefully controlling
the extent of chemical cleavage, overlapping fragments
of large proteins may be obtained. These may be isolated relatively quickly by HPLC to be used for functional, as well as structural studies. (Supported by NIH
Grants GM 28349 and RR 05386)

FUNCTION OF NAD AT THE ACTIVE SITE OF UROCANASE

Allen T. Phillips

Biochemistry Program, Pennsylvania State

University, University Park, PA   16802

Urocanase (EC 4.2.1.49) from Pseudomonas putida has one mole of tightly bound NAD per 110,000 daltons.   The enzyme is composed of two apparently identical subunits of 55,000 molecular weight each.   Although NAD is essential for catalytic activity, the reaction involves the addition of water across the unsaturated system of urocanate (imidazole acrylate).   Mechanistic evidence indicates that hydride ion transfer does not occur in this reaction.   The question thus arises as to what is the role of NAD in this situation.

Imidazole propionate (dihydrourocanate) binds to urocanase to produce a relatively stable intermediate, as evidenced by the formation of an absorption maximum at 328 nm and a fluorescence emission maximum at 420 nm.   To render the imidazole propionate-urocanase intermediate amenable to structural analysis, the complex was stabilized by oxidation with phenazine methosulfate and dissociated from the protein by treatment with sodium dodecyl sulfate. After purification by gel filtration, DEAE-cellulose chromatography and high voltage electrophoresis, analysis of the material was conducted by proton NMR with a 360 MHz instrument.   The spectrum indicated that NAD was attached through the 4-position of the nicotinamide ring to the τ nitrogen of the imidazole ring of imidazole propionate. Thus the material prior to oxidation was apparently a dihydropyridine-like addition product of NAD and imidazole propionate.

Urocanate does not form an analogously stable intermediate that can be isolated from urocanase but the existence of a similar complex between urocanate and NAD could
be seen by stopped flow fluorescence analysis. The kinetics
of the pre-steady state for urocanate and for imidazole
propionate were examined and the results indicated a
reaction sequence: $E + S \rightleftharpoons ES \rightleftharpoons ES^* \longrightarrow E + P$, where ES*
is the fluorescent intermediate. In the case of imidazole
propionate, ES* cannot decompose to a product P. Calculation of individual rate constants for each step with
urocanate as substrate revealed all steps prior to the
formation of ES* were faster than the decomposition of ES*
to products.

The isolation of a covalent NAD-imidazole propionate
complex, the evidence that urocanate forms a similar adduct
with NAD, and the pre-steady state rate measurements all
suggest that NAD functions in urocanase by virtue of its
formation of an addition product with the urocanate
imidazole ring. This novel role for NAD in catalysis is
consistent with all available mechanistic information but
a fuller understanding will require additional data on
protein groups actively participating during catalysis.

AMINO ACID SEQUENCE OF THE ALPHA AND BETA SUBUNITS OF OVINE PITUITARY FOLLITROPIN (FSH).

M.R. Sairam, N.G. Seidah[1] and M. Chrétien[1]

Reprod. Res. Lab. & [1]Protein-Pituit. Horm. Lab.

Clinical Research Institute of Montreal, Canada

Ovine pituitary follitropin (Follicle Stimulating Hormone, FSH) is an oligomeric glycoprotein with two non-covalently linked subunits, designated α and β. Both subunits contain carbohydrate and individual subunits are inactive. The objective of our investigation was to determine the primary structure of ovine FSH. The α and β subunits of purified ovine FSH were prepared by dissociation in 8 M urea and chromatography on DEAE-Sephadex A25 followed by filtration on Sephadex G-100. The α subunit was similar in amino acid composition to the α subunit of another gonadotropic hormone lutropin (luteinizing hormone, LH) isolated from the same source. The α subunits of these two hormones are freely interchangeable in the formation of the hormone (α + β) recombinants. The composition of tryptic and chymotryptic peptides and the terminal sequence of the subunit, was identical to that of LH α. Based on this, it is proposed that the 96 amino acid residues present in ovine FSH α are arranged in the same order as determined previously for ovine LH α.

The amino acid sequence of the hormone specific β subunit was derived from complete sequence analysis of all of the tryptic peptides from the reduced and aminoethylated β subunit. A total of 27 tryptic peptides including the glycopeptides were isolated by the conventional method of paper chromatography-electrophoresis. Thermolysin and chymotryptic peptides provided the necessary overlaps. In the Beckman sequencer with the spinning cup the intact but

579

SHEEP FOLLITROPIN

completely unfolded β subunit protein exhibited strange
behavior which resulted in extremely poor recoveries of
PTH amino acids. The N-terminal Serine as well as several
gaps along the sequence in the first 20 residues could not
be identified. Automated sequence analysis of the reduced
and ^{14}C alkylated β subunit confirmed the positions of the
first five -cys- residues and also indicated the existence
of amino terminal heterogeneity. In this run also the low
yield of PTH-cys* dropped by more than 90% after the first
cycle. The overall analysis of our results suggests that
in our β subunit preparations over 95% of the molecules
have the N-terminal sequence Cys-Glu-Leu ... There was
some evidence of heterogeneity at the C-terminus of the β
subunit which is yet to be completely defined. Based on
these data, it is proposed that the 111 amino acids in the
β subunit are arranged in the following sequence:

SER-CYS-GLU-LEU-THR-ASN(CHO)-ILE-THR-ILE-THR-VAL-GLU-LYS-
GLU-GLU-CYS-SER-PHE-CYS-ILE-SER-ILE-ASN(CHO)-THR-THR-TRP-
CYS-ALA-GLY-TYR-CYS-TYR-THR-ARG-ASP-LEU-VAL-TYR-LYS-ASX-
PRO-ALA-ARG-PRO-ASX-ILE-GLN-LYS-THR-CYS-THR-PHE-LYS-GLU-
LEU-VAL-TYR-GLU-THR-VAL-LYS-VAL-PRO-GLY-CYS-ALA-HIS-HIS-
ALA-ASP-SER-LEU-TYR-THR-TYR-PRO-VAL-ALA-THR-GLU-CYS-HIS-
CYS-GLY-LYS-CYS-ASP-SER-ASP-SER-THR-ASP-CYS-THR-VAL-ARG-
GLY-LEU-GLY-PRO-SER-TYR-CYS-SER-PHE-SER-ASP-ILE-GLU-ARG-
[GLX].

The two oligosaccharide moieties are linked to asparaginyl residues at positions 6 and 23. The amino acid
sequence of the β subunits of the sheep and human hormone
show 88% identity. (Supported by MRC of Canada).

REFERENCES

1. Sairam MR. Arch Biochem Biophys 194: 71-78, 1979.

2. Sairam MR. Biochem J 197: 535-539, 1981.

3. Sairam MR, Seidah NG, and Chrétien M. Biochem J 197:
   541-552, 1981.

AMINO ACID SEQUENCE DIVERSITY AMONGST CYTOCHROMES C-556

IN THE GENUS AGROBACTERIUM.

P. Tempst and J. van Beeumen

Laboratory of Microbiology, State University

of Gent, Ledeganckstraat 35, 9000 Gent, Belgium

In the study of bacterial phylogeny, the amino acid sequences of cytochromes c, isolated mainly from photosynthetic bacteria, nitrate reducing *Pseudomonads* and sulphate
reducing *Desulfovibrio's* have been determined over the last
years (Ambler and coworkers). We now focused our attention
to cytochromes c from species of the genus *Agrobacterium*.
The organisms are aerobe heterotrophes some of which are
known to cause plant cancer ('Crown gall'). Three strains
were studied : B2a, IIChrys and Apple 185, each one being
a representative of the 3 main genetic races of *Agrobacterium*. The 3 strains contain both a cytochrome c-522 and a
cytochrome c-556. Both types are soluble, acidic proteins
with a molecular weight around 12000 (1).
     We used classical sequence methodologies for the deter-
mination of the primary structures of the cytochromes c-556.
Automatic liquid phase sequencing was carried out on the
dehaemed proteins and on some tryptic and chymotryptic peptides, always in the presence of polybrene and with 0.33 M
quadrol buffer. PTH-amino acids were identified by isocratic HPLC analysis (5) and by TL-chromatography. Manual
sequencing was done according to the Hartley procedure
using 3-5 nmoles of peptide per amino acid residue.
     The sequencesof the cytochromes c-556 from the strains
B2a and IIChrys, determined independently without relying
on evidence of homology, have 102 out of the 122 residues
in common (2). Seventeen out of the 20 mutations involve
changes between residues of the same chemical nature; all
changes require only a single nucleotide mutation. The

581

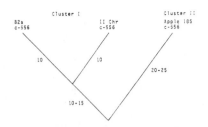

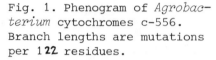

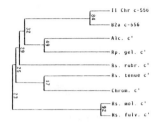

Fig. 1. Phenogram of *Agrobac-*
terium cytochromes c-556.
Branch lengths are mutations
per 122 residues.

Fig. 2. Unweighted average
linkage dendrogram of known
class II cytochromes c.

full evidence for the complete sequence of Apple 185 c-556
is not yet available, but it can already be said that this
protein has a maximal similarity with B2a and IIChrys c-556
of only 67 %. The sequence differences are visualised in
the phenogram of Fig. 1. The result is as expected if the
differences in phenotypical features and in the DNA : DNA
hybridisations are reflected in the sequence differences of
a single gene product. Strains B2a and IIChrys indeed
show some 50 % DNA homology, whereas each of them has only
10 % DNA homology with Apple 185 (3).

The single haem group of the cytochromes c-556 is bound
near the C-terminal end of the polypeptide chains. The proteins therefore belong to cytochrome c sequence class II.
In contrast to the 7 other class II cytochromes, the so-called cytochromes c' (4), the *Agrobacterium* cytochromes c-556
are of the low-spin type. The dendrogram constructed in
Fig. 2 shows the remarkable fact that the *Agrobacterium*
cytochromes are not more similar to the cytochrome of the
other aerobic organism than the latter is to the cytochromes c' of the photosynthetic bacteria. Without other molecular data, the dendrogram shown can not be concluded to
represent the evolution of the organisms in which they are
found.

1. Van Beeumen, J., Van den Branden, C., Tempst, P. and De
      Ley, J. (1980). Eur. J. Biochem. 107, 475-483.
2. Van Beeumen, J. (1980). in Protides of the Biological
      Fluids (H. Peeters, ed.) 28, 61-68, Pergamon Press.
3. De Ley, J. (1974). Taxon, 23, 291-300.
4. Ambler, R.P., Meyer, T.E. and Kamen, M.D. (1979). Nature
      278, 661-662.
5. Van Beeumen, J., Van Damme, J., Tempst, P. and De Ley, J.
      (1980) in : Methods in Peptide and Protein in Sequence
      Analysis (C. Birr, Ed.). Elsevier North Holland pp.521.